T0332376

Praise for

BANDWIDTH

"*Bandwidth* is a relentless ride through the peaks and valleys of an industry that's shaped our world. Like any great race, it's filled with high-stakes gambles, fierce competition, and the unwavering pursuit of the finish line. Some players crashed and burned while others found redemption, but the race never stopped. This book isn't just about technology—it's also about resilience in the face of overwhelming odds. *Bandwidth* is a reminder that even after the hardest falls, there's always a chance to get back up, rebuild, and press forward, making it a must-read for anyone who believes in pushing boundaries and defying expectations."

> —**Lance Armstrong,** professional cyclist and endurance athlete; managing director of Next Ventures

"*Bandwidth* is a business story for the ages, brilliantly told by a man who was intimately involved in the greatest multitrillion-dollar infrastructure race of all time. It is a story of personal achievement, of heroes and villains and crooks, of investment tragedy and rebirth, and of CEO guts on a level not seen since the Industrial Revolution and the transcontinental railroads.

Bandwidth is *Barbarians at the Gate* on steroids. And it continues to play out and impact human life everywhere."

> —**John J. Keller,** former award-winning reporter for *The Wall Street Journal*

"*Bandwidth* is an essential read for investment bankers, Private and Public Equity Investors, and technology consultants. The intertwining tales of hostile acquisitions, IPOs, industry consolidations, and bankruptcies are thrilling to read. This is a one-of-a-kind learning opportunity."

> —**Jason Rowe,** managing director at Goldman Sachs

"In *Bandwidth*, Dan Caruso lets us all in on the secret history of today's mega-scale tech companies. *Bandwidth* unwinds the tale of the pioneering telecom entrepreneurs in the 1990s and early 2000s who set in motion the golden age of tech startups and scaleups. Every subsequent tech wave—from the Dot-Com Boom (and bust) to the rise of US Webscale Giants to the flourishing of AI, blockchain, and, soon, quantum technology—has depended on fiber-optic networks and on bandwidth. Bandwidth is quite literally the foundation upon which every successful founder operates, and yet few know its story. Dan, a veteran executive of three digital infrastructure decacorns, is uniquely positioned to lead us along this rip-roaring ride of the creation, survival, revival, and triumph of dozens of the world's most storied tech players. *Bandwidth* is a must-read for scaleup leaders and their investors, as well as for anyone who enjoys a heart-pounding story!"

—**Linda Rottenberg**, co-founder and CEO of Endeavor

"This remarkable book recounts a hero's journey, wherein the hero is both the author and the bandwidth industry. Caruso takes us through the thrilling early days, where billion-dollar fortunes were created and lost. In classic fashion, rock bottom comes quickly when the Internet bubble bursts, to the surprise and ruination of many.

The hero's journey back to life follows with a select few visionaries (including Caruso himself) rebuilding bandwidth from the scraps of bankrupt companies. This classic storytelling allows us to follow a complex, labyrinthine web of subplots to an understanding of how the Internet came to be and to where it might be heading. Individually, the characters and their stories are fascinating. Together, they present a page-turning account of a moment in time when the world changed completely."

—**Steve Aranguren**, creative story artist, formerly of Disney and Creative Artists Agency

"With unparalleled access to industry titans and behind-the-scenes deal-making, Dan Caruso offers an insider's view into the business strategies, technological breakthroughs, and visionary personalities that have driven the exponential growth of our digital infrastructure. Dan delivers a fascinating and insightful firsthand account from the unique position of having worked for or started companies owned by the largest surviving US bandwidth providers. The book's engaging narrative weaves together personal anecdotes, technical explanations, and economic analysis, creating an essential read for anyone interested in the past, present, or future of global communications."

—**Steven Kaplan**, distinguished service professor of entrepreneurship and finance and Kessenich E.P.; faculty director at the Polsky Center, University of Chicago

"Like I was in *Molly's Game*, Dan Caruso is both the storyteller and a central participant—which allows him to tell the explosive story of the bold characters who created the infrastructure of the Internet. This always entertaining and sometimes chilling adventure will stay with you for years to come."

> —**Molly Bloom**, inspirational keynote speaker, entrepreneur, and bestselling author of *Molly's Game*

"I witnessed Dan successfully execute both a highly complex restructuring and launch a start-up that he grew into a $14.5 billion business. It is extremely rare for the same executive to have accomplished both. *Bandwidth* details Dan's extensive knowledge of unit economics, ability to examine product profitability, and relentless focus on value creation."

> —**Gillis Cashman**, managing partner at M/C Partners

"Ambition, greed, innovation, and resilience clashed on the grand stage of the Internet revolution. Dan 'the Bear' Caruso recounts the stories with amazing detail and brings this human drama to life, showcasing his fearless quest to push the boundaries of what was possible."

> —**George Nazi**, CEO of Saudi Company for Artificial Intelligence; former global VP and general manager of telecom at Google (2020–2023)

"A thrilling firsthand account of the development of the fastest-growing industry of all time by one of its leading protagonists. A wonderful read."

> —**Brady Rafuse**, former CEO of euNetworks; former CEO of Level 3 Europe

www.amplifypublishinggroup.com

Bandwidth: The Untold Story of Ambition, Deception, and Innovation That Shaped the Internet Age and Dot-Com Boom

For more information, please contact:
Amplify Publishing, an imprint of Amplify Publishing Group
620 Herndon Parkway, Suite 220
Herndon, VA 20170
info@amplifypublishing.com

Author photo on dust jacket by Zeb King

Library of Congress Control Number: 2024918722

CPSIA Code: PRFRE1024A

ISBN-13: 979-8-89138-459-0

Printed in Canada

My inspirations to write Bandwidth *were my wife Cindy,
my daughter Kailey, and my son Danny.*

*I needed to share the journey of my professional life, including my
vulnerabilities, struggles, and mistakes. The three of you gave me
the strength to complete my CEO chapter and then turn my life
focus on a broader purpose.*

Please share this with your children and grandchildren.

With enduring love and deep gratitude,

Dan

Bandwidth *is a celebration of innovators, entrepreneurs, and business professionals.*

Tens of thousands were directly part of the Bandwidth Revolution, as operators, accountants, engineers, sales and marketing professionals, financiers, and analysts in the companies featured in the story. Together, we built the fiber networks and created the Internet Backbones that powered the digital revolution. We were colleagues and rivals, usually friends, and always bursting with passion, creativity, relentless drive, and unfettered enthusiasm.

Many of you were my close colleagues—with paths that crisscrossed Ameritech, MFS, WorldCom, Level 3, ICG, and Zayo. We locked arms, overcame adversity, and had a whole lot to be proud of. I think of many of you every day. Thank you for your support and our shared memories.

I've come to appreciate that innovation and entrepreneurship are traits of humankind that transcend geographies, industries, and generations. I hope that Bandwidth *inspires leaders across the globe who are tackling new frontiers—including Artificial Intelligence, Quantum Technology, Sustainable Energy, and the Space Economy—to persevere. The world needs inspirational, industrious, and creative entrepreneurs now more than ever.*

BAND WIDTH

The **Untold Story** *of* **AMBITION, DECEPTION,** *and* **INNOVATION That Shaped** *the* **Internet Age** *and* **Dot-Com Boom**

DAN CARUSO

Entrepreneur, Tech Investor, and Founding CEO of Zayo Group

an imprint of Amplify Publishing Group

CONTENTS

In this book, individuals and companies that are not integral to the story are explicitly not named to declutter the story and streamline the narrative. The List of Names, located in the back of the book, provides the names of these individuals and companies.

FOREWORD

I moved from Boston to Boulder in November 1995, just as both the rise of the web and the commercial Internet were picking up speed. While I focused all my energy on software and Internet-related companies, I was aware of the rapid growth and evolution of the telecom industry, which was adjacent to and integrally intertwined with the Dot-Com boom.

In 1996 I co-founded SOFTBANK Technology Ventures (renamed Mobius Venture Capital in 2000), a Venture Capital fund sponsored by SOFTBANK, one of the most prolific investors in Internet-related companies. I rode the Internet boom up until its 2000 peak and then suffered mightily under the weight of the Dot-Bomb collapse. I regrouped and continued to co-lead Mobius Venture Capital while co-founding Foundry Group, which, over the past fifteen years, became one of the most impactful Venture Capital firms in the Mountain West region. I also co-founded Techstars, which became one of the leading start-up accelerators in the world.

I met Dan in 2003 as he was leaving Level 3. We gradually discovered that we had a shared interest in developing the Colorado entrepreneurship ecosystem. While we both had day jobs—me as a Venture Capitalist and Dan as a serial entrepreneur—we always found time to collaborate in support of Colorado's ascent as a place to innovate. Though I followed Dan's Zayo adventure, the depth of my understanding was at the surface level.

In April 2024, Dan asked me to look at his manuscript. I read the introduction and, within a week, the rest of the book. Reading it reminded me of places where my path crossed his, such as Exodus acquiring Service Metrics (a Mobius investment), which returned our entire first fund. Of course, several years later, after the result of a horrifying merger with Global Crossing, the combined Exodus-Global Crossing went bankrupt. Or, the day Howard Diamond, CEO of Corporate Software (another Mobius investment), walked into my office and said, "Level 3 is acquiring us for $125 million," to which I said, "Why the fuck would a telecom infrastructure company buy a software reseller?" And then there were stories Dan didn't even know, like the time I sat in Level 3's office in Broomfield, Colorado (five minutes from my office in Superior, Colorado), in 2001, praying Level 3 would buy Interliant (a failing public company I was co-chair of) for no other reason than Level 3 had bought other failing public companies. And, no, Level 3 didn't buy Interliant, and yes, Interliant did fail.

Before reading *Bandwidth*, I saw the Internet Revolution through the lens of software and Internet companies. I knew some of the stories of Qwest, Enron, Global Crossing, Level 3, and others—but didn't appreciate the details and magnitude of the Bandwidth Boom and Bust. The adventurous stories of business legends such as Phil Anschutz, Craig McCaw, Jim Crowe, and even Carl Icahn were exciting to read. The exploits of bad actors such as Joe Nacchio, Gary Winnick, and Bernie Ebbers were even more exhilarating. Epic battles of hostile takeovers, botched business plans, phoenixes rising from ashes, and entrepreneurs who persevered reminded me why I get so much fulfillment from helping start-ups become large successes and enduring companies. And why I am willing to fight for any company in which I'm involved.

As I read, I also learned a lot more about Dan. I now know why so many people have strong feelings about Dan—some think he walks on water, while others wish a big wave would knock him over. In this book, he openly shares the challenges he had to overcome and the pivots he had to make. At each step, he is delightfully self-deprecating and aware of what worked and what didn't. I, for one, enjoy watching him continually do extraordinary things. Put me in the "Dan walks on water" camp.

Brad Feld, venture capitalist, entrepreneur, and author of multiple books, including *Venture Deals* and *Startup Communities*

CAST OF CHARACTERS

360networks / Ledcor

Ron Stevenson, Ledcor executive

Greg Maffei, 360networks CEO

AboveNet

Stephen A. "Steve" Garofalo, CEO

John Kluge, founding investor

William G. "Bill" LaPerch, CEO (2004–2012)

Allstream / AT&T Canada

Michael "Mike" Strople, CEO

D. Craig Young, MetroNet CEO

AT&T / SBC Communications

Robert "Bob" E. Allen, AT&T CEO (1988–1997)

C. Michael Armstrong, AT&T CEO (1997–2002)

Edward "Ed" E. Whitacre Jr., SBC CEO

Bolt Beranek and Newman (BBN)

Leo Beranek, co-founder and president
J. C. R. "Lick" Licklider, computer scientist
George H. Conrades, CEO

Broadwing / Cincinnati Bell

Ralph Swett, Broadwing CEO (1985–1999)
Dr. David Huber, Broadwing CEO (2002–2005)
Richard G. Ellenberger, Cincinnati Bell CEO

Enron Broadband / Dynegy

Jeffrey "Jeff" Skilling, Enron President
Kenneth "Ken" Lay, Enron CEO
Scott Yeager, Enron Broadband executive
Charles "Chuck" Watson, Dynegy CEO

Genuity (prior GTE)

Charles R. "Chuck" Lee, GTE CEO (1992–2000)
Paul Gudonis, Genuity CEO (2000–2003)

Global Crossing

Gary Winnick, chairman
Robert "Bob" Annunziata, CEO (1999–2000)
Leo Hindery Jr., CEO (2000)
Thomas J. "Tom" Casey, CEO (November 2000–October 2001)
John Legere, CEO (October 2001–October 2011)

ICG Communications

J. Shelby Bryan, CEO (1995–2001)
Carl E. Vogel, CEO (2001)
Randy Curran, CEO (2001–2003)
Dan Caruso, CEO (2004–2006)
John Kane, COO
William "Bill" Beans, COO
James "Jim" Fleming, Columbia Capital
John Siegel, Columbia Capital
Peter Claudy, M/C Partners
Gillis Cashman, M/C Partners

Level 3 / Kiewit

James "Jim" Q. Crowe, Level 3
CEO
Walter Scott Jr., Kiewit
Corporation CEO
Dan Caruso, Level 3 executive
(1997–2003)
Kevin O'Hara, Level 3 COO
Sunit Patel, Level 3 CFO
Jeffrey K. "Jeff" Storey, Level 3
CEO (2013–2016)

Lightower

Robert "Rob" Shanahan,
Lightower CEO
David McCourt, RCN CEO
John Purcell, FiberTech CEO

Lumen (prior CenturyLink)

Glen F. Post III, CEO
(2009–2018)
Jeff Storey, CEO (2018–2022)
Sunit Patel, CFO
Kate Johnson, CEO
(2023–2024)

McLeodUSA

Clark McLeod, founder and
CEO
Chris A. Davis, CEO
(2001–2005)
Royce Holland, CEO
(2006–2008)
Theodore "Teddy" Forstmann,
Forstmann Little & Company

MFS Communications / Kiewit

Jim Crowe, MFS CEO
Walter Scott, Kiewit
Corporation CEO
Royce J. Holland Jr., MFS COO
Kevin O'Hara, MFS president
Dan Caruso, MFS SVP
(1991–1996)

PSINet

William "Bill" Schrader, founder
and CEO

Qwest

Joe Nacchio, CEO
Philip Anschutz, founding
investor
Richard "Dick" Notebaert, CEO

Salomon Smith Barney

Jack Grubman, analyst

TelCove / Adelphia

John Rigas, Adelphia founder
Bob Guth, TelCove CEO

Teleport

Robert "Bob" Annunziata, CEO
John Malone, TCI CEO
Brian L. Roberts, Comcast CEO

Touch America (prior Montana Power)

Robert "Bob" Gannon, CEO

tw telecom

Larissa Herda, tw telecom CEO
Glen Allen Britt, Time Warner
    Cable CEO
Gerald M. "Jerry" Levin, Time
    Warner Inc. CEO
John Warta, GST Telecom CEO
James "Jay" Monroe III,
    Thermo Group
James "Jim" Allen, Xspedius
    CEO
Anthony "Tony" Pompliano,
    e.spire founder

UUNET

John Sidgmore, CEO
Richard "Rick" Adams, founder

Velocita

Kirby G. "Buddy" Pickle,
    Velocita CEO
John Warta, PF.Net founder and
    CEO

Verizon

Ivan Seidenberg, CEO
Lowell McAdam, president,
    CEO

WilTel

Roy Wilkens, CEO

WilTel 2

Howard E. Janzen, CEO
Jeff Storey, CEO

WorldCom

Bernard J. "Bernie" Ebbers,
    CEO
John Sidgmore, vice chair, CEO
    (2002–2003)
Scott D. Sullivan, CFO
Dan Caruso (1997)

XO Communications

Craig McCaw, founding investor
Daniel F. "Dan" Akerson, CEO
    (1999–2002)
Theodore J. "Teddy"
    Forstmann, Forstmann Little
    & Company
Carl C. Icahn, Icahn Capital

Zayo Group

Dan Caruso, founder and CEO,
    Zayo
John Siegel, Columbia Capital
Gillis Cashman, M/C Partners
Michael Choe, Charlesbank
Phil Canfield, senior partner,
    GTCR
Marc C. Ganzi, DigitalBridge
    Group
Brian McMullen, Stonepeak
Jan Vesely, EQT Partners

INTRODUCTION

The Bandwidth adventure was a wild, multidecade roller-coaster ride—an epic story involving start-ups and scale-ups that experienced grand successes and devastating failures. The unbridled ambition of newly minted billionaires and high-profile tycoons created extraordinary value soon followed by unprecedented value destruction. It was a time of insight, creativity, and courage accompanied by deception, fraud, and prison sentences. The pace of industry consolidation was fierce, often involving hostile multibillion-dollar acquisition battles. Growth was unprecedented, and opportunities seemed endless. Then the tech bubble burst, leaving the industry in shambles and many investors devastated. The stage was set for a glorious rebirth.

The tale begins in 1983 with two world-impacting events. One was the dismantling of the world's biggest company, American Telephone and Telegraph Company (AT&T), an event that was noticed by all. The other was the birth of the Internet, which went entirely unnoticed. Further amping up these events were two technology breakthroughs: mobile communications and fiber-optic networks powered by lasers. The collision of these four forces ignited a revolution that would forever change how humankind worked, entertained and educated themselves, and how we communicated with one another.

The prehistory of the Internet began in 1957, when America's Cold War rival, the Soviet Union, launched the first-ever satellite, called Sputnik, into orbit. The United States needed a communications network that would function in the aftermath of a nuclear war. This led the Department of Defense to develop the Advanced Research Projects Agency Network (ARPANET). For years ARPANET was used only by the government, research institutions, and universities. In the 1980s ARPANET gradually evolved into the Internet. Specifically, the combined efforts of three companies led to this commercialization of the Internet. BBN, UUNET, and PSINet were the Inventors of the Internet Backbones.[1]

The AT&T breakup caught the attention of America's wealthiest eccentric entrepreneurs, most notably John Kluge of Metromedia, Cable TV titan John Malone, industrialist Philip Anschutz, and the scrappy telecom veteran Craig McCaw. Many of the nation's high-profile multigenerational family businesses—including the Kiewit construction behemoth of Omaha, oil giant Williams Companies of Tulsa, and the Roberts family of the Comcast media empire—also sniffed the aroma of opportunity. These tycoons partnered with ambitious entrepreneurs such as Robert Annunziata, who scaled Teleport; Jim Crowe, who founded MFS; Joe Nacchio, who became Qwest's first CEO; Roy Wilkens, who formed WilTel; and Steve Garofalo, who launched National Fiber Networks, which later was renamed AboveNet.[2] Together, they became the Renegades that built fiber networks to carry the traffic of the Internet.

---

1 The Internet Backbone is an expression that describes the physical links, electronic hardware, and operating software that serve as the on-ramps and pathways for the Internet traffic to flow between people and applications.

2 Another ambitious entrepreneur was Clark McLeod of Cedar Rapids, Iowa, who founded McLeodUSA. At the height of the Boom, McLeodUSA was worth $10 billion, making McLeod among the wealthiest people in Iowa. McLeodUSA went bankrupt in 2002 and again in 2005. See appendix 3, "More Bandwidth Stories," to learn more about McLeodUSA.

While the AT&T monopoly was dying, the Internet was coming to life. The Internet Backbones merged into the spiderweb of fiber networks, resulting in overnight fortunes being created for both the Renegades and the Inventors of the Internet Backbones. Some turned their paper riches into cold, hard cash. Others decided to hold, believing their piles of cash would continue to multiply.

The riches being racked up by the Renegades didn't go unnoticed. With wealth creation occurring at an unprecedented pace, a wave of business leaders, most notably Bernard J. "Bernie" Ebbers of WorldCom, did not want to miss the opportunity of their lifetimes. Frustrated executives from large companies—investor Gary Winnick launched Global Crossing, Greg Maffei left Microsoft to lead 360networks, and big company tech executive Michael Armstrong joined AT&T—made bold moves to get in on this action. Leaders of boring utilities—Cincinnati Bell, which became Broadwing; Montana Power, which transformed into Touch America; and GTE, which gave birth to Genuity—sought to transform their stagnant monopolies into new-age Internet giants.[3] Even those who sold out during the first wave, notably Crowe and Annunziata, fought their way back in.

The laser light shooting through the fiber networks was glimmering bright gold. All was playing out according to the Bandwidth players' grand plans. Ever bigger fortunes were being created. But like all stampedes for gold, the window of opportunity would not stay open for long. The late 1990s Bandwidth Boom was followed by the early 2000s Bust, the biggest industry crash in history. Over a trillion dollars of value disappeared. Recently minted billionaires saw their paper fortunes flame out. Criminal and civil probes were initiated, leading to jail time and multimillion-dollar fines for some Bandwidth players.

---

3 Enron and Dynegy also entered the Bandwidth industry, both ending up with disastrous outcomes. See appendix 3, "More Bandwidth Stories," to learn about Enron Broadband Services and Dynegy Global Communications.

The SEC investigated, leading to one of the most consequential securities reforms in history—the Sarbanes–Oxley (SOX) Act of 2002.

Like the phoenix from Greek mythology, the Bandwidth industry rose from the ashes to be stronger, smarter, and more powerful than before. A new breed of entrepreneurs scavenged the fiber mess and spearheaded the industry's rebirth. A strange optronics inventor named Dr. David Huber brought Broadwing back from its deathbed. Operator Larissa Herda received an unwelcomed battlefield promotion to rescue tw telecom from a premature death. Famed financier Carl Icahn reached into his hostile takeover bag of tricks to wrest control of bankrupt XO. And I, the teller of this grand tale, began my ascent to be the face of the industry by snagging a dying Boom-era fiber provider named ICG Communications. We made fortunes buying assets even as fear gripped the industry.

The bigger the mess, the longer it takes to clean up. It took two decades for the Bandwidth Infrastructure industry to be reconsolidated into four robust national providers.

The first is AT&T, but not the old AT&T that was broken apart to jump-start the Bandwidth adventure. Seeking transformation, the original AT&T launched a bold multifaceted acquisition strategy, which included buying one of the pioneering fiber companies and bankrolling three fiber upstarts. The transformation failed, leaving AT&T no choice but to sell itself when the Bandwidth Bubble burst. The buyer adopted the still resilient, more formidable AT&T brand as its own.

The second is Verizon, but only after surviving two multibillion-dollar debacles during the meltdown. Verizon acquired post-bankrupt WorldCom, the biggest Bust of them all, melding the MCI and WorldCom network assets into a new Verizon Business. Years later, Verizon acquired XO Communications, which traced its origins to the early fiber build-out.

The third is Lumen Technologies, a brand introduced in 2020. Lumen is the consolidation of many of the most storied companies of the Bandwidth Boom and Bust, most notably Level 3, Qwest, and Global Crossing.

The fourth is Zayo Group, the company I founded in 2007. Under my leadership, Zayo acquired over forty fiber companies. Unlike AT&T, Verizon, and Lumen, Zayo is in the enviable position of being the only national pure-play Bandwidth Infrastructure provider.

Like electricity, Bandwidth enables every aspect of our lives: mobile communications, emailing, texting, social media, analytics, GPS security, podcasting, and telemedicine. Without the Bandwidth Revolution, we would still meet friends and soulmates at parties and in bars instead of today's online social media and dating apps. We'd be watching linear TV, not streaming our content from Netflix, Google TV, and their dozens of rivals. Driverless vehicles would crash without the powerful Bandwidth. ChatGPT would not yet exist. Yet we take this networking resource for granted.

Bandwidth is the story of the characters who created this essential infrastructure of the Internet.

Bandwidth is my story.

Bandwidth is also your story because Boom and Bust cycles repeat themselves. In fact, the next one is already ignited, and it entwines the powerful forces of blockchain, generative AI, and quantum technology. Learnings from the Bandwidth Boom and Bust will prepare you well for navigating the new era.

CHAPTER I

THE STORYTELLER

You only have to do a very few things right in your life.

—Warren Buffett

I am in a unique position to tell the Bandwidth story, having been perched inside the tent from its exciting beginning to its glorious end.

I was part of all four of the major surviving players—AT&T, Verizon, Lumen, and Zayo. AT&T's buyer acquired my first employer, Ameritech Corporation. My second and third employers—MFS and WorldCom—are now part of Verizon. I was a co-founder of Level 3, Lumen's largest acquisition. Before the acquisition, Level 3 had acquired my fourth company, ICG Communications. My fifth and final employer was Zayo, which I founded in 2007 and sold in 2020 for $14.3 billion.

I am the Forrest Gump of fiber—an unwitting average guy who survived and then thrived on the multidecade Bandwidth roller-coaster ride. I excelled, failed, learned, and emerged as the industry's most successful value creator.[1]

---

1 In the ICG subsection of Chapter IV, I will provide the financial metrics that substantiate this claim.

My uneventful childhood was spent in a blue-turning-white-collar suburb south of Chicago, in a family with a strong Italian heritage. One of my great-uncles was Guido Marconi, a bread maker, and another was Nicky Ripani, a butcher. My godfather, Jimmy Zeranti, was a barber. My grandparents on my dad's side owned Dog n Suds, a fast-food burger joint that was eventually wiped out by McDonald's. Marino Lale, my maternal grandfather, was a factory worker whose friend and neighbor was Mike Costabile, the enforcer for the Chicago Heights Italian mob.

Costabile pulled a shotgun on me when I was ten years old. I was playing alone in my grandfather's backyard when I heard, "Hey, kid," from the yard that abutted my grandfather's. I looked up to see Costabile with the shotgun pointed from his hip. He looked like a body double of the short and round Peter Clemenza, the fictional caporegime, or captain, in *The Godfather.* As if mimicking Clemenza, Costabile laughed loudly as he stared me down. I beelined, tears flowing, into my grandfather's house. Five minutes later my grandpa sat me down on his outdoor bench, fetched Costabile, sat him next to me, and ordered him to apologize. "Kid, I was just trying to toughen you up" was the best he could do.

My dad was the first in his family to get a college degree. He spent his entire career as an engineer at AT&T.

Brookwood was my junior high, located seven miles south of Chicago and four miles from the Indiana border in my hometown of Glenwood.

When I entered sixth grade, Brookwood adopted a new method for assigning students to a classroom. Students took a test. Depending on how well you did, you'd get assigned to one of six cohorts. For sixth-graders, the smart kids were in cohorts 6:0 and 6:1. The average kids were in 6:2 and 6:3. The not-so-smart kids were in 6:4 and 6:5.

My brother, who was eleven months older and in the seventh grade, made it to 7:1. He was a smart kid.

I didn't do as well. I was assigned to 6:2. I was average, according to Brookwood Junior High.

I must have shown childish disappointment to my friends, as they started calling me "Av," short for "average." This infuriated me. My friend Jeff Frigo poked at me the most, and finally, I had enough. In the classroom, after lunch and just before class started, I charged at him. We exchanged punches, though neither of us was a fighter. We were sent to the principal's office, and each of us had a choice: the wooden paddle or detention for a week.

Frigo chose the paddle. I chose detention.

I always regretted this moment. The fight wasn't a regret—after all, Frigo was my friend, and we barely touched each other. My regret was choosing detention instead of the paddle. I was embarrassed that my friends knew I feared a smack on my ass.

Reflecting, I also dreaded a life of being average. I wanted to be somebody.

In my family it was a given that we'd go to college. Only later in life did I realize this didn't have to be the case. Without this implied expectation, I might not have gone to college.

My brother went to the University of Illinois in Champaign-Urbana to be an accountant. The conventional wisdom of the time was that being an accountant was the path to a good life: a country club membership, a three-car garage in the suburb of Naperville, and an annual vacation to Florida.

With my brother picking the best major at the best public university in Illinois, how would I one-up him?

I learned that engineers made starting salaries of around $28,000, while accountants lagged at $22,000. So, I decided to apply to the University of Illinois School of Engineering. I made this decision without really understanding what engineers, including my father, did. Yet I was so convinced this was the right decision that I didn't apply anywhere else.

The University of Illinois' engineering school was among the top rated in the nation. They rejected my application because accepting applicants whose junior high nickname was "Average" was not in their bag of groceries.

This was a wake-up call for me and a foreshadowing of my life to come. I could have let it go, and if I had, I wouldn't be writing this book. Instead, I applied to the University of Illinois' liberal arts school, with the intent of backdooring into engineering via a transfer.

My counselor warned me against this: "To transfer, you will need a 3.6 out of 4.0 grade point average for two years while taking predominantly math and science courses.[2] Highly unlikely, don't you agree?" I ignored him and his know-it-all smile. Stubbornness is another one of my admirable traits.

Two years later, despite a C in Spanish that pushed my GPA just below 3.6, I transferred into mechanical engineering. For the first time, I felt that maybe I was better than average. Maybe I would be somebody.

Fortunately for me—this kid who yearned to be more than average— the Bandwidth Revolution was on the cusp of being ignited.

I was in the middle of college in 1983, the year of both AT&T's breakup and the Internet's birth. Judge Harold H. Green, a United States district judge for the District of Columbia, was assigned to a 1974 anti-trust lawsuit filed by the US Department of Justice to blow apart AT&T's regulated monopoly. On August 5, 1983, Green approved the consent decree, thereby breaking up the largest corporation in the world.

The judge's move required the divestiture of AT&T's local telephone operations, resulting in the creation of seven Regional Bell Operating

---

2 The University of Illinois was on a five-point scale instead of the more standard four-point scale. My GPA was 4.6 out of 5.0.

Companies—Ameritech, Bell Atlantic, BellSouth, NYNEX, Pacific Telesis (PacTel), Southwestern Bell (SBC), and U S West—known as the "Baby Bells." These new companies were required to provide equal access to their networks, enabling competition in local telephone services, but the Baby Bells were prohibited from providing long-distance services and information services.

During this time AT&T was fighting off a pesky new competitor named Microwave Communications, Inc. (MCI), which offered an alternative for long-distance phone calls. A few years later, Sprint joined in on the fun. The local phone business, however, did not yet face competition.

In the early 1980s, phones hung on walls or sat on tables. If you wanted to use one, you had to walk up to it, pick up the handle, and press it to your ear and mouth. You couldn't roam around too far while talking because the cord tethered you to the phone outlet on the wall.

Life was simple in that way.

The idea of a portable communication device wasn't new. Walkie-talkies had been around since World War II. Yet the vision of mobile phones for everyday calls began its path to fruition in the early 1980s. To enable wireless phones, the Federal Communications Commission (FCC) awarded spectrum licenses—initially through hearings and lotteries and, beginning in 1994, through auctions—with the hopes of sparking a new era of mobile communications.

Motorola's 1985 brick phone marked the beginning of the wireless phone era. We stretched our imaginations to envision a world where most everyone would carry a phone in their pockets and a wearable à la Dick Tracy on their wrists.

Clearly, the underlying infrastructure that transmitted phone calls was on the verge of a disruptive transformation of massive proportions. Copper was used by Alexander Graham Bell when he invented the phone call in 1876. Two thin copper wires, twisted together and called a "copper pair," carried a single phone call. The signals were analog, meaning information was converted into electromagnetic waves that were transmitted through the copper pair, with variations of the waves' amplitude, frequency, and phase being used to code and decode voice conversations. Like music records played on a turntable, static was an annoying part of most phone conversations.

In the 1960s AT&T standardized on the Digital Service 1 (DS1). This breakthrough allowed a single copper pair to carry 24 voice calls at a time. Digital technology—the sequencing of 1s and 0s to encode information—replaced analog as the communication modality. It improved the clarity of communications, and gone was the static.

In the 1970s AT&T advanced to the Digital Service 3 (DS3) standard, which multiplexed 28 DS1s—meaning 672 phone calls—over a single copper line. The DS3 was state of the art when I was in high school.

The capacity of a DS3, though revolutionary at the time, is extraordinarily tiny by today's standards. Most individual homes surpass a DS3 worth of capacity, and larger businesses use the equivalent of dozens of DS3s. Data centers and cloud facilities each use more than ten thousand times the capacity of a DS3. If we still relied on copper, nothing we experience today would be possible.

In 1977 AT&T deployed the first experimental fiber-optic telecommunications system, connecting a Bell Laboratories facility to a nearby office building. Fiber-optic systems use strands of glass that are as thin as hair to transmit light waves. The digital 1s and 0s are represented as pulses of "light on" and "light off." Relative to sound transmitted through copper, the light, fired by lasers and carried by the fiber, enabled gigantic increases in the amount of information—or bandwidth—that could be transmitted over a single signal.

Quality was also better on fiber than copper. Sprint highlighted the quality advantages through its pin drop advertising campaign. Sprint proclaimed its newly activated fiber network provided better user experience over AT&T and MCI, both of which relied heavily on microwave. "You can hear a pin drop" was Sprint's memorable tagline.

Fiber was destined to replace copper as the medium for communications, gradually gaining momentum throughout the 1980s. As this became clear in the 1990s, entrepreneurs envisioned an infrastructure buildout opportunity on the scale of the early 1900s' Golden Age of Railroads and the oil pipelines in the mid-twentieth century. The first movers of fiber deployment would own and control the most important infrastructure of the twenty-first century. Thus fiber deployment accelerated, and, it appeared, the Fiber Barons of the twenty-first century were being anointed.

The answer to the question, "What would data be used for?" was hiding in plain sight. The technological breakthroughs that became the Internet began in 1957, as the United States became locked in the Cold War with the Soviet Union. Russia had launched the first-ever satellite, called Sputnik, into orbit, further presaging an intense geopolitical era between the United States and USSR. America was determined to stay ahead of its principal rival. The Department of Defense sought a communications network that would function in the aftermath of a nuclear war. This led to the formation of ARPANET in the 1960s and 1970s, a closed network for researchers closely connected to Department of Defense projects and associated work.

January 1, 1983, is considered the birthday of the Internet, marking the date when ARPANET adopted the Transmission Control Protocol/Internet Protocol (TCP/IP), a programmed set of rules for transferring data and by which computers could communicate on a network. The inventors of TCP/IP—Dr. Bob Kahn and Dr. Vint Cert—are honored as the fathers of the Internet. Taking things to a higher level of specific data communications in 1989, Sir Tim Berners-Lee introduced the World Wide Web, the Hypertext Markup Language, the URL system, and the hypertext transfer protocol (HTTP). Marc Andreessen released the Mosaic web browser while he was a student at the University of Illinois, my alma mater, and followed up with Netscape web browser in 1994. These advancements aided the transmission power, usability, commercialization, and rapid growth of the Internet.

So there you have it. Break up the AT&T monopoly. Give birth to the Baby Bells and open the local telecom markets to competition. Untether telephones and fit them in pockets. Use fiber optics to juice up capacity. Commercialize the Internet.

Mix all this together in a firepit, light a match, and pour a giant pile of money on it to stoke the flames. In less than ten years, the world had become overwhelmed by competition. The Bandwidth bonfire was erupting, all serendipitously timed to coincide with the launch of my career.

———————

My first job out of college was with the Bell System—in my case, Illinois Bell, which was a division of Baby Bell Ameritech. I accepted without hesitation when offered a position in the management development program for engineers. I was intrigued by the AT&T breakup and the impending competition. I was aware of, though didn't fully understand, the looming revolution of mobile, data, and fiber. I also craved the management development program as a way of supercharging my career.

I would be remiss if I didn't mention the other reason for joining Illinois Bell: it was the only employment offer I received. The job market, which was weak during my graduation year, was unenthusiastic about engineers who backdoored into the program and took five years to get a four-year degree.

My rotational program began with a field supervisory position at an Illinois Bell garage in Chicago's fast-growing, upscale suburb of Naperville. It ended with a position in Corporate Development at Ameritech's headquarters in the heart of downtown Chicago.

One of my assignments was network planning, where I was tasked with evaluating fiber to the home. I worked on variations of this for over a year and found my passion. How much would it cost to replace all the copper with fiber? How would we architect the network, and what technology would we deploy? What services would be enabled, and how much new revenue would be generated? Where should we start? At what pace should we invest? Would it be a good investment?

This was fun. I built Lotus 1-2-3 spreadsheets, plotted colorful pictures using Harvard Graphics presentation software, and shared my insights in a storytelling fashion. The thirty-year veterans I worked for—most of whom were comfortable only with their Texas Instruments TI-81 calculators—didn't know what to make of me.

In the Bell hierarchy, the Seventh Level was the big boss; he ran Ameritech. The Sixth Level ran Illinois Bell, and the Fifth Level managed large departments such as network, sales and marketing, finance, and corporate strategy. The Fourth Level was a division head—a big deal.

Rarely was a First Level, which I was, allowed to speak to a Fourth Level and almost never to be in the presence of a Fifth Level. That's just the way the bureaucracy worked.

My boss gave me the assignment of creating a long-range plan to replace all the copper serving homes with fiber. I did my analysis and put together a slick presentation. However, no one followed up with me, so I didn't know what to do with it.

Months later, the Fourth Level sent down the word. His boss, the Fifth Level, needed the work immediately because he was presenting to the board the following week. "Is it done?" my boss asked. I showed it to him. He shrugged and, with no time to change anything, told me to drop it off at the Fourth Level's office.

The Fourth Level reviewed it, and he, too, was perplexed. He told me to go to the Fifth Level's office first thing the next morning. Strangely, neither my boss nor the Fourth Level joined me. I went to the hallowed floor of the executives, checked in with the secretary, and waited outside.

After a long wait, the door opened, and he invited me in. I presented my study, which I knew was good work. He asked many questions, and frankly, I could tell he was absorbing a lot.

He then saw the results of the spreadsheet, which showed an investment of $6 billion over a ten-year period to wire most homes with fiber. He was astonished and said he could not present it to the board. He would lose all credibility. There is no way the regulators would let Illinois Bell spend that amount of ratepayers' money. "Are you sure it is right?" he asked me.

It was. The Fifth Level escorted me from his office. I never heard what happened next.

I went on to my final rotational assignment as an analyst in Ameritech's Corporate Development group. This opened up a whole new world to me. My colleagues all had MBAs from top universities, including Harvard, Yale, and Northwestern. Nearly the entire team was hired from outside the Bell system. All my colleagues grew up in upscale suburbs, or so it seemed. I was blue collar to them.

We met with investment bankers and management consultants. We supported Mergers and Acquisitions (M&A). We studied new lines of business, including Cable TV.

John Malone of Tele-Communications Inc. (TCI) was leading the charge in rewiring America for Cable TV service. It was an unregulated monopoly that had the glam of content and the promise of getting rich. Should Ameritech overbuild its networks to offer a competitive service? If so, should it deploy the architecture of the Cable TV industry, which was the heavily touted hybrid fiber/coax?[3] Or should Ameritech leapfrog the cable industry with fiber to the home? Alternatively, Ameritech could just buy Cable TV companies.

Those were big questions. I created numerous Lotus 1-2-3 spreadsheets and Harvard Graphics presentations. I learned about various architectures and technologies. More fun for me. No clear answers were

---

3 Coax, short for coaxial cable, is a type of transmission line used to carry high-frequency electrical signals with low losses. Among the use cases was the Cable TV industry because it had advantages over the twisted pair copper cables used by telephone companies for transmitting television signals.

Fiber/coax architecture referred to the combination of fiber-optic cables being used in the central portions of the network, where traffic was consolidated, and coax being used to feed a relevant portion of the traffic to customers in a neighborhood cluster.

self-evident. With seven Baby Bells all sorting through their Cable TV strategies separately, variations of each possibility were pursued.

In all cases the Cable TV industry got the better of the Baby Bells. The Cable Cowboys, as the Cable TV pioneers were branded with Malone the lead cowboy, were simply better entrepreneurs. The Baby Bells were forced to retreat from each expensive mistake, while the Cowboys got rich.

My tour through Ameritech unveiled my purpose in life. Though I chose engineering for the wrong reason, I discovered my passion was assessing the financial outcomes associated with new technology. Business, not engineering, was my true love. I loved using spreadsheets to gain insights into how to leverage disruptive changes to create wealth.

While at Ameritech, I applied to Northwestern and University of Chicago's MBA programs. I chose Booth, Chicago's MBA program, for two reasons. First, I was analytical, which made Booth and me well suited for each other. Second, Northwestern rejected me. Are you detecting a pattern?

The MBA experience is when I realized I was more than just average.

I purchased a book on the Graduate Management Admission Test (GMAT) preparation and spent a weekend or two breezing through it. I enjoyed problem-solving. When I took the GMAT, I was surprised by how easy it was and thought I got every answer correct. I scored in the top 2 percent, so I clearly got a few wrong.

I unlocked quantitative skills, financial discipline, and business acumen, all hallmarks of a University of Chicago MBA, that provided me with the capabilities that catapulted my career journey.[4] I also gained confidence that I could hold my own with the smart kids.

---

4 My most impactful professor was Steven N. "Steve" Kaplan. His use of case studies, which were rare at Chicago in the early 1990s, shaped my career. Kaplan prompted his students to share their thoughts on the business decisions in play while requiring that rigorous financial analysis be used to analyze the cases and defend conclusions.

Kaplan emerged as one of the most important professors at Chicago and made a global name for himself for his contributions in both finance and entrepreneurship. I invited Kaplan to join my Zayo board, where he helped us navigate the 2020 exit.

In the mid-2010s, Kaplan wrote a case study on Zayo's IPO and featured it in his course curriculum for many years thereafter. Each year I am a guest to provide firsthand color to the dynamics that led to the Zayo IPO.

As I navigated my MBA, I knew I belonged on the other side, the one attacking, not defending the Bell System monopoly. Soon after I completed my MBA, I left Ameritech to join founder Jim Crowe's MFS Communications, which was rapidly emerging as the biggest rival to the Baby Bells.

CHAPTER II

THE INVENTORS OF THE INTERNET BACKBONES

Predicting rain doesn't count, building the ark does.

—Warren Buffett

W hen I joined MFS Communications, I knew nothing about the Internet. Neither did anyone else. Our chairman and CEO, Jim Crowe, held a meeting in MFS's Chicagoland headquarters. Located twenty minutes straight west of downtown Chicago was Oak Brook's twenty-story blue glass structure. In this shiny blue landmark, a meeting called "The Internet" was held.

I arrived at the meeting perplexed, as I didn't realize why I was included as one of MFS's twenty-five most senior executives in the invitation. The meeting room was set up like a classroom, with all the chairs in rows facing the front, where a computer, an overhead projector, and a white movie screen were arranged for a presentation. I sat near the back, as I thought it best to blend in.

Today, projecting a computer onto a big screen in meetings is commonplace. In 1994 this was done far less frequently. Those of us who gathered were curious.

Once the senior executive team settled in, Crowe called the meeting to order. He sat in the front of the room, next to the computer, and typed while he spoke. Again, I need to stress that this setup, though common today, was very unusual in 1994. More common was the use of slide transparencies. If a senior executive needed a computer for a presentation, an underling was doing the clicking. However, Crowe wasn't clicking through a PowerPoint presentation. *He was going online.*

Online? Huh? What? In those days employees used corporate networks for email. To the extent they used the Internet at home, it was America Online (AOL) or Netscape via a dial-up connection. As hard as it is to believe, the Internet wasn't really viewed as a business tool. Crowe, a true visionary, knew this was changing, and he decided his executive team needed a wake-up call.

"The Internet is going to be a powerful force in all of business," Crowe told the group. "To show you what I mean, I am going to log onto the Internet."

In general, the senior team saw Crowe as out of touch with the day-to-day goings-on at MFS. As such, his declaration triggered a few smirks from some in the room. "Doesn't he know we have work to do?" said someone as we watched Crowe log onto the Internet.

"We've all used corporate data networks for a long time," Crowe said. "Primarily, these were used for email. The Internet is something entirely different. Corporate networks are closed systems, not much more than speeding up intracompany mail." The group grew quiet as we listened. Some in the room looked skeptical.

Crowe continued, "The Internet is an open network. But to think of it solely as a faster way to exchange email would be a mistake. It will change the way people live and work throughout the globe."

Crowe added, "The Internet will speed up business. Using File Transfer Protocol, large files can be sent over vast distances both rapidly and cheaply.[1]

---

1 The File Transfer Protocol (FTP) is a standard communication protocol used for the transfer of computer files from a server to a client on a computer network. FTP was commonly used before web browsers became commonplace.

Whereas FedEx would charge $20 to overnight a document, the Internet will deliver the same document in a few minutes at less than 5 percent of the cost."

Crowe then used the Internet to send a document, making his point.

"Information is power. The Internet will speed up the pace at which people get access to information," Crowe said. Then he asked a question that would look silly in today's world: "Who here has heard of Yahoo!?" Eight hands went up; mine was not one of them.

"Who has actually used Yahoo!?" prompted a couple of the hands to drop.

"Yahoo! is a web portal," Crowe explained to the perplexed crowd. "Internet users get information from a portal. Some of the information is prepackaged by Yahoo! itself, like an online magazine. But a portal is also used to find information that could reside anywhere on the Internet."

I seriously doubt Crowe's audience followed his "anywhere on the Internet" comment.

"Who knows what a search engine is?" Crowe asked the room. Two or three hands rose proudly. Crowe demonstrated a search engine, perhaps Yahoo!'s or another pre-Google version such as Ask Jeeves. He would type in words, and links would appear on the big screen. Wow! Powerful. There were some startled looks around the room.

Yet it was what Crowe said next that made that day so important in my career.

"I have come to believe that the Internet will fundamentally change the way people live their lives and conduct business," Crowe told the group in a voice that was even deeper and more serious than his usual deep and serious voice. "I have concluded that MFS Communications will shape its business plan, and everything we do, around the Internet."

Then as that statement began to sink in, Crowe said, "I have called each of you to this room to offer you a choice. One possibility is to leave this meeting as a skeptic. You can view the Internet as a trend, a fad, or a toy. You can be slow to understand the Internet. You can be an old dog that is not ready to learn a new trick. If this is your choice, you will need to find an employer other than MFS Communications."

Then Crowe offered an alternative, though not necessarily a reassuring one. "Your other choice is to embrace the Internet. Learn what it is and how it works. Use it frequently and leverage it to make our business better. Understand how others are using it, and use this knowledge to guide our sales, marketing, and investment decisions. If you do this—if you embrace the Internet—I can offer you only one thing: you will have met the minimum qualification for being an executive at MFS Communications."

Yep. No special bonuses. No promotions. In fact, not even assurance of continued employment. The gist of Crowe's message was to embrace the Internet, and perhaps you will keep your job. Shun it and do not let the door hit you on the ass on the way out.

I picked door number one—I embraced the Internet.

BBN—Leo Beranek and J. C. R. Licklider

Long before the birth of the Internet in 1993, computer scientists spread across several government research and higher education institutions invented the technologies that became the backbone of the Internet.

The oldest Internet pioneer was a company named Bolt Beranek and Newman (BBN). The company's Internet journey began in 1957, when it hired a computer scientist named J. C. R. Licklider. After only a few months on the job, "Lick," as he was known, insisted BBN buy a $30,000 computer. Leo Beranek, BBN co-founder and CEO, asked Lick what he would do with it.

"I don't know," Lick responded. "But if BBN is going to be an important company in the future, it must be in computers."

The next year a large industrial company developed a prototype of its first computer. The company's president was impressed by Lick's understanding of digital computation, which led to BBN becoming a test site for the prototype computer.

In 1960 BBN leased a $150,000 computer. In search of customers, Lick and Beranek visited several government agencies, including the National Science Foundation (NSF) and the Department of Defense.

Lick hired John McCarthy, a computer science researcher from MIT, who envisioned time-sharing of computers. This concept resonated with Lick, leading him to publish "Man-Computer Symbiosis," which foreshadowed the Internet.

> Man-computer symbiosis . . . will involve very close coupling between the human and the electronic member of the partnership. The main aims are 1) to let computers facilitate formative thinking as they now facilitate the solution of formulated problems, and 2) to enable men and computers to cooperate in making decisions and controlling complex situations without inflexible dependence on predetermined programs. In the anticipated symbiotic partnership, men will set the goals, formulate the hypothesis, determine the criteria and perform the evaluations. Computer machines will do the routine work that must be done to prepare the way for insights and decisions in technical and scientific thinking.

In 1962 BBN demonstrated computer time-sharing with one operator in Washington, DC, and two in Cambridge. Later that year BBN installed a time-sharing information system at Massachusetts General Hospital, allowing several nurses and doctors to create and access patient records stored on a BBN computer. For the first time, users could interact with computers in real time from a terminal—a major advancement relative to submitting a stack of punched cards and, hours later, getting back a readout.

In 1968 BBN responded to the Defense Advanced Research Projects Agency's (DARPA) request for proposals (RFPs) to build a network, later named ARPANET. BBN was awarded a $1 million contract to develop a small electronic unit, which evolved into the modern-day Internet router. BBN shipped the router to the University of California in September 1969.

Beranek elaborated on what happened next.

> [T]his project required us to make a computer network with five stations on it first. And then if that worked it would be 19. So the idea was to hook together 19 computers and these would be located in universities and at a couple of the government laboratories. . . . [W]e were . . . hooking together these very expensive computers.

And so, ARPANET became the Internet's precursor.

Beranek explained the bigger need of smaller universities to gain access to big computers spawned internetworking.

> [I]f we make a network then [the small universities] can hook into . . . these big computers, maybe after midnight, when they're not in use in the big schools. . . . [W]e invented this sort of cabinet. [The IMP,] the size of a refrigerator, . . . would hook into their computer. . . . So there are probably four different [computers], maybe five, that were in these different universities . . . connected together to form the network. [T]his became the big network, which is like the Internet today, only it was just one network and started off with only two people. And the first two went on the air, I think, in 1969. . . . And then every month, [BBN built and shipped one] to another university. And then . . . we had 19 of these out. . . . And the government financed the ARPANET as it grew up until 1982. At which time they said, well, look, we think that this can be handled by people like Comcast. . . . They can handle it and they can charge people to use it. Whereas [the government had] been paying for the whole thing [allowing users to get] . . . free use of it. . . . They stopped financing it in 1982.

In 1972 BBN veteran Dr. Bob Kahn joined DARPA to tackle the problem that the three emerging government packet networks, including ARPANET, couldn't talk to one another. The three networks used different interfaces, packet sizes, labeling conventions, and transmission rates.

Kahn enlisted Stanford professor Dr. Vinton Cerf. Their 1973 solution was to invent the Transmission Control Protocol (TCP), which led to their historic May 1974 Institute of Electrical and Electronics Engineers publication titled "A Protocol for Packet Network Interconnection." They later combined TCP with the Internet Protocol (IP), forming TCP/IP. TCP addressed how to collect and reassemble packets of data, while IP ensured the packets were sent to the right destinations.

Dr. Cerf and Dr. Kahn achieved a major demonstration in 1977 by "internetting" ARPANET with the other networks. Messages were relayed from a van in the Bay Area across the United States on ARPANET, then to University College London and back via satellite to Virginia, and finally to the University of Southern California's Information Sciences Institute via ARPANET.

Six years later ARPANET adopted TCP/IP—the specific event that led to January 1, 1983, being marked as the birth of the Internet and Dr. Kahn's and Dr. Cerf's anointing as fathers of the Internet.[2]

From the late 1970s through the early 1990s, BBN leveraged its early Internet expertise and its invention of the router by focusing on selling network hardware to the government and government-related contractors. BBN did well as a hardware company during this period. However, the early 1990s were a different story. The ending of the Cold War resulted in a cutback to government defense spending. Moreover, Silicon Valley

---

2 BBN was awarded one of the three contracts to implement the Kahn-Cerf TCP protocol. In doing so Ray Tomlinson, who led BBN's implementation of TCP/IP, wrote a program to enable electronic mail to be sent over the ARPANET. Tomlinson chose the @ sign for the email address convention, and "user@host" convention stuck as the worldwide standard for email addresses.

upstarts, notably Cisco Systems, developed more technically advanced IP network products.

Despite their contributions in developing Internet technology, BBN found itself on the sidelines as it witnessed its two main Internet rivals develop commercial Internet Backbones. We will share how BBN responded after first telling the stories of those two rivals, UUNET and PSINet.

UUNET—John Sidgmore and Rick Adams

Richard L. "Rick" Adams had the appearance of an Internet tech pioneer—scruffy with a beard, disheveled, and typically sporting a wrinkled polo shirt at work. Adams preferred to toss around a Frisbee during the day and do his programming at night. He described himself as "the quintessential computer nerd."

Adams's first employer was not keen on his work habits or attire, so he quit in 1982 to join the Center for Seismic Studies in Northern Virginia. There, he became an ARPANET user and, within two years, invented SLIP, enabling individuals to connect their personal computers to the ARPANET via dial-up modem.

Adams was a member of a nonprofit organization for Unix software programmers, whose mission was tied to innovative research. Adams noticed that Unix developers wanted access to ARPANET. "Everyone wanted this wonderful resource, even though it was mostly just electronic mail and not much else," he explained.

In 1987 Adams convinced the organization to provide $250,000 to buy hardware from Cisco and rallied programmer volunteers to build out the network software. He named his nonprofit organization UUNET (pronounced "you-you net"), which meant nothing to most people, but the "Unix-to-Unix Network" underpinnings appealed to the tech crowd. Though Adams's goal was to charge only for what was necessary to recover costs, he was soon collecting $1 million of revenue.

"If people were begging you to take their money for this, that's the business I want to be in," Adams said.

True to his words, in 1989 Adams left the seismic center and moved the UUNET hub from a USENIX facility to his home in Northern Virginia. The following year he turned UUNET into a for-profit company and launched Internet connectivity services for enterprise customers. UUNET had four employees on its first day.

Al Gore, who was Bill Clinton's vice president and, but for a few votes in Florida, might have been Clinton's successor, claimed to have invented the Internet. He did not. However, in 1991 he championed a $1.75 billion spending package to support his vision of an information superhighway. While others were vying for this funding, UUNET was growing in popularity with the user community. Observing this, the NSF allowed UUNET—and a few of its rivals—to interconnect with ARPANET. With that, the commercialization of the Internet was underway.

Other innovations conspired to cause the popularization of the Internet: Berners-Lee's 1989 invention of the World Wide Web and the Hypertext Markup Language; Andreessen's 1992 launch of Mosaic and, soon thereafter, Netscape; and AOL launching its Internet service in the early 1990s. Traffic on UUNET's network was skyrocketing.

Augmenting capacity on the UUNET network was expensive. Dozens of computer servers were deployed in data centers across the nation. These servers were connected using dedicated lines leased from fiber network companies. The exponential growth made network design and implementation a daunting undertaking. I know this firsthand because, as the story unfolds, much of this responsibility eventually fell into my lap.

"The project was swallowing cash by the trash-can full," recalled UUNET's chief scientist. "We had to put hardware everywhere. There were these big-boy pants that we had to grow into real quick."

The need for cash prompted Adams to raise money for the first time in 1992. Mitch Kapor, best known for his spreadsheet software Lotus 1-2-3, befriended Adams and offered to invest in UUNET. "He told me that there was a new poker game going on and he wanted to get some chips," Adams said. Kapor invested several hundred thousand dollars. "I liked him, trusted him, and I also wanted him to stop bugging me [to let him invest in UUNET]," explained Adams.

Kapor, knowing his investment wouldn't fund UUNET for very long, turned to Silicon Valley. John Doerr, a top Venture Capitalist at a leading Venture Capital firm Kleiner Perkins, wasn't interested in meeting Adams, as he viewed UUNET as capital intensive and lacking durable advantages over deep pocket rivals.

Kapor next reached out to another major Venture Capital firm, Accel Partners, which focused on the telecom industry, and pitched to them that UUNET could "make this into a thing where everyone talks to each other." Despite multiple attempts, Accel would not bite.

While Sand Hill Road struggled to see the UUNET opportunity, my bosses' boss—MFS CEO Crowe—was ready to invest.[3] UUNET's consumption of bandwidth to interconnect its servers—and to connect its customers to the UUNET network—was resulting in UUNET becoming one of MFS's biggest customers. In February 1993, just months after the Internet presentation in Oak Brook, Crowe offered Adams $8 million for his company. "I wanted $12 million, and they wouldn't consider spending that much. Everyone thought I was crazy to walk away, but I was having fun and I thought the Internet business was a good bet." So were the 1985 Bears.

Kapor used MFS's offer to nudge Accel, which led to an invite for Adams to pitch to Accel's investment committee. Accel remained on the fence. Adams emailed Kapor: "They are currently waffling around deciding the size of the market. They are convinced that UUNET can become a $30 million company but aren't (yet . . .) sure it has the potential for $100 million."

---

3 Sand Hill Road is a street in Palo Alto where most of the highest profile Venture Capitalists had their offices in the 1990s. The expression "Sand Hill Road" is a reference to the Venture Capital community.

In October 1993 Accel and Menlo Ventures, along with another firm, invested $1.5 million in UUNET at a valuation just over the $8 million offered by MFS.[4]

Soon after the investment, $750,000 of unpaid invoices were discovered, swallowing half of the new $1.5 million investment. Facing insolvency, the Venture Capitalists provided an additional $1 million in exchange for more control and greater ownership. "I've got a gun to my head," Adams emailed Kapor, explaining why he was accepting the onerous terms.

The investors convinced Adams that UUNET should hire GE veteran John Sidgmore as CEO. At first Sidgmore was dismissive of the opportunity: "Why in the world would I want to go to this little company, YoYo Net, WeWe Net, or whatever you call it?" Ultimately, he acquiesced.

Sidgmore explained why he accepted the job. "I respected [the investors] and they thought UUNET had a lot of potential that just had not been realized and wanted a professional manager to make that happen."

With Sidgmore at the helm, UUNET's investors were content. "John brought [UUNET] credibility that said we were serious because we had brought in serious management . . . [I]t said we were ready to play," said John Jarve of Menlo Ventures.

Sidgmore, who became my boss a couple of years later, was both disarming and intense. He was shorter and heavier than he looked in his online profile. Sidgmore wore expensive Canali suits and trendy Ferragamo shoes, giving him the sheen of a new-generation Internet executive versus the old-time phone company leader derisively called a "42 long" for the typical cut of a CEO's suit from that era. Sidgmore was easy to respect, fun to be around, and, given the insatiable capacity requirements

---

4 Accel doubted Adams's ability to lead the fast growing company. Accel approached New Enterprise Associates (NEA), which had an office in Maryland. Peter Barris, a new partner at NEA, had executive-level experience at GE Information Services and two successful software start-ups. Barris's background and NEA's proximity to UUNET de-risked the investment. This led to NEA's becoming UUNET's third Venture Capital investor.

of the UUNET network, difficult to please. He focused hard on the business of the Internet, not so much the technology.

UUNET chose well, as Sidgmore proved to be a perfect leader for the Internet pioneer. Sidgmore convinced Microsoft to become a 17 percent owner of UUNET and use UUNET as its underlying network backbone, supporting Microsoft's quest to be a global leader in Internet access.

Sidgmore shared his excitement at this to the *New York Times*: "It's going to enable new classes of applications and users." He also recognized the significance to UUNET: "I never dreamed there was one killer opportunity that could make the company. I have a very happy board." Adams added, "If Microsoft's successful, we'll be successful." Luckily, Microsoft did just fine.

One month later Sidgmore struck a deal with AOL, which meant that UUNET would carry most of the Internet traffic for the two largest Internet brands. "We're seeing almost out-of-control growth in the industry. It's been a rocket ride," Sidgmore exclaimed.

In May 1995 UUNET raised $68 million in an IPO. Its stock price doubled in its first week of trading, which put the Market Value of UUNET at $900 million. By November the stock reached its peak of $93.25, valuing UUNET at close to $3 billion.

To fuel its rapid growth, UUNET was burning cash. In early 1996 Sidgmore announced plans to raise $285 million by selling additional shares to the public. In March 1996 he withdrew the offering, citing UUNET's then mid-$20s stock price as the reason.

Meanwhile, MFS continued to be the beneficiary of UUNET's growth, and Crowe's interest in owning UUNET grew stronger. Knowing Microsoft's support was critical, Crowe visited Microsoft CEO Bill Gates and received Gates's blessing, but only after reinforcing his commitment to retaining Sidgmore.

By May Crowe's MFS affirmed its courtship by agreeing to acquire UUNET for $2 billion. Sidgmore explained, "This is a scale game, and to prevail you have to get big fast. We could have stayed independent, but we could not be the leader if we did that."

MFS was also seen as a good cultural fit based on its entrepreneurial and aggressive corporate style. "They are just like us, not some Bell-heads we have to reshape," said UUNET board member Daniel C. Lynch. "We couldn't find a better match."

Adams did well. His take was $138 million. Reflecting on his 1993 decision not to sell to MFS for $8 million, Adams said, "It looks like I made the right decision. But you know, it was never about the money." Sure it wasn't.

Explaining his rationale for the combination, Crowe saw "a fundamental shift in the nature of communications, from the voice-oriented network that has been around for 70 years or so, to a whole new type of network based on IPs. More and more, data, voice and even video will be moving to Internet-based platforms. It's a force that is almost unstoppable, with tremendous cost advantages."

Sidgmore was onboard. Still in reflection, he said, "It is a little bittersweet to hand over the company, because, really, what we are giving up is our baby." I felt the same way when I sold my baby, Zayo Group, in 2020.

If Sidgmore knew the nightmare that was soon to transpire, he certainly would have done something else with his baby. Stay tuned.

PSINet—Bill Schrader

William "Bill" L. Schrader had an audacious vision: "I would like to build a system that can deliver all the media to all the customers on earth."

Schrader was without question one of the pioneers of the Internet. He remained at Cornell University after his 1974 graduation. NSF launched NSFNET, which funded several regional data networks that collectively became the initial Internet Backbone. Schrader built the first NSFNET backbone, which linked the national supercomputer centers. He then created NYSERNet, a nonprofit organization to connect New York State's universities, research institutions, medical centers, and libraries.

In 1989 Schrader became one of the pioneers intent on commercializing the Internet when he co-founded PSINet. He financed the company the old-fashioned way, using credit card loans and proceeds from the sale of a family car. Later, he received backing from Venture Capitalists, including Bessemer Venture Partners, Matrix Partners, and Sigma Partners.

To get started PSINet purchased NYSERNet, the nonprofit launched by Schrader, for $34 million. Then in 1991 PSINet partnered with UUNET to co-found a trade association of Internet providers. The nonprofit established the Internet's first exchange point, which exchanged traffic between providers via a router operated by PSINet. This router was moved from its original location in Santa Clara to Palo Alto. This site, dubbed the Palo Alto Internet eXchange (PAIX), became the early Internet's most important interconnect.[5]

On May 1, 1995, PSINet went public at a valuation of $362 million. Schrader used his equity as currency to acquire bandwidth to support PSINet's network. In exchange for 20 percent of PSINet's equity, he secured long-term credits to purchase $240 million worth of bandwidth services from the fiber upstart Broadwing Communications, which will be introduced in the next chapter of this story.[6]

Schrader was bold as he predicted the future of communications. When discussing Bell Atlantic, he likened the Baby Bell to the long-lost island of Atlantis, declaring, "Bell Atlantis is dead. They cannot exist in [the future]."

Schrader told an executive as he was interviewed to join PSINet as COO, "We're going to kill the phone companies, and we're going to be bigger than AT&T, and it will be a lot of fun."

---

5 PAIX was co-founded with Internet pioneer Paul Vixie, whose contributions included the development of the Domain Name System (DNS). AboveNet, before being acquired by MFN, acquired PAIX from Digital Equipment Corporation. When MFN acquired AboveNet in 1999, PAIX came along with it. Vixie was the CTO of the combined company from 1999 to 2001.

6 The original name of Broadwing was IXC Communications. The Broadwing name was adopted when the company was sold in 1999.

Schrader was known for speaking his mind. Sidgmore joked, "Bill's report card in sixth grade said 'Doesn't play well with others.'"

In 1998 Schrader rejected an offer to buy at least a 51 percent stake in PSINet for $10 a share in cash, which was a 19 percent premium to its stock price.

Sidgmore also provided Schrader the opportunity to cash out, saying, "I've tried to buy PSINet at least 10 times over the last few years. Right after MFS bought us we tried . . . a bunch of times, and [Bill] would never sell. . . . I'm a big supporter of Bill, as wacky as he is. Whatever else you say, good or bad, he is steadfastly committed to his vision for the company."

Schrader was a gambler. He wasn't ready to walk away from his dream. When we pick up the PSINet story later in "The Bandwidth Boom Gold Rush," we'll see whether Schrader made the right bet by holding rather than selling PSINet.

BBN Continued—George Conrades

When we left the BBN story, the Internet pioneer was on the outside looking in as UUNET and PSINet launched the commercialization of the Internet. "We were a company whose time had come, and yet we didn't have a presence," observed George H. Conrades, the thirty-one-year veteran of IBM who was named BBN CEO in 1994.

Conrades's vision was to reposition BBN from a hardware company to a leading provider of Internet network services. To fund the transition, BBN sold its router business to Cisco for $120 million, deploying the proceeds to acquire Internet service providers (ISPs).

BBN spent $25 million to buy three of the original nonprofit NSF-NET networks.[7] In 1995 BBN rebranded itself as BBN Planet to signify its intent to be a global Internet provider. Conrades announced a $55

---

7 BBN purchased New England Academic and Research Network (NEARNET) from MIT, Harvard, and Boston University in 1993; the Bay Area Regional Research Network (BARRNET) from Stanford University in 1994; and SURAnet, which was bought from the Southeastern Universities Research Association in 1995.

million contract to provide Internet Backbone services to AOL, which grew over several years to an annual revenue of $400 million.

Next, Conrades formed a strategic relationship with AT&T, making BBN Planet the exclusive provider of Internet access and AT&T a minority owner of BBN Planet for $8 million. BBN's stock doubled in the month following the AT&T announcement.

In its 1995 article titled "Innovator Is Leaving the Shadows for the Limelight," the *New York Times* took note of the BBN transition and the enthusiastic support of its investors: "Until recently other companies took [BBN's] inventions and turned them into multimillion-dollar businesses. With its stock now trading at a record high, Conrades's instincts apparently proved correct."

BBN Planet grew rapidly in the 1990s, achieving a revenue run rate that approached $200 million in 1997, more than double the prior year. To keep pace with the revenue growth, substantial investment was needed for bandwidth and equipment. Like Sidgmore, Conrades concluded BBN needed a fiber network partner. "It's obvious that we don't have global scale or reach or the necessary financial resources to compete going forward," he said.

Given the AT&T strategic relationship, Wall Street assumed AT&T would be the acquirer of BBN. Much to the Street's surprise, however, a different suitor approached BBN. In the spring of 1997, BBN announced it was selling to the regional phone company GTE, led by CEO Charles (Chuck) R. Lee, for $690 million. "Together we are an almost perfect strategic fit—our technology and their resources," explained Conrades.[8]

---

8 Conrades's biggest accomplishment of his fifty-year career occurred after BBN. In 1999 he was named CEO of the start-up Akamai Technologies, which became the global leader in content distribution networks. Though I never had the opportunity to meet Conrades, Akamai was among Zayo's largest customers, and Conrades and I were both recipients of the University of Chicago's Distinguished Alumni award, his in 2014 and mine in 2016.

"Almost perfect." The reference reminds me of an exchange between Jack Singer, played by Nicolas Cage, and Betsy, played by Sarah Jessica Parker, in *Honeymoon in Vegas*.

> JACK: Do you know what a straight flush is? It's like . . . unbeatable.
> BETSY: Like unbeatable is not unbeatable.
> JACK: Hey. I know that now. Okay.

Conrades's almost perfect strategic fit proved to be far from perfect, as we will learn when we cover Lee leading GTE's swoop into the Bandwidth fray during the Gold Rush and the flaming downfall of GTE's spinout named Genuity during the meltdown.

CHAPTER III

THE RENEGADES THAT LAID THE FIBER

What the wise do in the beginning . . .

**—Warren Buffett, on savvy investors who jump in early
on transformational opportunities**

UUNET's Sidgmore and BBN's Conrades sold because the capital requirements to fund their Internet Backbones were daunting. Schrader bartered 20 percent of PSINet to secure bandwidth on Broadwing's network. Yet constructing the fiber networks to support these Internet Backbones was even more capital intensive. Venture Capital, which was a tiny industry in the early 1990s compared with the 2020s, was in no position to fund the fiber build-out.

The source of the early funding came typically from a high-net-worth individual or family. These tycoons backed ambitious entrepreneurs to build out their fiber networks. Within a few years, the Internet Backbones were merged into larger fiber platforms, and when combined, a new industry—Bandwidth Infrastructure—would emerge.

The Cable TV titans, including the leading Cable TV Cowboy John Malone, partnered with Bob Annunziata to scale Teleport, which AT&T acquired to anchor its local fiber strategy.

Walter Scott, the CEO of the largest private construction firm Kiewit Corporation, backed Jim Crowe to launch MFS Communications. WorldCom's acquisition of MFS was its entry into local fiber networks.

Philip Anschutz used his billions and his land alongside his railroads to give birth to SP Telecom, which became Qwest, and enticed Joe Nacchio from AT&T to accelerate Qwest's growth. Qwest would be one of the two fiber titans acquired by Lumen.

Ralph Swett of Broadwing did not have deep pockets or a sugar daddy partner. After some bootstrapping, he turned to the public markets for funding but then hit a wall. Ownership dramatically changed a couple of times. Eventually, Broadwing also found its way to Lumen.

The Williams Companies of Tulsa, Oklahoma, led by division president Roy Wilkens, blew fiber through empty gas pipelines. Their company, WilTel, now sits inside Verizon.

Eccentric billionaire Craig McCaw, whose backstory has much in common with Anschutz, followed the lead of Teleport and MFS by creating XO Communications. It, too, now belongs to Verizon.

John Kluge, the megarich media mogul, sponsored the construction pioneer Steve Garofalo to fund AboveNet Communications.[1] which today is part of Zayo.

Teleport—Bob Annunziata

Ironically, Teleport, which would be acquired by AT&T, was built by a seventeen-year veteran of AT&T, Robert Annunziata, who resigned from Ma Bell after his lack of a college degree caused his career to stall.[2]

---

1 When Steve Garofalo started AboveNet, the company's name was National Fiber Networks. The name was changed a few years later to Metromedia Fiber Network (MFN), reflecting the Metromedia brand that John Kluge used to brand many of the companies he controlled. The AboveNet name was adopted when the company emerged from bankruptcy, using the name of a data center company that was acquired by MFN.

2 "Ma Bell" is a slang reference to AT&T as the mother of the Baby Bells.

The *New York Times* headline on October 17, 1982, read, "Staten Island Teleport Is Seen as a Link to the Future." Merrill Lynch and Western Union announced a partnership to build a satellite farm—a facility that housed a collection of satellite receivers—to send and receive information via satellites.[3]

A fiber network would be constructed to connect New York City and New Jersey to the satellite farm. "New Jersey will be linked [to the Teleport] by a fiber-optic cable system that will begin at the World Trade Center, cross the Hudson River to Journal Square in Jersey City by way of the PATH tunnel and continue south through Bayonne to Staten Island," the *New York Times* explained. It added, "The Teleport will relay information in the form of telephone conversations, data transmitted from one computer to another, television and so-called 'facsimile' transmission, in which a printed document is fed into a machine and then reproduced in a receiving machine." We've come a long way since 1982.

A New Jersey commissioner stated, "The amount of data that will be flying around in space is just unbelievable." A New York City representative added, "It's like an airport. Every major city needs an airport. Well, this is an airport for communications."

Merrill Lynch and Western Union invested $18 million to turn the project into a start-up business named Teleport Communications. Soon thereafter, Western Union then lost interest in the project and sold its 40 percent Teleport ownership stake to Merrill Lynch for $4.7 million.

Merrill Lynch's next step was to find experienced leadership, which led them to name Annunziata CEO in 1984. Under Annunziata's supervision, network construction was completed in 1985.

Annunziata soon realized that the original intention of Teleport—to connect to the Staten Island satellite farm—was insufficient to support the cost of building the network. Teleport turned its focus to linking its customers to their long-distance providers in competition with New York

---

3 The Port Authority of New York and New Jersey was also involved in the partnership.

Telephone, a subsidiary of Baby Bell NYNEX. Customers benefited from having a diverse second connection, which improved the resiliency of their corporate communications networks. Moreover, Teleport touted better service and more flexibility.

By 1990 Teleport added networks in Boston and San Francisco, and construction was underway in Chicago, Houston, and Los Angeles. In some of these expansions, Teleport partnered with local Cable TV companies. In exchange for use of its fiber, the Cable TV provider would receive a revenue share. These partnerships allowed Teleport to serve customers who otherwise would be outside the reach of Teleport's fiber networks.

To support these capital-intensive buildouts, Merrill Lynch sought a financial partner. When Merrill opened its door, nearly the entire Cable TV industry marched through it. Atlanta Cable TV giant Cox Communications and its CEO, Jim Kennedy, led the charge.[4] In 1991 Cox acquired a 12.5 percent ownership stake in Teleport, prompting an expansion to forty additional cities. In early 1992 Merrill Lynch sold the rest of its Teleport stake. Cox increased its ownership in Teleport to 50.1 percent, and its six board seats put Cox in control.

The other 49.9 percent was purchased by Cable TV provider TCI, which gained the right to name the remaining five board seats. TCI's CEO was John Malone, the granddaddy of Cable TV. Malone joined Colorado-based Cable TV provider TCI in 1973, and over the next two decades, he spearheaded the consolidation of the industry. Malone was credited with developing innovative programming, including many new specialized channels, pay-per-view, and interactive services. By the early

---

4 The origin story of Cox dates back to 1898, when James M. Cox purchased the *Dayton Daily News*. Cox later became the governor of Ohio and, in 1920, won the Democratic Party nomination for president of the United States. He lost despite having Franklin D. Roosevelt as his running mate.

His son, James M. Cox Jr., followed in his father's footsteps by also acquiring media properties, including, in the early 1960s, Cable TV providers. In 1988 Jim Kennedy, the grandson of the founder, was named CEO. Kennedy led the investment in Teleport. By 2020 the Cox family was worth $34.5 billion.

1990s, TCI was the biggest Cable TV provider, and Malone was the industry's gun-slinging leader.

With Malone as the industry's champion, Cable TV companies were friendly rivals. Since their cable franchises served different geographies, they didn't compete for subscribers. As such, they sought ways to collaborate on content (such as HBO and QVC), services (pay-per-view and broadband), and technology.

This collaborative industry culture led two more Cable TV leaders—Comcast Corporation and Continental Cablevision—to each purchase a 20 percent stake in Teleport. Comcast Corporation, which would eventually become the largest Cable TV provider, was led by Brian L. Roberts, the son of founder Ralph J. Roberts. The other, Amos Hostetter Jr.'s Continental Cablevision, was the nation's third-largest Cable TV provider.

Teleport promised to solve a major challenge faced by the Cable TV industry. Each Cable TV operator served a mosaic of geographies. In New York one provider might serve lower Manhattan, another northern New Jersey, a third Long Island, and a fourth Brooklyn. A top five Cable TV operator might have a dominant presence in some cities, such as Boston and Philadelphia, but no operation in other markets, such as Connecticut and New Jersey.

To compete with the telephone companies for voice and Internet services, the Cable TV industry needed a more ubiquitous platform, which led Cox, TCI, Comcast, and Continental to become the owners of Teleport.

The Teleport platform opened two opportunities for the Cable TV industry. One was to use Teleport's fiber networks to reach enterprise customers. The second was to leverage Teleport's voice and Internet access to offer a bundled package of Cable TV, voice, and Internet services to both their consumer and enterprise customers.

In 1996 the Cable TV Cowboys—Malone, Roberts, Kennedy, and Hostetter—turned to the public markets to fund their prized Teleport, raising $400 million in the IPO that valued Teleport's equity at $2.8 billion. The 15 percent of Teleport that was sold to the public came with only 3 percent of the voting rights, leaving the Cable TV owners in firm control.

Teleport used the IPO proceeds to fund a nationwide expansion, including multiple acquisitions of smaller fiber providers. By 1998 Teleport served sixty-five US markets.

On paper the ownership of Teleport by the largest Cable TV operators made strategic sense. Teleport's network reach was extended by aligning with the properties operated by its Cable TV owners. Teleport-enabled voice and Internet services were offered to Cable TV customers. Cable TV services were added to the services offered by Teleport to its enterprise customers.

In practice, execution was enormously difficult. Each cable property operated with fierce independence. Local general managers (GMs) controlled investments, product offerings, and strategic alliances. TCI, Cablevision, Cox, and Comcast thrived by holding their GMs accountable for their financial performance. Cash flow mattered, so capital spending was tightly constrained by the GM. Implementing top-down initiatives was hard, as buy-in was required from each GM. Doing so with consistency across geographies was impossible.

Malone and crew turned what could have been a disastrous and costly mistake into a great financial outcome. On January 8, 1998, Teleport announced its sale to AT&T for $11.3 billion.

Strategically, it was logical for AT&T to acquire Teleport. However, AT&T fell into a honeypot from which it would fail to escape.

MFS (Chicago Fiber Optics)—Jim Crowe and Royce Holland

Teleport's archrival was my second employer, MFS. Just as Teleport is now embedded into AT&T, MFS is now buried in Verizon. Crowe, the founder of MFS, whom we first met through his landmark Internet presentation to my colleagues and me at MFS, will be a main character throughout *Bandwidth*, as will his financial sponsor, Kiewit Corporation of Omaha, led by CEO Walter Scott.[5] Warren Buffett was a childhood and lifelong friend of Scott, a connection that would prove critical when the industry was in shambles.[6]

In 1986 the start-up Chicago Fiber Optics acquired the rights to deploy fiber cable in the freight tunnels that ran underneath the financial district of downtown Chicago. The fiber was used to connect financial trading firms directly to their long-distance carriers, bypassing Illinois Bell, and thereby reducing the cost and accelerating the speed of stock transactions. Chicago Mercantile Exchange, the Chicago Board of Trade, industrialist Amoco Corporation, and retailer Marshall Field's were among the first customers.

The Chicago start-up hired Kiewit to construct its fiber network. Crowe, a division president at Kiewit, was assigned to lead the project.[7] When the start-up ran out of money in 1987, Crowe convinced Kiewit to buy a controlling investment.

Crowe was tall and had pronounced facial features that made him stand out with presence in a room full of suits. Crowe spoke not as a corporate leader drone but as a compelling intellectual, connecting the worlds of engineering and business and captivating his audience. Crowe enjoyed framing and sharing his futurist visions. He carved out a reputation for making bold strategic investment decisions well in advance of

---

5 Kiewit Corporation was founded in 1924 as Peter Kiewit Sons' Company by Peter and Andrew Kiewit. It was led by Peter Kiewit Jr. until 1979 and became the largest private construction company in the nation and iconic to its hometown of Omaha, Nebraska.

   When Kiewit Jr. passed away in 1979, and his successor died soon thereafter, Walter Scott Jr. was named CEO. Scott was consistently cited as one of the wealthiest Americans. When Scott died in 2021, his net worth was estimated at $4.2 billion.

6 Walter Scott sat on the board of Warren Buffett's Berkshire Hathaway, and Buffett's office was located in Kiewit Plaza in Omaha.

7 Before Kiewit, Jim Crowe was an executive at the construction and engineering firm Morrison–Knudsen Corporation, where he gained experience overseeing the construction of portions of Sprint's intercity fiber network.

other business leaders. Crowe would describe the future of the Internet, fiber, and regulatory disruption in the same tone and manner as President Obama articulated his *Audacity of Hope*. As Barack Obama became "Obama," Jim Crowe in his world became "Crowe," and he could hold and mesmerize his audiences with his compelling and hope-inspiring messages.

However, as an operator, Crowe's record was mixed. He aced corporate strategy, fundraising, and M&A, but he wasn't an operator. He did not like details and was annoyed by facts that contradicted his far-reaching conclusions. And perhaps he was better at talking than he was at listening.

Crowe needed a partner who would complement his strengths and cover for his weaknesses. For this, Crowe turned to one of his former executives, Royce J. Holland Jr. In the first several years of MFS, Holland became Crowe's perfect wingman.

Regulatory strategy was a primary focus for Holland. The Bell System had a one-hundred-year monopoly, owned a cash flow machine, and was a regulatory powerhouse. Sure, the Baby Bells were open to competition, so long as the Bells were permitted to write the rules. Armed with endless resources to spend on lobbyists, the Bells were arrogant, hidebound competitors certain of controlling outcomes and their destinies. The Bells would let what they viewed as lesser competition nibble around the edges while they gained the pricing and regulatory freedom to have the times of their lives.

In comparison their upstart rivals had almost no money and practically no political or market influence. Building networks was expensive. Convincing customers to replace their trusted local phone company was difficult. The dependency on the monopoly competitor for access to conduit and telephone pole infrastructure could be massively problematic. The competitive carriers' budgets for lobbying were miniscule compared with the incumbent, established Bells.

The fiber start-ups needed a showman, and MFS's Royce Holland played it to perfection. Holland was fun-loving and humorous, a personality accentuated by his constantly messy white hair and wrinkled white collared shirts, which more often than not had brown stains at the armpits. The larger-than-life Holland was the megaphone that undressed the Baby Bells.

Whereas Crowe was introverted and intellectual, Holland was gregarious and freewheeling. He was featured in the publications of every trade rag spouting off on the Bells, repeating his greatest hits: "Their prices are too high; their service horrible. The Bell Heads refuse to innovate, and they toy with the regulators. They are fat, dumb and happy monopolists living off of ratepayers' utility bills, while they play country club golf near their big homes in the nicest suburbs."

Holland shamed the regulators into pushing forward with the changes that enabled competition. He made MFS appear as if it was a gigantic company with the resources to take on all the Baby Bells. It worked.

Holland caught my attention. Not only was he bold and charismatic, but he also sounded right. Holland inspired me to believe it was my moral, ethical, and spiritual obligation to take down the evil monopoly, notwithstanding the fact that they launched my career and paid for my expensive MBA.

Working under Holland, I learned the power of being a vocal industry leader, of establishing a brand with little to no marketing budget and thereby setting the course of an industry. This would come into play when, fifteen years later, I launched Zayo. Holland remains one of my iconic heroes.[8]

With Crowe and Holland at the helm, MFS was ready to grow its fiber network beyond the means of construction. On the East Coast, in Washington, DC, a 1987 start-up was also constructing a local fiber network to bypass the Baby Bells. When the start-up ran into financial

---

8 Following MFS, Holland remained an important telecom executive for years to follow. He founded Allegiance Telecom, eventually selling it to Carl Icahn. Holland became CEO of McLeodUSA after its second bankruptcy, and was later chairman of data service provider Masergy. He passed away at the age of seventy-five in March 2024, when I was finishing the first draft of this book.

challenges, ownership passed onto its creditors, who sold it to MFS in 1991. Their CEO, Ron Beaumont, joined MFS as president of the business unit that provided dedicated connections to its customers. Beaumont would remain important throughout MFS and during the WorldCom saga, where he was briefly my boss.

By 1993 MFS had fiber networks in fourteen major markets, including New York City, Los Angeles, San Francisco, and Dallas. These NFL markets were seen as a starting point for a far more grandiose expansion.

With MFS established as the leading competitive fiber company, Crowe and Scott decided it was time to monetize MFS's investment through a public offering. The first sentence in the business overview of the 1993 IPO prospectus described MFS as "the largest provider of local competitive access telecommunications services in the United States, based on an analysis of total fiber miles, buildings served, customer locations and total number of employees."

MFS went public in May 1993, achieving a Market Value of $1.4 billion. The public markets loved MFS, and its valuation rapidly soared to multiple billions.

With the success of the IPO, MFS continued to develop its internal talent. Kevin O'Hara was MFS's rising young executive. He began his career at Kiewit and moved to MFS. O'Hara was a hockey-playing fighter from Philadelphia with a chip on his shoulder. He chose construction management as a career and Kiewit as his first employer. A particularly loyal guy, O'Hara worked for Kiewit and its portfolio companies, MFS and its successor Level 3, until he almost turned fifty. By then O'Hara had become a son to Crowe and a feisty elder brother to me.

In 1991 O'Hara was named president of MFS Development, the group that led expansions into new markets. During the transition O'Hara's predecessor handed him my résumé, along with the comment,

"The last thing MFS needs is a Bell head."[9] O'Hara ignored the advice and hired me to lead the metro network expansions. If not for O'Hara, I probably would still be living in the south suburbs of Chicago.

O'Hara was my boss on and off again for the next fifteen years. He fired me at least once, sort of. I'm sure he considered firing me multiple times. I would have if I were him—I wasn't the easiest guy to manage.

The group at MFS Development was raw, with blue-collar talent, absurd naivety, and unfettered enthusiasm. Our energy came from Holland. We were told to open up new markets. Through brute force and youthful ignorance, we overcame the obstacles the Baby Bells and local regulators placed in our way. I was reminded of my days at Ameritech Development when I toiled as the square peg in a round hole. Then I was Gulliver, abandoned on the shores of Brobdingnag. Not so at MFS Development. I was the opposite—a round peg in a square hole, a Gulliver in the land of Lilliput. In an ironic twist, I was viewed as the sophisticated and educated outlier in a blue-collar, construction-driven group of misfits. They were entrepreneurial. They got shit done. It was such fun!

I was proud of the reference in MFS's 1994 annual report, which highlighted that MFS had expanded from fourteen networks at the end of 1993 to forty-two US markets. Why? I drove most of this expansion.

I remember with fondness the group that was there when I arrived. One team member claimed to have a side job as a male stripper. Once when he joined me on a business trip, he noticed our plane was light on passengers. As we boarded my colleague told me to follow his lead. He plopped down in first class and motioned me to sit next to him. He whispered, "If we act like we belong, they will let us sit here." The flight attendant came by, papers in hand,

---

9 Rick Kozak was the executive who was replaced by Crowe. Kozak later partnered with Tony Pompliano, the founder of Chicago Fiber Optics, to create MFS's rival, American Communications Services Inc. ("ACSI"), which later was renamed e.spire Communications.

But e.spire ran into problems and went bankrupt in 2001. The company's assets changed hands a few times before eventually ending up within Lumen. The subsection on tw telecom, in Chapter VI, provides additional detail on e.spire.

asking for our names and seat assignments. My coworker played dumb and turned up his charm. It didn't work. After a few minutes, we got up and went to our coach seats. Such was MFS Development's early culture.

While I was at MFS Development, we expanded the team—most of whom were hired by me. The majority of my MFS Development colleagues became part of Level 3 and Zayo. Others joined our rivals.

As MFS grew, expanding our product offering to include voice services became a priority. In the early years, Baby Bell competitors offered direct connections, called private lines, between a corporate customer's private network and the network of their long-distance provider.

Private lines were a great initial product to attack the local phone monopoly. The technical requirements were straightforward. The only necessary coordination with the Baby Bell competitors was shared infrastructure, such as conduits and poles.

Unfortunately, private lines were a tiny portion of the Baby Bells' market. The big bucks were in voice-switched services. How could competitors penetrate the gigantic market for carrying local phone calls? MFS had a regulatory team that was grappling with this question.

In 1994 I received a shoulder tap from O'Hara that led to a make-or-break assignment from Holland. "We need you to be the business leader to spearhead the initiative opening up local competition for voice traffic." The combination of my Bell company experience, business development background, and financial acumen made me uniquely qualified. However, my tenacious reputation for execution to drive outcomes was perhaps the most important factor.

The technical name of the pursuit was co-carrier, meaning that competitors and Baby Bells would be equal providers of voice traffic. Today, we take interoperability for granted. Whether dialing from an office phone, iPhone, or computer; whether using T-Mobile, Comcast, or Verizon; and whether calling someone in your building, your home city, or the other side of the world, you dial the number, and the call goes through.

In 1994 no one knew how this would work in a post monopoly environment. If a customer wanted to switch to a new carrier, could they keep their phone number? How does the company that completes the call get paid? Who sets the per-minute charge?

Does the originator of the call cover all the costs? If so, how will they know what they will be charged? What if the call traverses through a network that is owned by a third party? Does this provider also get paid? How much? By whom? How and where would the competitors interconnect their network to the phone companies, and who would pay for the costs of these interconnections?

How do 911 calls get completed in this environment? What about 411 directory assistance calls, especially considering that these platforms were owned by the Baby Bells? If a customer switches to a Baby Bell competitor, will their number get delisted from the Yellow Pages, which also were owned by the Baby Bells? If so, why would any business shift providers and lose their primary advertising platform?

The United States was the first to go through this quagmire. No precedent or guidelines existed. The path was littered with business, regulatory, and technical hurdles.

Luckily for me, I didn't understand what I was signing up for when I said yes. We pursued co-carrier with a vengeance.

I was able to bring along a few members of my prior team alongside several technical employees who were already engaged in the regulatory effort. With the heavy involvement of MFS's regulatory team, my team negotiated the nation's first-ever co-carrier agreement with Ameritech as the counterparty. Numerous agreements with the other incumbents followed. We implemented all the interconnections, ancillary services, and financial arrangements to be a local telephone company.

The Telecommunications Act of 1996, which had a stated intention to "to let any communications business compete in any market against any other," was signed into law by President Clinton on February 26. In

May, with the heavy involvement of MFS's regulatory team, my team negotiated a comprehensive agreement with Ameritech and MFS became perhaps the first competitive local exchange carrier in the United States.[10] The *Wall Street Journal* summarized as follows:

> MFS will be able to lease Ameritech's copper phone lines to reach prospective customers without also having to lease other facilities owned by the Bell, including expensive switching centers. Ameritech customers who choose to use MFS's service will be able to retain their phone numbers. The two carriers also agreed on rate parity: Ameritech and MFS will pay each other the same amount for local calls made on one company's network but ending in the other's.

Numerous agreements with the other incumbents followed. We implemented all the interconnections, ancillary services, and financial arrangements to be a local telephone company.

Despite our success in pioneering competitive local voice services, the experience taught me that providing voice services was enormously complicated, with the need for lots of disparate systems and a vulnerable dependency on regulation. Customers, reluctant to leave their trusted Bell company given their reliance on voice, required steep price discounts. The voice market was a zero-sum game, as revenue was relatively flat. Add this all up and it was nearly impossible to make money as a provider of competitive voice services.

A clear lesson emerged: stay away from voice.

Thankfully, we were able to apply our success in building out the competitive voice networks to the termination of Internet traffic. UUNET and other ISPs relied on MFS's co-carrier platform and fiber network to

---

10 Andy Lipman was the regulatory pioneer who spearheaded this effort both as senior vice president of regulatory for MFS and in his overlapping role as a partner in the Washington, DC, telecommunications law firm, Swidler & Berlin. Lipman and his right-hand person, Alex Harris, were instrumental in the co-carrier effort.

terminate Internet modem traffic and keep pace with the rapid growth of AOL, MSN, and others. MFS's role, enabled by the co-carrier initiative as key supplier to UUNET, led to Sidgmore accepting Crowe's $2 billion bid to buy his company.

The MFS-UUNET marriage was a match made in heaven. To understand why, it is essential to know that in the early days of the Internet, consumers accessed it through dial-up phone connections. At busy hours in the evening, customers would try over and over again to log in. Once in, they would never log out, which put a huge strain on the phone system.

Co-carrier regulation at the time resulted in Bell companies paying their rivals by the minute to terminate the calls. As a result, modem connections were draining the Bell companies of their cash flow while enriching MFS.

UUNET had an initiative named Project DAN, for dial-up access network, that was supercharging the augmentation network capacity.[11] Responsibility for executing Project DAN was assigned to my team, leading most to believe the project was named after me. The undertaking involved network planning, network implementation, and endless pestering of the Bell companies to increase capacity between our network and theirs. It was invigorating as we turned up capacity throughout the United States and glorifying to watch as the revenue poured in. John Sidgmore, though never expressing satisfaction, appreciated our progress.

Despite the excitement of the UUNET acquisition, the ramping up of revenue, the soaring stock price, and the global praise for MFS, something was amiss. The cause of the ruckus was related to MFS's transition from a simple to a complex organization.

---

11 The Internet was initially accessed through modems attached to phone lines. The expression "dial up" referred to launching a modem connection through a phone line. Dial-up access network was the collection of equipment and data links that supported these modem connections to the Internet.

For the first several years of MFS's existence, its approach was simple, as its only product was private lines.

The organization structure at MFS was centered on business units: MFS Telecom, MFS Intelenet,[12] and MFS Datanet.[13] They each had their own sales team, product group, and network organization.

As MFS grew it became clear that having three standalone business units was nonsensical. Customers bought transport, voice, and data services. The sales team needed to be aligned with the customers. Moreover, the higher-layer business units—Intelenet and Datanet—used the network resources of MFS Telecom, yet no mechanism existed for the use of these valuable and expensive assets.

Should the network remain separated by product layer, as is currently the case? If so, should the higher layers purchase services from the lower layers? Or should the network be viewed as a factory that is available for use by all the products? The latter would match up well with the traditional Bell System.

If the network was treated as a factory, the majority of the organization would be reassigned to this cost center. The roles of the presidents of

---

12 MFS created MFS Intelenet in 1992. Kirby "Buddy" G. Pickle, a sales executive with Sprint, MCI, and AT&T, was named president. Two years later MFS Intelenet acquired two voice and voice equipment resellers to become one of MFS's largest units.

After MFS was acquired by WorldCom, Pickle became COO of Teligent, a company similar to MFS Intelenet. He left just before its collapse.

Next, Pickle became CEO of Velocita, the company funded by Koch Industries under the name PF.Net. Interestingly, Bob Annunziata was his chairman. Velocita went bankrupt in June 2002. Velocita's primary customer was AT&T, which bought the assets out of bankruptcy for pennies on the dollar. To learn more about Velocita, see appendix 3, "More Bandwidth Stories."

13 When MFS acquired Network Communications in 1989, its co-founder, Scott Yeager, was named city manager of MFS Houston. Yeager developed a new type of data service using customer premises equipment that ran on the Ethernet protocol. He shared his ideas with Rick Adams of UUNET, which sparked the creation of MAE-East, the important interconnect location in the early commercialization of the Internet. Sparked by Yeager, MFS created a new business unit named MFS Datanet to focus on data connectivity.

Yaeger left MFS and became a leader at Enron Broadband, which played a major role in the collapse of Enron. Yaeger struck a deal with the SEC to avoid prosecution.

the three business units would be unclear, as most of what would remain were the sales, marketing, and customer service organizations.

Crowe asked Holland to come up with a recommendation. Holland, after consulting with his presidents, recommended only modest changes to the current model. They wanted to preserve the business unit structure while allowing the sales teams to cross-sell products among the business units.

Crowe overrode this recommendation. He decided the network was a factory, and it would be most efficient to collect all network assets, as well as the engineering and operations personnel, in a newly created Global Network Services organization. With that, 1993 marked the year of MFS's big reorganization.

Crowe put O'Hara in charge, despite O'Hara having no experience in running a large network organization. The three business unit leaders retained their president titles, but their roles were dramatically narrowed.

In early 1995, with this organization in place for less than two years, I received an out-of-the-box request to travel to Omaha for my first one-on-one interaction with Crowe, outside of larger group meetings. Though the topic was unclear, the request was urgent. When I sat with Crowe, he explained the company had an issue, and he couldn't get his arms around it.

"Revenue is growing, yet the profitability is shrinking," he fretted. "The business unit executives say they are hitting their revenue numbers, but the network company doesn't know how to run the network efficiently."

Crowe continued, "The network executives argue it is the other way around. The business unit executives are making their numbers by selling products that are unprofitable—either because of low prices or misalignment with MFS's network assets. The network organization blames the shrinking margins on misguided sales execution."

Crowe told me he couldn't figure out which perspective was correct. "The only thing both sides agree on is you. They think you have the analytical skill, intellectual honesty, and tenacity to get to the bottom of this." He told me to do whatever was necessary to figure it out. As a public company CEO, Crowe was under a lot of pressure.

Getting to the bottom wasn't hard. Doing it without getting fired was the tricky part.

The business units weren't really business units managing profits and losses. They were revenue units. They were accountable for beating their revenue targets. Most of their costs came from Global Network Services, the network factory outside their control.

The network factory wasn't really a company; it was a cost center. Global Network Services built and operated networks, spending most of the company's capital and managing most of the company's people and expenses. It was accountable for beating its expense and capital budgets, but its workload was driven by the business units. If Global Network Services cut costs to make a budget, it would directly affect the revenue growth and customer experience of the business units.

Everyone knew this, and that was why the riddle was easy to decipher. However, no one acknowledged the source of the problem. To do so, the business unit leaders would have to admit they were driving their revenue without knowing the underlying costs. The network company would have to admit it didn't know whether the investments it was making were creating value.

That was why not getting fired was the trick I needed to master. How do you tell the CEO and CFO that the organizational setup of the company was at the core of the problem? How do you tell the business unit leaders that they have no clue whether they were selling the right products to the right customers in the right locations at the right prices? How do you tell the network executives that they have no way of knowing if they are running the network appropriately?

Well, I told them my opinion. I didn't get fired. Instead, Crowe and Scott decided to take their chips off the table by selling the company to WorldCom in 1996 for the staggering price of $14.3 billion. WorldCom's CEO, Bernard J. "Bernie" Ebbers, who was rapidly emerging as a darling of Wall Street, bailed us out.

I knew that Crowe was committed to building MFS for the very long term. The decision to sell to WorldCom was only because he knew he was about to disappoint his investors. Losing financial control of the company was the reason. The impression this made would weigh on me throughout my career, eventually leading me into redefining how a large telecom company should be organized and managed. I attribute the future success I had as CEO to the financial discipline I would instill in an organization, which originated with my unique insight into why MFS waved the white surrender flag.

Qwest (SP Telecom)—Philip Anschutz and Joe Nacchio

While Teleport and MFS were building local fiber networks, Qwest, bankrolled by Philip Anschutz and led by hired gun and CEO Joe Nacchio, took a different approach. They focused on building out intercity fiber networks. With a unique business plan, they got off to a fast start. Qwest attracted many copycats, which, along with undisciplined ambition and industry-shattering missteps, led to a historic collapse. Eventually, the Qwest assets would become part of Lumen.

By the late 1990s, Philip Anschutz had established himself as a Wild West entrepreneurial legend. Already a billionaire and listed as the wealthiest person in Colorado, he turned his attention to fiber networks. His business journey frames how he gave birth to Qwest.

Anschutz's plans to attend the University of Virginia School of Law were aborted when his father Fred's illness left the Anschutz family in financial straits. Anschutz the senior was a dealmaker who thrived and dived on risk-taking in the oil market. Fred would make a small fortune when the oil market soared but lose it just as fast as the market plummeted. At times the Anschutz family would struggle to pay their bills. Anschutz recalled, "You never knew what was going to happen next."

Anschutz characterized the first five years of taking over the family's troubled business as the hardest of his life. "As a wildcatter, 95 percent of everything you do is a failure. Most holes are dry."

At the age of twenty-seven, Anschutz's flirtation with bankruptcy became an early chapter of his legendary story. An exploratory oil well blew up at a site that Anschutz was drilling near Gillette, Wyoming. Anschutz promptly chartered a plane, rented a pickup from a neighboring farmer, and visited the site. The land was ankle-deep in crude oil, and natural gas was thick in the air. Armed with the inside knowledge of the discovery of a big oil field, Anschutz embarked on a buying spree, acquiring oil leases for surrounding land on thirty days of credit.

When Anschutz returned to Denver, the TV news was covering a major oil field fire in Gillette, Wyoming. The land associated with his oil rights was in flames, putting Anschutz at risk of failing to meet his obligations related to his newly acquired leases.

Anschutz tracked down the famed oil firefighter Red Adair. Lacking the money to pay Adair in advance, Anschutz's first attempt to persuade him didn't go well. "Kid, I checked you out, and you don't check out," Adair responded.

Despite Adair's initial hesitation, Anschutz persisted and eventually succeeded, though Adair issued a warning: "If you don't pay me, don't ever have another oil field fire."

Serendipitously for Anschutz, Universal Studios was making a movie about Adair, starring John Wayne. Armed with this knowledge, in true entrepreneurial folklore, Anschutz sold Universal the rights to film the crews fighting his Wyoming fire for a fee of $100,000. This windfall provided him the runway to secure bank loans needed to close on the oil rights he had purchased on credit. Anschutz made millions.

Anschutz learned from the financially devastating close call: "It's important to have your back to the wall. It teaches you how to think outside the box."

In 1978 the next chapter of the Anschutz legend unfolded when Amoco Corporation discovered a large natural gas reservoir adjacent to an Anschutz ranch. Instead of immediately selling to Amoco, Anschutz waited. Four years later he sold half of his interest in the property to Mobil Oil for $500 million in a $1 billion transaction that also included rights to additional Anschutz properties. This move allowed Anschutz to reduce his debt ahead of the oil market crash of the 1980s.

Following his exit from the oil industry, Anschutz shifted his focus to railroads. He purchased Denver and Rio Grande Western Railroad in 1984, which bolstered his confidence to purchase the significantly larger Southern Pacific Railroad in 1988. Anschutz's contribution to the $1.8 billion purchase price was only $90 million, with the majority funded through bank loans and junk bonds. His $90 million investment was obtained by merging the Rio Grande with Southern Pacific. Soon thereafter, he sold 25 percent to Morgan Stanley.

Southern Pacific was a distressed railroad that was burning cash and in need of substantial investment to improve its operations. Anschutz capitalized on underlying land rights, monetizing the rights for $2.2 billion, which provided the necessary funds to support the operations and upgrade the facilities. He took the Southern Pacific public in 1993.

Anschutz's railroad investment thesis centered on consolidation, anticipating that rail properties would increase in value as the number of rivals decreased. The 1985 merger of Burlington Northern with Santa Fe left Union Pacific and Southern Pacific as the only other major railroad operators west of the Mississippi. As a result, Union Pacific was eager for a merger, providing Anschutz the leverage to negotiate a $5.4 billion purchase price in 1996, with Anschutz's stake worth more than $1 billion.

As it so happens, railroads sell communications companies the rights, known as rights-of-way,[14] to use the land alongside their railways.

---

14 Rights-of-way are legal agreements that allow others permission to use the owner's land for specific purposes. Typically, utilities leverage right-of-way agreements to construct pipelines, bury power or telecommunications cables, and erect cell towers.

Long-distance telephone companies such as MCI and Sprint acquired right-of-way from railroads to install their microwave towers.[15] These towers were instrumental in expanding their intercity microwave-based communications networks, enabling them to compete with AT&T. As fiber began to replace microwaves, railroad rights-of-way became even more valuable, as fiber could be laid alongside rail lines more quickly and less costly than most construction alternatives.

Southern Pacific had a small division named SP Telecom, which focused on monetizing its rights-of-way. Additionally, SP Telecom offered fiber construction services to its communications customers. SP Telecom had a long-term agreement with its railroad sibling to lay fiber alongside Southern Pacific railways.

In 1995, one year before the Southern Pacific sale to Union Pacific, Anschutz separated SP Telecom from Southern Pacific. He then acquired Qwest, a small microwave communications company in Dallas, merged it with SP Telecom, and rebranded the combined entity as Qwest. Leveraging his insights and relationships from within the railway industry, Anschutz negotiated rights-of-way agreements with other major railroads.

In its December 1996 article titled "Qwest Sees Itself Rising to Top with Its High-Speed Network," the *Wall Street Journal* unveiled Anschutz's ambitious plan: "In the mad scramble by the nation's large telecommunications companies to find more network capacity, tiny Qwest Communications seems poised to deliver. Qwest . . . is building a $2 billion high-speed network to transmit voice, video, and data."

---

15 Southern Pacific deployed a communications network that it used to provide telegraph services. In the 1970s it began to use this network to offer long-distance calling services that competed with those of AT&T. In the mid-1970s, Southern Pacific selected Sprint, an acronym for "Southern Pacific Railroad Internal Networking Telephony," as the new name for its communications company.

In 1983 Southern Pacific sold Sprint to GTE. Starting in 1986 and culminating in 1990, GTE sold its stake in Sprint to United Telecom, which eventually changed its own name to Sprint.

Southern Pacific's history with Sprint prepared Anschutz well for the fiber-optics revolution and his formation of Qwest.

Anschutz said, "There are probably those who would say this isn't the sexy part of the business. But this is the underlying platform upon which strategies and companies are built." Fiber networks were sexy to me.

Salomon Smith Barney analyst Jack Grubman proclaimed that, because of the Internet, bandwidth capacity requirements were now doubling every four months, compared with doubling every twelve months just one year earlier.[16] Grubman was emerging as the highest-profile analyst of the Bandwidth industry, leading to mischievous behavior and ultimately dire consequences.

To offset the costs of the network, Anschutz sold 25 percent of the fiber on the Qwest network in 1997—twenty-four fibers—for $500 million.[17] With this funding in hand, Anschutz sought a seasoned communications industry leader to execute his vision for Qwest. In late 1996 Grubman introduced AT&T executive Joe Nacchio to Anschutz.

Nacchio described himself as "an entrepreneur trapped inside a big, bureaucratic organization," which seemed like an odd self-reflection given that he had spent twenty-six years in the AT&T bureaucracy. Perhaps Nacchio's departure had more to do with him being passed over in the AT&T CEO succession plan.

In any case Nacchio was the executive Anschutz was seeking. Anschutz explained, "Nacchio is certainly one of the best-known guys in the business, plus he's got a lot of the discipline and background I was looking for," referring to Nacchio's engineering, marketing, and sales

---

16 Salomon Brothers was the name of the firm when Grubman joined. The name was changed to Salomon Smith Barney in 1997 as a result of a merger. Travelers Group purchased Salomon in 1997 and retained the brand as its banking division. Citigroup merged with Travelers in 1998 and dropped the Salomon Smith Barney brand in the early 2000s.

17 The fibers were sold to Frontier Corporation, the rebranded name of Rochester Telephone. In 1999 Global Crossing purchased Frontier and took possession of the fibers.

experience. Nacchio was firmly in the Anschutz companies' family. On his first day at Qwest, Nacchio found a pair of cowboy boots on his desk. Next to the boots was a note from Anschutz: "Welcome to Denver. Horses and guns to follow."

Nacchio was excited as well. "This is an industry at an historic inflection point," he declared.

Shortly after Nacchio joined, Qwest struck two deals to sell fiber on its under-construction network. The transactions infused an additional $600 million of cash on top of the $500 million from Anschutz's earlier fiber sale.[18] In 1997 Qwest acquired SuperNet, a Colorado Internet Backbone provider, for $20 million, giving Qwest a chance to dip its toe into the Internet.[19]

Nacchio's vision for Qwest was to rival AT&T as a full-fledged telephone company that offered a full stack of communications services, including voice, data storage, consulting, and website hosting. To fund this vision and reduce debt load, Qwest raised $321 million in a June 1997 public offering. The initial stock price was $22 per share, which valued Qwest's equity at $2 billion. By the end of 1997, Qwest's Market Value increased to $6 billion. Anschutz's 84 percent ownership was valued at $4.9 billion, representing a ninety-fold increase on his $55 million investment.

Nacchio was amazed at how easy it was to raise money. In the fall of 1998, he claimed to have completed a large fund raise in a ten-minute call with bankers while driving to his son's soccer game. "Before the game was over, we had subscribed for a billion bucks at 8.9 percent interest," bragged Nacchio.

Nacchio was ready to make his first move toward his big vision; he just needed the right acquisition target. In 1998 Qwest executed a $4.4

---

18 The transactions were with WorldCom and GTE.

19 SuperNet was a nonprofit ISP owned by Colorado Advanced Institute of Technology, the University of Colorado, the Colorado School of Mines, Colorado State University, and the University of Denver.

billion stock swap to purchase LCI International, turning Qwest into the fourth-largest, long-distance company behind AT&T, MCI, Sprint and WorldCom.[20] The acquisition provided Qwest customers and traffic to redirect onto its network as it was completed.

H. Brian Thompson, a former MCI executive, was named vice chairman. Reflecting on the sale, Thompson remarked, "In some respects it's a little like watching your mother-in-law drive over a cliff in your new car. You're not sure what you're going to feel about it." Perhaps Thompson was just being nostalgic. Or maybe he was peering into a crystal ball.

Either way, Qwest stock price continued to rise. In April 1999 Bell-South, the Baby Bell of the Southeast, purchased 10 percent of Qwest for $3.5 billion, pushing Qwest's Market Value to $35 billion. In an SEC filing, BellSouth reported it would eventually consider purchasing all of Qwest.

The rise of Qwest brought attention to the reclusive Anschutz. In 1999 CNN Money published "Billionaire Next Door Philip Anschutz May Be the Richest American You've Never Heard Of."

> The intriguing thing about Philip Anschutz is not that he's worth well over $10 billion. It's that he's a genuinely nice guy worth more than $10 billion. He didn't make his money by being a nasty, grasping, miserly bastard. Once he got rich, he didn't turn

---

20 LCI International, which was born as LiTel in 1983, was founded by telecommunication industry veteran Lawrence McLernon. LiTel built out a 1,500-mile fiber network in the Midwest. By 1990 LiTel, though tiny in size at $200 million in revenue, was in the top ten of US long-distance companies. However, profitability wasn't sufficient to keep pace with its debt service.

Warburg Pincus, LiTel's biggest investor, replaced McLernon with H. Brian Thompson. Under Thompson's leadership, LiTel shifted its focus from business to residential customers and honed in on more profitable international traffic. In 1992 the company was renamed LCI International. In 1999 Thompson took LCI public, raising $200 million. Throughout the 1990s LCI grew. By 1996 Thompson tripled the revenue to $600 million. Despite its early focus on fiber networks, LCI mostly relied on the networks of other carriers to deliver its service.

into a twisted, weirdo billionaire like Howard Hughes. . . . As billionaires go, Anschutz is abnormally normal.

Anschutz wasn't entirely under the radar. After all, he had been recognized as the richest person in Colorado for the previous fifteen years.

The *Los Angeles Times* piled on in its article titled "One Word for Phil Anschutz: Private." The focus of its article was not fiber but on the grand opening of LA's new sports palace, the Staples Center, which Anschutz had built.

Anschutz and Nacchio reached an inflection point. Would they follow the leads of Malone and Annunziata of Teleport, Scott and Crowe of MFS, and Conrades of BBN by taking their billions of dollars off the betting table?

Or would they gamble, like Schrader of PSINet?

In June 1999 *Forbes* published an article titled "Merger Rumors Surround Qwest," which opened with "Like a character in a French bedroom farce, Qwest is running from room to room amidst the slamming of doors, looking for love. [Qwest] has been portrayed in various reports as both the suitor . . . and as an object of desire. What is Qwest doing? Buying or being bought out?"

The company that Qwest was rumored to be courting was U S West, the Baby Bell headquartered in Qwest's hometown of Denver. However, U S West was already committed to selling itself to a different fiber provider—a new upstart named Global Crossing, now run by Annunziata, another castoff of the 1996 AT&T succession debacle, soon to be covered. The Global Crossing-U S West deal was priced at $32 billion. Qwest would have to top that bid and convince U S West to terminate its merger agreement.

The headline in the *New York Times* on June 15, 1999, was "At Qwest, a Chance to Become Really Big." Two days earlier, on Sunday,

June 13, Nacchio fired off hostile bids to acquire both U S West and Frontier, another company that had agreed to sell itself to Global Crossing. The combined value of the two bids was $55 billion.

The *New York Times* honed in on the face-off between the AT&T veterans.

> In recent years, Mr. Nacchio and Mr. Annunziata have come to be rivals in almost every sense, competing for the same investors, and many of the same customers. Now they are competing for a place at the table with the moguls of communications. . . . So, while Global Crossing is by no means out of the game for U S West and Frontier, given the sharp thumbs-down the market gave Qwest yesterday, a winning deal by Qwest would catapult it into the ranks of the world's biggest communications companies. That would be an impressive feat for Mr. Nacchio, a personable, engaging, fast-talking salesman who is widely admired for his smarts.

Clearly, Anschutz and Nacchio were ready to double down.

Broadwing (IXC)—Ralph Swett

The poor man's version of the Qwest story is the tale of Broadwing.[21] Its founder, Ralph Swett, had the ambition of Anschutz but not the billionaire bank account. His focus, like Qwest's, was on the construction of an intercity fiber system. But he didn't have the railroad land assets to lower his costs. He built out an early intercity fiber network anyway and did quite well for himself. Today, his assets are part of Lumen.

---

21 The prior names of Broadwing were IXC Communications and Communications Transmission Inc. (CTI).

Swett's story traces its roots to the Times Mirror Company, which was born in 1884 when the Mirror Printing and Binding House was combined with the *Los Angeles Times*. The media company owned *Popular Science*, *Outdoor Life*, *Golf Magazine*, and *Field & Stream*; a Cable TV business; and a microwave communications network that, like MCI and Sprint, was an early competitor to AT&T. Times Mirror decided microwave infrastructure no longer fit its business strategy and sold the division in 1985 to a group led by its Cable TV president, Swett, for $175 million.

Through acquisitions and organic buildouts, Swett expanded his Communications Transmission (CTI) microwave network to more than ninety cities. However, CTI lost Western Union, its most important customer, in 1990, causing default on several loan agreements.

Swett rebounded, using loans secured by his personal balance sheet, and restructured the business, now rebadged as IXC Communications. Throughout the first half of the 1990s, IXC remained focused on selling private connections between cities on its hybrid microwave and fiber network. As momentum grew with fiber and the Internet, IXC seized the opportunity through acquisitions, fiber build-out, and the rollout of nationwide data networks.

By 1995 IXC's revenue jumped to $203 million, enabling the company to raise $93 million in a Nasdaq public offering with its shares trading at $16. Swett described IXC's competitive position as a wholesale provider to other service providers, stating, "[IXC's] role in the data communications industry is to supply capacity to companies that resell broadband services to the commercial market. . . . Unlike other carriers, we do not compete with our customers for data communications business." Though true at the time, this changed quickly in the next two years as acquisitions pushed into providing retail services to commercial customers.

In 1997 and 1998, IXC continued its acquisitions, several of which strayed from its primary business of selling bandwidth. Swett even secured 20 percent ownership of PSINet in exchange for $240 million of capacity on the IXC network.

Yet Swett was frustrated with the company's stock performance—with IXC's valuation being 10 percent of rival Qwest's value—and concluded it was time to sell. In November 1999 Cincinnati Bell announced its purchase of IXC for $3.2 billion.[22] The combination then changed its name to Broadwing, now consisting of the Cincinnati Bell local phone operations and the former IXC intercity fiber system.

Cincinnati Bell would soon regret this decision.

WilTel—Roy Wilkens

Kiewit wasn't the only multigenerational industrial company from heartlands that saw fiber as its next big frontier. The Williams Companies of Tulsa, Oklahoma, mobilized behind Roy Wilkens, the president of its oil pipeline division. Like Anschutz, Williams planned to connect NFL cities, and they, too, had a shortcut. Whereas Qwest had railroad rights-of-way to leverage, Williams had empty oil and gas pipes as its edge. Williams's brass called their fiber sortie WilTel. Today, Verizon is home to those assets.

Wilkens, who is twenty years my senior, grew up fifteen miles from my hometown. When I met him in 1996, I was living in Orland Park, Illinois, the town immediately adjacent to his hometown of Tinley Park.

Wilkens left Illinois to attend University of Missouri-Rolla, known now as Missouri University of Science and Technology, and received an electrical engineering degree. His career path led to Tulsa, where he became an executive of Williams Companies, which was among the largest transporters of natural gas and petroleum in the country through its network of pipelines.

In the early 1980s, Wilkens, then president of Williams Pipeline Co., attended a Harvard University program for midcareer executives. Inspired, he returned to Williams with a proposed solution to its vexing problem.

---

22 $2.1 billion of the $3.2 purchase price went to IXC shareholders; the other $1.1 billion was associated with IXC's debt.

At the time, we were heavy into the energy sector, real estate, and production. All three of those sectors of business were on the down side. . . . So what Williams did is try to look for an additional industry to get into that was either non-cyclical or counter-cyclical. And we found communications had no cycle at all. . . .

. . . the telecommunications industry [was] exploding, and Williams Pipeline had a vast network of decommissioned oil and gas pipelines doing nothing.

Wilkens's idea was that Williams should use the pipeline infrastructure to build out a network of fiber cables and offer wholesale bandwidth services to long-distance carriers such as AT&T, MCI, Sprint, and their emerging rivals.

Wilkens pitched the idea to Joe Williams, the chairman and CEO of Williams, who was the son of Williams's co-founder, David Williams.[23] The business plan was approved, $50 million funding was allocated, and the business unit named WilTel Communications was created, with Wilkens named CEO.

A 1990 article in the *Oklahoman* titled "Firm Combines Fiber Optics with Pipelines" described the process WilTel used to build out the fiber network.

In 1985, Williams Cos. shot pipeline "pigs" through a sector of tubing in the Midwest, much of it with an eight-inch diameter. The plastic balls acted like bullets in a gun barrel, cleaning the pipes.

---

23 Williams was founded as Williams Brothers in 1908 by Miller and David Williams. In 1949 John H. Williams, a nephew of the founders, teamed up with his brother, Charles Williams, and David's son, David Williams Jr., to buy out his uncles. They took Williams public in 1957 and, in 1966, purchased Great Lakes Pipe Line Company, the largest petroleum pipeline network at the time. In the 1970s they rebranded to the Williams Companies. In 1982 they expanded into natural gas transport with the purchase of Northwest Energy Company.

Then the crews blew a thousand gallons of alcohol through the lines, flushing the residue.

Fiber optics followed. The glass filaments, sheathed, and bundled, slid inside pipes that were nestled beside as many as a half-dozen other tubes in the Williams Cos. rights of way.

The elapsed time to build out the fiber network was reduced from more than four years to less than two years. The pipelines reduced the capital investment by more than 80 percent. Moreover, the oil and gas pipes provided a hardened environment that provided protection to the fiber that was far superior to competitors' fiber, which often was directly buried into the ground. These factors provided WilTel enormous advantages in cost structure, speed to market, and service quality.

By 1989 WilTel had eleven thousand miles of fiber network that stretched from the Atlantic to the Pacific, covering most major US cities.[24] The business posted revenue of $300 million and an operating profit of $60 million.

In 1990 MCI announced plans to invest $1.1 billion to upgrade its network from an analog to a digital system, with WilTel selected as a major supplier. In a powerful endorsement of WilTel's carrier strategy, Bert C. Roberts Jr., MCI's COO, said, "The agreement with WilTel . . . will permit us to provide fully digital communications while giving us the flexibility to most efficiently invest our own capital."

WilTel was seen as an acquisition target by MCI and others. Wilkens enjoyed the attention: "I think we are probably one of the most

---

24 WilTel constructed most of its network. However, it also acquired companies with fiber networks. In 1987 WilTel acquired LDX NET, which owned 1,295 miles of fiber in Texas, Louisiana, Oklahoma, Missouri, and Kansas. In 1989 WilTel purchased LIGHTNET, which had 4,500 miles of fiber network, linking thirty-eight major cities in the eastern United States for $365 million.

sought-after telecommunications businesses in the country. I think we make the Big 3 a little nervous."

Surprisingly, the bidder for WilTel was not one of the Big 3—AT&T, MCI, or Sprint. It was a fast-growing reseller of long-distance services—WorldCom. "We have a very large customer base that would fit very well on their network," said WorldCom CEO Ebbers. In May 1994 Ebbers offered Williams $2 billion to buy its WilTel subsidiary. In an attempt to pressure the bid's acceptance, Ebbers pointed out that the $2 billion was 76 percent of Williams's total Market Value.

An analyst at Merrill Lynch endorsed the WorldCom bid, noting that WorldCom did not own a nationwide fiber network: "It's an excellent fit—more than most other mergers."

Williams cavalierly rejected the WorldCom offer. "They met, they talked, they considered," explained a Williams spokesperson. At the same time, the Williams board indicated that WilTel was in play, directing management to explore several other alternatives, including possible alliances with one or more partners.

In July 1994 Ebbers increased his offer to $2.5 billion, and the offer was accepted. "It was a time of enormous opportunity," reflected Wilkens. "Technology was moving extremely fast, and lots of people jumped into the industry. It was the wild, wild west for five to ten years."

This transaction hinted that the industry might soon enter a new stage. Was a Bandwidth Boom Gold Rush beginning to build momentum?

Nice job, Wilkens! You sold at the right time.

XO (Nextlink)—Craig McCaw

Craig McCaw, similar to Anschutz, overcame adversity at a young age. He, too, had extraordinary prowess as a tech entrepreneur in infrastructure businesses. Like Anschutz, his love of business kept him going even after achieving incredible financial success. Instead of oil and railroads, McCaw's pre-Bandwidth exploits were in Cable TV and mobile communications.

Whereas Anschutz's fiber focus was on connecting cities, McCaw vied with Teleport and MFS on metro fiber networks. The company he created was initially named Nextlink and later rebranded as XO Communications. The company bobbed and weaved until eventually becoming part of Verizon. Along the way, the famed financier Carl C. Icahn would jump into the XO drama.

Craig McCaw was nineteen years old when he walked into his parents' bedroom to find his father dead from a massive stroke. Elroy, his entrepreneurial old man, left behind a significant estate, including several radio and TV stations, investments in fifty-four companies, money in twenty-five bank accounts, a seven-bedroom mansion on five wooded acres in the Highlands community near Seattle, go-karts, motorcycles, vintage cars, a Learjet, and a sixty-three-foot yacht.

Elroy also left the McCaw family with a ton of debt. Bill Gates's father, William Gates Sr., served as the McCaw family estate attorney for over ten years. When all the dust settled, only $20,000 of value remained from the original estate valuation of $12 million. The only asset that survived was a small cable company in Centralia, Washington.

No doubt, McCaw gained a lifetime's worth of entrepreneurial lessons by sorting through his family's collapsed empire.

"My father was a visionary who did not hire great people. . . . I'm picky about the people I do business with. He wasn't picky. I learned to be fastidious about accounting, to be careful about the ethical nature of people. And to keep things simple, to focus on quality," McCaw reminisced.

Craig McCaw was one of four brothers. Though all became wealthy, it was only Craig who focused long term on work. A 1993 *Seattle Times* article characterized the contradictions that described McCaw:

> He's a fierce competitor with no ego. A dyslexic who sees things others miss. A visionary who communicates clumsily. A fitness nut who looks frail. A wallflower bold enough to wager billions

on risky business deals. A person who avoids people, yet has powerful insights into personal motivations. A man who is worth half a billion dollars, owns planes, and yachts, yet wears a cheap digital watch. A strategist who plots his steps but tries to be spontaneous. A man with the subdued persona of a seminarian who shoots Nerf balls at visitors to his office.

To some, a genius. To others, mainly unwitting rivals later left in his wake or outsmarted in a deal, a crazy fool.

McCaw was a student at Stanford when Elroy passed. During his senior year, McCaw became CEO of the Centralia Cable TV company. Using junk bonds, he spent the next fifteen years buying cheap, fixer-upper cable systems; raising quality and prices; and generating cash flow. By the 1980s McCaw Cablevision was the twentieth-largest cable carrier in the United States. In 1987 he sold it for $755 million, of which $250 million stayed with McCaw.

In 1979 the FCC established its cellular policy and began the process of issuing spectrum licenses. In 1980 AT&T hired McKinsey & Company to assess the opportunity. McKinsey estimated the market would grow to only nine hundred thousand subscribers by the year 2000, leading AT&T to conclude that the cell phone market was simply too small for the company. McKinsey's estimate turned out to be less than 1 percent of the actual subscriber number. This massive miscalculation blows my mind. Did AT&T get a refund from McKinsey for this multibillion-dollar mistake?

McCaw's conclusion was the opposite of McKinsey's, which led him to enter FCC's lottery for spectrum licenses. He was granted licenses in six of the top thirty markets in the United States in 1983. He used his proceeds from the sale of the Cable TV company along with debt from the junk bond king Michael Milken to gather up additional licenses from other FCC spectrum awardees.

On a 2020 podcast, Milken reflected (paraphrasing), "As hard as it is to believe today, of all the industries that I financed, the most difficult was mobile. People couldn't understand why they needed mobile. I had seen Captain Kirk tell Scotty to beam him up. Both Craig and I believed the future was mobile."

Milken and McCaw convinced the capital markets of the value of the spectrum licenses, which they used as collateral to raise $1.25 billion in junk bonds for their new venture, McCaw Cellular. Additional cash was raised by selling 22 percent of McCaw Cellular to British Telecom. His acquisition spree culminated in 1989 with a $3.5 billion acquisition of another mobile provider. By the early 1990s, McCaw Cellular was the largest mobile phone network in the United States.

Like his father, Elroy, McCaw accumulated a significant amount of debt to the tune of $5 billion. Unlike his Pop, McCaw cashed out. AT&T purchased McCaw Cellular for $12.6 billion, making McCaw a billionaire.

The value of McCaw Cellular, rebranded as AT&T Wireless, rose dramatically. Six years after the sale, AT&T took AT&T Wireless public in the then-biggest IPO in US history, resulting in a valuation of $73 billion. McCaw may have compounded his earlier mistake of holding investments too long by prematurely selling his cellular business. Perhaps he'd get another bite at the apple.

McCaw launched his family office in 1993, with a grand ambition around global communications. McCaw explained in 1996,

> The revolution is here. People need high-capacity networks to homes, to remote locations, to farms, to villages that have never had fiber and won't for a long time. It's pretty obvious that there's a need, and that demand will build.
>
> You arrive at moments in time when an entrepreneur, a technology, and the needs of people coincide. You get

serendipity every once in a while. You try to be willing to accept it when it works on your behalf.

McCaw invested $1.1 billion in another wireless venture to turn a mobile dispatch company into a unique new entrant into the mobile provider space. Nextel partnered with Motorola to provide a "push to talk" phone feature that was well suited for truckers, taxi cabs, construction workers, and other service industries. Nextel grew to serve 3.6 million customers. In 2005 Sprint purchased Nextel for $35 billion.[25]

In 1994 McCaw also launched XO, under the original name Nextlink, with a mission to mimic MFS and Teleport by building local fiber networks to compete with the Baby Bells. Initially, XO focused on second-tier cities, such as Nashville and Spokane. Soon thereafter, it targeted the Tier-One markets.

By 1997 XO was operational in twenty-three markets. The company announced plans to expand to the MFS stronghold of Chicago and the Teleport home turf of New York—perhaps in an attempt to lure World-Com and AT&T into acquisition offers. Neither took the bait. In September 1997 XO went public, raising $258 million at $17 per share. By the end of its first day of trading, the stock rose to $28. McCaw retained 80 percent of XO's super-voting Class B shares.

Salomon analyst Grubman characterized XO as "the best opportunity ever in telecom history. . . . All the ducks are aligned for success. . . . XO has superb people running it and has already shown that it can get markets up and running with paying customers."

---

25 In 1994 McCaw partnered with Bill Gates to form Teledesic, a satellite venture aimed at operating a constellation of three hundred low-Earth-orbit broadband communications satellites. He made the cover of *Fortune* magazine with the headline "Craig McCaw's Cosmic Ambition." The venture failed; Teledesic halted satellite production in 2002 and sold its spectrum licenses in 2003. However, the success of SpaceX's Starlink and perhaps Amazon's Project Kuiper validate that Teledesic's concept was the right idea, just way ahead of its time.

In stark contrast to Grubman's beaming optimism, *Forbes* expressed skepticism. In a June 1998 feature titled "Craig McCaw: The Wireless Wizard of Oz," *Forbes* grappled with the uncertainty of XO's prospects: "[XO] is no investment for the faint-hearted either. Building a competitive fiber network in the most populous sectors of the country is very expensive. In the past year XO has raised a billion dollars in bonds, and another $191 million through an initial public offering, all to be used for the buildout. And this fact hasn't escaped the market's eyes: The stock is trading . . . well below its 52-week high of $38."

Clearly, McCaw was doubling down.

AboveNet (MFN)—John Kluge and Steve Garofalo

AboveNet Communications, like MFS, Teleport, and XO, focused on metropolitan fiber. However, AboveNet's strategy was unique, as it focused squarely on fiber as a shared infrastructure, and company leaders stayed true to this focused approach for many years. When they strayed from it through untimely and poorly conceived acquisitions, they dug a deep hole that had bankruptcy as the only way out.

The AboveNet story begins with two key individuals, media mogul John W. Kluge and Steve Garofalo, the CEO of a New York City construction contractor. Let's start with Kluge.

John W. Kluge's fortune building began in the 1940s when he launched the radio station WGAY in Silver Spring, Maryland. He added radio stations in cities such as Buffalo, Dallas, Nashville, Orlando, and St. Louis. In 1959 Kluge acquired a controlling 24 percent interest in Metropolitan Broadcasting Corporation, which owned both television and radio stations, from Paramount Pictures for $4 million. Soon thereafter he took the company public and became the largest network of independent television stations in the country.

In the early 1980s, Kluge focused on the emerging mobile phone industry. Instead of buying spectrum licenses, Metromedia spent $300 million to acquire paging companies, which then applied for cellular spectrum licenses.

When the reaction from Metromedia's public shareholders was unsettled, Kluge took Metromedia private at a price of $1.6 billion, funding the buyout with $1.3 billion in junk bonds. However, instead of building up the business, he decided to sell. The *Washington Post*'s 1986 article titled "Parts Worth More Than a Whole Metromedia" detailed how the exits generated over $4.5 billion in cash.[26]

"There's not anything left but a big fat bank balance," commented an analyst.

Metromedia's exit tsunami resulted in Kluge replacing Ben Walton, the heir of Walmart, in the top slot of *Forbes*'s richest person in America. Kluge, now a septuagenarian, was in no mood to retire. So he bought more businesses.

Metromedia built up a controlling interest in Orion Pictures. Under Kluge, Orion produced *Platoon, Hoosiers, RoboCop, Dances with Wolves,* and *Silence of the Lambs*. Alongside these successes were a larger string of disappointments. Kluge gave up in 1992, allowing Orion to declare bankruptcy.

Metromedia Restaurant Group acquired Bennigan's, Steak and Ale, and Ponderosa Steakhouse in 1988 and Bonanza Steakhouse in 1989. This business also went bankrupt, resulting in the closure of Steak and Ale and many of the Bennigan's.

In 1994 Metromedia backed the formation of a professional soccer team named MetroStars. This, too, ended badly. In the team's first home game, a MetroStars defender handed the opponent a 1–0 victory by diverting a cross kick into his own net. The play was dubbed the "Curse

---

26 The $4.5 billion included $2 billion from Rupert Murdoch's purchase of the television stations, which became the backbone of the Fox News Network; $285 million from a management buyout of nine radio stations, funded by Morgan Stanley; a $710 million prize from the sale of the outdoor billboard advertising business; and the sale of the entertainment division, which owned the Harlem Globetrotters and the Ice Capades featuring Olympic Champion Dorothy Hamill, for $30 million. Finally, SBC bought the paging assets for $1.65 billion.

of Caricola," and despite much hype, the MetroStars never won a title. Pelé played his final professional game for the team. MetroStars was eventually sold in 2006 and renamed the New York Red Bulls.

In its 1997 article titled "Rich, 82, and Starting Over," the *New York Times* characterized Metromedia's communications division as the crown jewel in Mr. Kluge's empire. They noted that "Kluge's strategy rests on a beguilingly simple premise: More than half the people in the world have not yet made their first telephone call."

Kluge formed partnerships to provide communications services in China, Russia, and several eastern European countries such as Latvia and Georgia. The article questioned the risky nature of Kluge's strategy, asking, "Why would one of the world's richest men choose this forbidding business landscape for what is very likely to be the last stand of his career?"

Some of Kluge's explanations made sense: "Every country needs two things to develop economically: transportation and communications. Communications is not a luxury in emerging markets; it is a necessity."

Though it also seemed to be another chapter in Kluge's fondness for dice throwing, who once said, "I don't really get comfortable when I haven't got something at risk."

Clearly, the stock market wasn't buying it, as Metromedia's stock dropped 33 percent of its value. Foreign ownership restrictions in these historically communist countries did not bode well for a US businessperson.

Domestically, Kluge continued to focus on communications. When Kluge did his 1986 fire sale, he retained ownership of a regional long-distance communications service called Metromedia Long Distance. In 1993 he merged Metromedia Long Distance into WorldCom for a value of $440 million. The resultant entity became the fourth-largest, long-distance company behind AT&T, MCI, and Sprint. Kluge was named chairman.

Kluge was playing multiple Bandwidth hands. He found one of those hands in the sewers of Gotham City, in the form of a true New Yorker named Steve Garofalo.

I owe a chunk of my fame and fortune to my paisano Garofalo, an individual I have yet to meet. To explain, as the calendar approached 2020, I was best known for two things: Zayo's acquisition of AboveNet and being the inventor of the product referred to as Dark Fiber. Without Garofalo, there would not have been an AboveNet for me to acquire. And the truth is I didn't invent Dark Fiber. Garofalo did. Thank you for both!

Forbes's 2000 article titled "Sewer Rats and Billionaires" reported Garofalo "grew up working beneath the streets of Manhattan working for his father's company, a Brooklyn-based electrical contractor. There he learned the tricks of the trade: negotiating manholes, placating regulators, wooing unions." Garofalo served as CEO of F. Garofalo Electric Company for over fourteen years.

The Garofalos placed cables for power companies, phone networks, and Cable TV. By the early 1990s, placing fiber cable emerged as a new opportunity. And it wasn't just communications companies requesting fiber. One of Garofalo's enterprise clients paid him to install fiber to link up its nearby offices, allowing the client to treat its multiple locations as if they were all in the same building. Once these enterprises had fiber links, they could bypass the phone companies for interbuilding communications.

Fiber construction, particularly in highly congested geographies, wasn't easy. "If anyone knows the difficulties of laying optical fiber," *Forbes* noted, "it's the Garofalo family."

Nor was it cheap. Often, streets had to be cracked open and then restored.

Yet Garofalo knew there was money to be made, as *Forbes* described, "No question, laying fiber-optic cables under city streets . . . is dirty work.

But glittering riches lie beneath the subterranean muck. Urban fiber networks are one of the least-developed pieces of Internet infrastructure, a market of vast potential."

Garofalo was among the first to apply when New York City announced it would grant a limited number of construction franchises to communications companies looking to build fiber networks within Manhattan. When awarded a franchise in 1993, he launched National Fiber Networks, which changed its name to Metromedia Fiber Network (MFN) before later settling on the name AboveNet. Garofalo raised $3 million by mortgaging his house and borrowing against Garofalo Electric. The Manhattan fiber network, which was built largely in ducts owned by the Empire City Subway (ECS), was open for business in 1995.

Garofalo saw the opportunity differently than others. He did not seek to offer competitive voice and data services like the other communications upstarts. In fact, he avoided investing in electronics altogether. Instead, his plan was to simply construct fiber networks and lease out fiber strands to incumbents such as Verizon and AT&T; their rivals WorldCom, XO, and others; and Cable TV companies. Enterprise customers, such as large media and financial institutions, were also targeted.

Garofalo's product was Dark Fiber.

"You could buy anything else in New York—anything. The one thing you couldn't get was a fiber-optic cable under a public street," Garofalo explained.

When I oversaw network development for MFS, we would typically put between 48 and 144 fiber stands in each of our fiber cables. Nearly all our rivals deployed similar fiber counts. The engineering justification was that optical electronics (optronics) allowed vast capacity on each fiber. As such, the smaller fiber counts seemed more than sufficient.

AboveNet took a different approach and maximized the number of fiber strands in each of its fiber cables—864 fiber strands or more. The cost of building a fiber network was mostly tied to construction. The incremental cost associated with running higher fiber strands was modest in comparison. As such, for a 20 percent increase in overall costs, AboveNet ended up with ten times the number of fibers.

Garofalo would carve out these fiber strands for sale to other carriers. "When carriers decide whether to build or buy their own fiber," *Forbes* explained, "Garofalo wants to offer a price that's just a bit less than it would cost to build a smaller network from scratch." This was AboveNet's pricing strategy and differentiation. "The key is not to get too greedy," Garofalo said.

Howard Finkelstein, who later became AboveNet's COO, explained the strategy: "AboveNet [competes] against infrastructure alternatives, rather than other companies." Before AboveNet, communications companies had only two choices: build their own networks or lease bandwidth capacity from the telephone company. Constructing a fiber network is expensive, time consuming, and hard. Leasing bandwidth from a telephone company makes it difficult to earn a profit or differentiate the service.

Finkelstein added, "A typical competitive telecom carrier builds their fiber because there was no other alternative. There wasn't Dark Fiber for them to purchase. AboveNet gives these companies the opportunity to focus not on being in the construction business, which most of them don't really want to be in."

The conventional wisdom was that once a customer purchased Dark Fiber, they would use optronics to avoid the need to purchase additional fiber. "Construction is just a one-time business. Selling services provides you with a recurring monthly business," said Teleport CEO Annunziata, which echoed the critique of most in the industry.

AboveNet's experience was different. If Dark Fiber was priced appropriately, customers would buy more—to reach new locations and to reduce their need for expensive optronics.

Beyond that, AboveNet recognized it wasn't just carriers that would purchase Dark Fiber. Enterprises were also interested. Hospitals and libraries desired to connect their various locations. Financial institutions wanted connections to the trading hubs. Tech companies sought direct links to their servers located in data centers.

Dark Fiber had a dark side, especially for the incumbent telephone companies. Nicholas Tanzi, Metromedia's president, explained to *Forbes*, "It breaks—absolutely destroys—the pricing of bandwidth. You buy the cow instead of the milk." This cannibalization was a great way for newcomers to create value for themselves, but it was undoubtedly harmful to the incumbents.

Kluge was drawn to AboveNet's Dark Fiber strategy. "We understood [Garofalo's] vision. You build it and they'll come, and they'll come for more, and then they'll come for more," Finkelstein explained.

Metromedia invested $33 million for 26 percent of the company, along with enough super voting shares to control the board.[27] Craig McCaw, who formed a relationship with Kluge when McCaw Cellular purchased Metromedia's paging business in 1990, invested alongside Kluge.

In October 1997 AboveNet raised $125 million in its IPO. It announced plans to expand to Boston, Chicago, Philadelphia, and Washington, DC, with additional targets, including Atlanta, Dallas, Houston, Los Angeles, San Francisco, and Seattle. To fund these expansions, $650 million in debt was raised.

The stock soared. Kluge's stake reached a value of $2.7 billion, second only to Garofalo's $3.8 billion.

Enwrapping Earth in a communications web of optical glass threads is "the largest infrastructure project in the history of the world," declared Garofalo. When *Forbes* pondered whether AboveNet might get stuck with more capacity than it can sell, Garofalo scoffed, "We'll never run out of customers."

Garofalo likely didn't consider the scenario that his customers would run out of money.

---

27 John Kluge was in his eighties when Metromedia invested in AboveNet. Stuart Subotnick, who represented Kluge in all aspects of Kluge's business, led the AboveNet investment.

CHAPTER IV

THE BANDWIDTH BOOM
GOLD RUSH

What the wise do in the beginning, fools do in the end.

—Warren Buffett

As the twentieth century was winding down, the riches being racked up by the Renegades didn't go unnoticed. A wave of business leaders feared they were missing out on the opportunity of their lifetimes. Frustrated executives from large companies made bold moves, while leaders of boring utilities sought to transform their stagnant monopolies into new age Internet giants. Investment bankers became Bandwidth entrepreneurs. Even those who sold out in the first wave fought their way back in.

The laser light carrying communications traffic through the fiber networks was shining bright gold. Fortunes were to be made.

The sale of MFS and Teleport marked the beginning of the Bandwidth Boom Gold Rush. The next three years were euphoric, with the stock market achieving record highs and Bandwidth providers achieving valuations in the tens of billions, often seemingly overnight. Tracking newly minted billionaires was the watercooler chatter.

For a short while, all was playing out according to their grand plans. But like all stampedes for gold, the risk of being crushed lingered in the air. The stampede that started in 1998 would continue into the new millennia. The wild ride was thrilling while it lasted.

The companies discussed in this section are no longer in operation. However, with one minor exception, their assets now belong to one of our four consolidators. The Gold Rush covers their brief years of glory, beginning with the guy who dove in first and deepest, Bernie Ebbers of World-Com. He ignited the Boom with his $14.3 billion acquisition of MFS.

WorldCom (LDDS)—Bernie Ebbers

After WorldCom acquired MFS, I spent a week working in its Jackson, Mississippi, headquarters. After days of meetings working on integrating the businesses, we trudged to the local watering hole for an after-work beer.

Bernie Ebbers was there, standing alone despite a number of his WorldCom employees at the bar. The 6'4" Ebbers had long gray hair, paired with a matching silver beard and accented with piercing blue eyes. He was intimidating, even to me. "He likes to drink alone while his team admires him from a safe range," I was told. I regret that I didn't walk up to him, introduce myself, and strike up a conversation. Now that I know his story, I suspect he had an unusually large thirst for attention.

Ebbers grew up in the Canadian city of Edmonton and attended the University of Alberta. He lasted only one year, departing because, as he put it, "my marks weren't too good." Ebbers transferred to Calvin College but only lasted two years, citing similar reasons: "my marks were not as good as they should have been, and I didn't know exactly what I wanted to do at that time, or just a little confused about my major."

Prompted by a basketball scholarship, Ebbers moved near Jackson, Mississippi, to attend Mississippi College. However, his college basketball career was cut short. "In between the first and second year I was mugged by a bunch of hoodlums and in the incident my Achilles tendon was cut," Ebbers explained. "And when you have your Achilles tendon cut, you can't jump anymore." Bernie had a way with words.

After graduating from college in 1967, Ebbers lived in a trailer and settled into the motel business, managing twelve properties while also coaching a junior high school basketball team. During this time, he encountered a tiny reseller of long-distance voice services, Long Distance Discount Services (LDDS), an encounter that would put him on the unlikely path to being a top dog in corporate America. In 1983 Ebbers convened with his three partners at a Day's Inn coffee shop in Hatties-burg, Mississippi, to brainstorm the first of many expansions and merger transactions the business would make. On this day they decided to extend LDDS service into the rest of southern Mississippi, thus igniting the massive growth years of the business Ebbers would come to rename WorldCom.

At the outset one of the others in that coffee shop held the position of CEO. However, in 1985 WorldCom was $1.5 million in debt and burning $25,000 a month. With confidence lost in the CEO, the investors turned to Ebbers.

WorldCom scraped by on thin margins, profiting from the difference between the retail rates charged to its business customers and the whole-sale rates it paid for bulk capacity from AT&T and the local Bell companies. Ebbers kept the overhead costs low while ensuring superior customer service compared with the phone companies. Within six months Ebbers was earning a profit and doubling revenue year over year.

Ebbers saw the opportunity to increase margins by becoming bigger, and he did so by buying other long-distance resellers. He started with small rivals in third-tier markets. Revenue grew to $100 million with profits of $4.5 million.[1]

In 1989 WorldCom backdoored its way to public company status by merging with a public company listed on the Nasdaq exchange. This move

---

1 WorldCom purchased Telesphere Network in 1987, Com-Link 21 and Telephone Management Corporation in 1988, and Inter-Com Telephone in 1989. The $35 million spent on these acquisitions expanded WorldCom's geography to include Missouri, Tennessee, Arkansas, Indiana, Kansas, Kentucky, Texas, and Alabama.

provided Ebbers with publicly traded stock to use as currency to continue his rollup.

In the early 1990s, WorldCom continued its aggressive expansion by completing over ten acquisitions, including its first major acquisition of a public company.[2] By late 1992 WorldCom revenue was approaching $400 million, and its stock price hit $40 a share.

In 1993 Ebbers elbowed his way into the three-way merger that included Metromedia Communications Corporation, which was part of Kluge's Metromedia empire. WorldCom shareholders owned 68 percent of the combined company. To get Kluge to agree to the transaction, Ebbers agreed to name Kluge chairman and also altered the name of WorldCom to incorporate Metromedia's brand.[3] WorldCom's network now reached forty-eight states, with revenue that eclipsed $1.5 billion and a ranking as the fourth-largest, long-distance provider, trailing only AT&T, MCI, and Sprint. Ebbers remained the all-powerful CEO.

Ebbers continued his rollup of long-distance resellers in 1994, spending $900 million to acquire two companies that provided him reach to 150 countries, including a large revenue base in his home country of Canada.[4] "This is an important acceleration of our strategy to penetrate

---

2 In 1990 Ebbers orchestrated the acquisitions of TeleMarketing Corporation of Louisiana and Mercury for a combined $26 million. The following year National Telecommunications of Austin, Phone America of Carolina, and MidAmerican Communications were purchased for a combined value of $90 million. Subsequent acquisitions in 1992 included Automated Communications, Prime Telecommunications, TFN Group, Telemarketing Investments, Dial-Net of Sioux Falls, and Touch-1 Long Distance in Alabama.

Also in 1992 WorldCom made its first major acquisition of a public company— Advanced Telecommunications Corporation, which served southern Florida and Texas.

3 WorldCom-Metro Communications was the new name. It didn't last long.

4 In early 1994 WorldCom acquired ACC Corporation for $204 million. Headquartered in Rochester, New York, ACC had significant revenue in Canada. In mid-1994 World-Com added IDB Communications Group to its rollup. The $700 million acquisition added a strong international voice business that reached 150 countries to WorldCom's product portfolio.

the lucrative segments of the international market," declared Ebbers. "But it is only a step."

———

Despite dozens of acquisitions, WorldCom had minimal fiber network. Consequently, Ebbers altered his acquisition strategy dramatically over the next few years to build a Bandwidth empire. In late 1994 WorldCom purchased WilTel for $2.5 billion, adding the freshly built national fiber network built in and along the gas pipelines of Williams Group.

In 1996 Ebbers added metro fiber networks throughout the United States by paying a stunning $14.3 billion to acquire MFS. UUNET, the world's largest Internet Backbone, came along with the acquisition.[5] MFS's CEO, Crowe, would replace Kluge as chairman.

Crowe jumped at the opportunity to sell, reaching agreement in only one hour. "I didn't expect him to be as bold and forceful as he was," Crowe said. "[Building MFS] was a labor of love, and I can't sell it without having a bittersweet taste. But it's only a 30-second feeling," he added, hinting that his $150 million of stock was a more than adequate consolation prize. Chairman Walter Scott's take was $425 million, and Richard "Rick" Adams, the founder of UUNET, received stock valued at $350 million.

"We are creating the first company since the breakup of AT&T to bundle together local and long-distance services carried over an international end-to-end fiber network owned or controlled by a single company," Ebbers said.

WorldCom's stock hovered in the low $20s per share after the deal was announced.

———

5 Interestingly, "Ebbers did not understand UUNET and initially did not want UUNET to be part of WorldCom. He considered putting it up for sale. However, the positive accolades from Wall Street analysts convinced him to retain UUNET."

Ebbers wasn't done. In 1997 WorldCom outlaid $2.4 billion to acquire MFS-copycat Brooks Fiber Properties, adding significantly more metro fiber markets to its portfolio.[6] "The merger will greatly expand our total number of local networks and advance our entrance into more secondary cities," Ebbers explained in a press release.

The additions of WilTel, MFS, and Brooks Fiber made WorldCom the largest owner of Bandwidth Infrastructure in the United States and, via UUNET, the largest Internet Backbone. On paper, he looked like the winner of the Bandwidth Revolution. The press saw the combination of MFS and WorldCom as particularly noteworthy, as reported by the *Wall Street Journal*: "The planned merger unites two of telecom's more respected management teams and their CEOs." MFS's head of corporate communications added, "We believe we're bringing together a dream team."

As I reflect back, this sentiment captured our mindset as we entered Bernie's world, though we quickly learned that we were delusional. The first glimpse of something amiss was during the period after the merger

---

6 Robert Brooks launched Brooks Fiber Properties in 1991, initially under the name Brooks Telecommunications. Brooks was able to fund the company using the profits he made as a Cable TV pioneer. The company he founded, Cencom Cable, became the twenty-first largest Cable TV company. He sold it in 1991 to Hallmark. Later, the company became Charter Communications, which was purchased in 1998 by Microsoft co-founder Paul Allen for $4.5 billion.

Brooks Fiber's strategy mirrored that of MFS, Teleport, and XO but focused on third-tier markets. It expanded mostly by constructing greenfield networks, though its growth was accelerated by a few key acquisitions. Brooks Fiber acquired Michigan-based City Signal and Phoenix Fiberlink. In 1996 Brooks Fiber went public on Nasdaq and, by 1997, operated in forty-four US cities. In October 1997 WorldCom purchased Brooks Fiber for $3 billion, which included $2.4 billion for the equity and the assumption of $600 million in debt. The combination of MFS and Brooks Fiber provided WorldCom with local fiber networks in eighty-six US markets.

Three Brooks Fiber executives would move on to other key industry roles. Brooks Fiber's co-founder and CEO James "Jim" Allen would be chairman of Xspedius. COO D. Craig Young would become CEO of Canada's Metronet Communications. Operations executive Mark Senda, an MFS veteran, would team with Allen to co-found and become CEO of Xspedius and team with Young to serve as COO of MetroNet.

Centennial Ventures—Colorado's oldest Venture Capital firm, whose final investment was Zayo—invested in Brooks Fiber and did exceptionally well in the $2.4 billion exit. M/C Partners, which backed me at both ICG and Zayo, was also a Brooks Fiber sponsor.

transaction was signed but before it closed. MFS executive Kevin O'Hara led a delegation of MFS's network team to Tulsa to meet with legacy WilTel executives who transitioned to WorldCom. The WilTel team was led by their CEO, Roy Wilkens. I was surprised by how casual and lightheaded they were, poking at our naivety and smirking at our enthusiasm.

"Buckle your seat belts," they warned us with smiles. "You have no idea what you've gotten yourselves into." This was one of Wilkens's final meetings as a WorldCom executive.[7]

WorldCom completed its purchase of MFS on New Year's Eve of 1996, naming the combined company "MFS WorldCom." MFS's top fifty executives were invited to a kickoff meeting at the posh Ocean Reef Club in Key Largo, Florida.

During our Tulsa visit, we were told that Ebbers would close and lock the entry door to his meetings at precisely the top of the hour. "Get to an Ebbers's meeting on time or you will be locked out," we were told. "If you leave to go to the bathroom during the meeting, don't bother coming back." We watched the doors to see if this was folklore. Nope, the doors were slammed shut just as the clock hit the meeting's start time of 9:00 a.m.

Ebbers started out, with Crowe the only person on the stage with him.

---

7 Soon after Roy Wilkens left WorldCom, he became a board member of a start-up named Splitrock, which was founded by WilTel veterans Kwok Li and Bill Wilson. Clark McLeod was also on the board. Splitrock purchased the network of Prodigy, a rival of dial-up Internet providers AOL and MSN.

In 1999 Splitrock went public, raising $100 million at a valuation of $700 million. They used the proceeds to purchase sixteen dark fibers on the Level 3 network.

In January 2000, just five months after the IPO, Splitrock sold to McleodUSA for $2.1 billion, comprising $350 million in assumed debt and the rest in McleodUSA stock. Wilkens joined McLeodUSA's executive team to lead the newly created data services business unit.

The ill-advised Splitrock acquisition contributed to McLeodUSA's bankruptcy.

Today, we will hear from our new Chairman, Mr. Jim Crowe, who will tell us why he is so excited about the launch of MFS WorldCom. Before I turn the mic to Mr. Crowe, I have a story to share.

Years ago, I took over as CEO of WorldCom, a small reseller of long-distance voice services. I bought a couple dozen small long-distance resellers. I had my eye set on Metromedia Communications Corporation, which had a great presence in the northeast. So I made a generous offer, but Metromedia owner John Kluge declined. I visited him again a few months later with an even higher price. He declined that as well.

Where was Ebbers headed with this story? It was out of context. Ebbers did not leave us hanging for long.

I flew back to Jackson and gave it some thought. I realized it was his ego getting in the way of the deal. He wanted to be in charge. So I flew back to their headquarters for another visit.

"What's your new offer?" Kluge asked me. I told him the offer remains the same. However, I'd like you to be Chairman of the combined company.

He said yes. So we did the deal and he got to call himself Chairman. I'm not sure what it meant, because he'd still be working for me.

So, getting back to today's agenda, allow me to introduce you to MFS WorldCom's new Chairman, Mr. Jim Crowe.

Crowe gave his canned investor speech about the future of Bandwidth but without the passion we had become accustomed to. The attention of the audience was on the Ebbers's introduction. The MFS delegation was stunned. The WorldCom insiders were smirking as if they saw this coming from a mile away.[8]

That evening, we joined Bernie on his 150-foot yacht named *Aquasition*. The boat ride was the last time I saw Crowe in his role of chairman of MFS WorldCom.[9] And that day was the last time Ebbers referred to the company as MFS WorldCom.

As it stood WorldCom was well positioned after its acquisitions of WilTel, MFS, and Brooks Fiber. Its collection of freshly built metro and intercity fiber was the strongest in the nation. Combined with World-Com's long-distance services business and UUNET's Internet platform, Ebbers could have turned WorldCom's focus to execution.

Instead, he wanted more—a lot more! He saw his opening when the takeover of MCI by British Telecommunications (BT) hit a snag.

In 1994 BT—the UK's equivalent of AT&T—acquired 20 percent of MCI for $4.3 billion. The two companies formed a joint venture branded as Concert, which would offer multinational customers end-to-end network services managed by one global operator.

Two years later BT agreed to purchase the rest of MCI for $21 billion, which equated to $36 per MCI share, a 44 percent premium to MCI's

---

8 Bill LaPerch recalled a similar experience that occurred a year later, when Ebbers introduced MCI's Bert Roberts as WorldCom's new chairman: "At the first big meeting post-acquisition, with Roberts on stage, Ebbers showed an organization chart with Roberts as Chairman. The only thing reporting to Roberts was MCI's flight organization. Ebbers told Roberts that his only responsibility was to get rid of the private jets and the pilots ASAP."

9 LaPerch shared a story on how Ebbers used *Aquasition* to torment executives who fell out of favor: "One of Ebbers favorite pastimes was to invite senior executives who recently sold their WorldCom stock for a trip on his yacht—in many cases, no doubt, to pay for their children's college education or perhaps family illnesses. Once on *Aquasition*, Ebbers would fire them, leaving them to spend unwanted time on the yacht, waiting unemployed to return to port with no job."

stock price. "If BT's shares hold up, this is a great price," commented a Merrill Lynch analyst. "It's an offer that can't be refused."

"The companies promised a communications powerhouse with annual revenues of $42 billion and 43 million business and residential customers in 72 countries," reported the *Wall Street Journal.*

"This merger creates the premier telecommunications company of the new millennium," declared Bert Roberts, who succeeded MCI's iconic founder, William "Bill" G. McGowan, as CEO.

Sir Iain Vallance, the BT chairman who was knighted by Queen Elizabeth and years later would sit in Britain's House of Lords, would be co-chairman with Roberts. BT's CEO, Sir Peter Bonfield, would be Concert's CEO.

"All mergers have surprises, and I'm not going to tell you, because you wouldn't believe me, that this merger won't have its challenges," Sir Peter told reporters at a press conference. He was correct—surprises were percolating.

The first came in the summer of 1997 with the regulatory approval process not yet complete. MCI reported huge losses, causing a 17 percent fall in its stock price and a $5 billion drop in Market Value. "This is a bomb, a major bomb" was how a Lehman Brothers' analyst characterized MCI's disclosure.

BT pressed for a price reduction. In August a revised agreement was announced with the $21 billion price being lowered to $18.9 billion. "We've survived the jolt and moved forward," Sir Peter said in a news conference. Sir Iain piled on: "It could well be best for the MCI shareholders to have a smaller slice of a bigger cake."

The second surprise was foreshadowed by a faulty phone connection.

US journalists participated in BT's London news conference via a phone connection. "It never worked," reported the *Wall Street Journal.* "One reporter tried three times, but her voice came out hopelessly garbled, with red-faced executives struggling to piece together the question. On a fourth attempt, there was nothing but silence from across the Atlantic."

"That's not an omen at all," Sir Peter insisted. He was soon to be proven wrong.

MCI shareholders approved the higher-priced deal. However, for the recut deal to proceed forward, shareholder approval was again required. This created an opening for Ebbers, and he drove his muddy pickup truck through it. WorldCom launched a $30 billion unsolicited bid, causing unwelcome noise to the once-harmonious transatlantic Concert. The $41.50 a share easily topped BT's $34 agreement. Though WorldCom's $7 billion of revenue was far less than MCI's $20 billion, WorldCom's shareholders would own more than 50 percent of the combined entity.

BT was in a pickle. The British telecom giant already publicly acknowledged its first bid was too high. Responding with an even higher bid would be hard to explain to BT's public investors.

MCI's board also faced tough decisions. The merger agreement prohibited the MCI board from entering discussions with WorldCom until and unless it concluded that WorldCom's offer was more favorable from a financial and strategic point of view. Though WorldCom's offer appeared materially higher, the form of currency was WorldCom stock, not cash. BT's offer as well was mostly in the form of stock. This meant that the board needed to evaluate which offer was truly superior, with due consideration to the relative strengths of each merged entity.

Would WorldCom-MCI or BT-MCI be a better company? Did one combination have more equity upside than the other? Which was riskier?

A unique complication was BT's 20 percent ownership stake in MCI and the duo's Concert alliance. Would BT elect to sell its 20 percent to WorldCom and end the Concert alliance with MCI? How should MCI's board factor this into their valuation comparison?

The merger agreement spelled out the process that was to be followed if an unsolicited offer was made. The MCI board was required to inform BT whether it planned to terminate their deal and recommend the alternative offer to MCI shareholders. This presented another dilemma to MCI's board. MCI would owe BT a $450 million breakup fee if the MCI board pulled out of the deal. However, if the BT merger was rejected by share-holders, the breakup fee would not apply.

As if Hollywood were writing the script, bad blood was also top of mind. Less than a year before the takeover battle, MCI shut down services in Jackson, Mississippi, over a billing dispute with WorldCom, leaving Ebbers and many others unable to make long-distance calls. The service was restored only after WorldCom paid a $41.5 million settlement to MCI. "How could WorldCom be more valuable than MCI?" must have been the angry comments ringing in the hallways of MCI's headquarters.

Ebbers called Sir Iain to support WorldCom's bid, roll BT's MCI ownership stake into WorldCom, and invite WorldCom into the Concert alliance. Sir Iain's cold British reply was that BT would respond at an appropriate time. A BT spokesperson called the conversation "brief but cordial." Two days later the same BT spokesperson said, "This has a hell of a way to go before this thing is over."

The following month brought another quake: GTE, made renegade by its purchase of BBN in 1997, lobbed in a $28 billion all-cash offer for MCI, along with expressing a willingness to substitute stock for cash if MCI or BT so preferred. If accepted, the transaction would be the largest all-cash acquisition in the history of business.

GTE's CEO, Chuck Lee, explained his strategic reasoning: "This is a great target of opportunity. MCI jump-starts our strategy. We can still go it alone, but with MCI, we can move to where we wanted to be four to five years sooner."

GTE's all-cash offer commanded the full attention of MCI's board. A $28 billion cash bird in the hand could be better than WorldCom's all-stock $30 billion bird in the bush.

Lee said he would be thrilled if BT became a partner. "They're an excellent institution. There are tremendous opportunities for us in the global arena," Lee told the press. BT was believed to be favorable to GTE but not so keen on WorldCom.

Ebbers responded by raising his bid, this time to $37 billion. It was poised to be the largest merger in the history of business. Roberts would chair the board, and the merged company would be renamed MCI World-Com. MCI's stock price rose to $41.50.

Lee considered responding but chose not to. WorldCom's stock price fell to $31, reflecting its investors' disappointment that the bidding war was ending. Conversely, GTE's stockholders were pleased that Lee backed down, as evidenced by GTE's stock price jumping up to $45. BT was happy as well, as it collected $7 billion for its 20 percent MCI ownership, plus an extra $450 million for the breakup fee.

"We have aligned ourselves with a management team and employees who share our entrepreneurial spirit and continue to pioneer competition in our industry," declared Ebbers.

Roberts chimed in, "In combining our unique strengths—our agility, innovative approach and competitive skills—we will be a new era communications company."

The deal required the approval of both the US Justice Department and the European Commission. To satisfy regulators, WorldCom agreed to sell MCI's Internet business to Cable & Wireless for $1.75 billion, a low price for one of MCI's most valuable assets.

When the acquisition finally closed in September 1998, WorldCom's vice chairman, Sidgmore, lamented, "It has been a very long struggle." Was it worth it? The price paid was a gigantic premium to MCI's share price. The Concert joint venture was terminated, opening the door for AT&T to knock out MCI as BT's global partner. The highly valued MCI Internet business was sold.

Did Bernie win the bloody battle but lose the war?

When the FCC approved WorldCom's acquisition of MCI, the FCC chairman warned that the telecommunications marketplace is "just a merger away from undue concentration." Ebbers heard the message, telling investors, "The regulators told me not to come back with another big one."

But Ebbers couldn't resist.

WorldCom's stock ascended significantly in the two years that followed the MCI merger. In June 1999 WorldCom's shares peaked at $62, translating to a Market Value of $115.3 billion.

Despite warnings from the FCC to lie low, Ebbers lusted for an even grander transaction, and that October, a mere four months after indicating he would heed regulators and stand down on further big deals, he used WorldCom's bloated stock as currency to interlope on an in-process acquisition of Sprint by Baby Bell BellSouth.[10]

Sprint was the third-largest, long-distance provider and was one of the largest wireless carriers. WorldCom's bid of $115 billion, three times the amount WorldCom paid for MCI, was successful. All that remained was for Ebbers to convince the regulators to approve the transaction.

"Telecom is about scale, scale, scale," said an analyst. "It's no longer a national playing field. It's a global playing field. . . . The game is 'Clash of the Titans.'"

In WorldCom's case it became "Crash of a Titan."

Analysts were stoked: "[WorldCom/Sprint] would be the largest of any telephone company in the world," said one. "It's the most attractive set of assets that any carrier has." Not to be outsold: "We believe that as in past deals, now is the best time to buy WorldCom shares," coaxed another analyst, "after they drop due to merger news, but before merger benefits are given weight in the market." Shilled a third, "We believe

---

10 WorldCom pursued an acquisition of the mobile carrier Nextel in 1999. John Sidgmore confided to John J. Keller, a former *Wall Street Journal* reporter turned CEO headhunter, that Ebbers pulled out of the acquisition on the day the purchase agreement papers were to be signed. Sidgmore said he had put together the deal. Recounting the story in 2024, Keller relayed to me: "Sidgmore told me that he and Ebbers were due to go to Nextel and meet with Nextel leaders to sign the papers. He went to Ebbers's office and said, 'Hey, Bernie, c'mon, we have to get going. [Nextel's] expecting us soon.' Sidgmore said Ebbers answered, 'I'm not doing that fucking deal. Forget it.'"

WorldCom's current valuation level represents an attractive entry point into one of the premier telecom companies."

Regulators begged to differ: In June 2000 they followed through on their stern warning to Ebbers, and the US Justice Department filed suit to prevent the takeover. The European Commission announced it, too, would block the merger on anticompetitive grounds.

Ebbers's feistiness began to ebb. He had spent nine months pursuing approval from the government. Fighting the regulators would take at least another year when WorldCom's stock had already fallen by 30 percent from its high. The companies pulled the plug on the WorldCom-Sprint merger. WorldCom's shares climbed 8 percent on the news, but its $48 a share price was well off its $72 high two years earlier.

Despite the failed merger, Ebbers the gambler was unbowed. In fact, he was steaming. And for those not familiar, steaming is a Texas Hold'em expression for a player who overreacts to a cold streak by betting recklessly on subsequent hands. Ebbers, fatigued by the Sprint failure, forced a deal with regional long-distance company Intermedia Communications,[11] which included a majority stake in the web hosting business Digex.[12] Ebbers's final acquisition was perhaps his worst.

---

11 Intermedia Communications provided local and long-distance services to ninety thousand customers in fourteen cities over its two-thousand-mile network in the Southeast. The debt-laden Intermedia put itself in play earlier in the year when it revised revenue guidance and was exploring strategic options.

12 Intermedia owned a majority stake in the web hosting company branded as Digex. The spark for WorldCom's acquisition of Intermedia was a call from Global Crossing CEO Gary Winnick to Bernie Ebbers, letting him know he was about to buy Digex. For reasons that were entirely baffling to Sidgmore, Ebbers determined he needed to outmaneuver Winnick. Ebbers steamed into buying Intermedia for $6 billion, with the transaction closing in early 2001. WorldCom's stock dropped by 17 percent when the deal was announced. Intermedia gave Ebbers majority control of Digex, which he intended to use to force the rollup of Digex into WorldCom. Digex minority shareholders challenged the transaction. Their case was strong, as a special committee of Intermedia's board that was set up to protect Digex's minority shareholders voted against the WorldCom deal. To settle the lawsuit, WorldCom gave Digex shareholders an additional $165 million in WorldCom stock while also providing Digex a $900 million loan.

Regulators remained feisty. They forced WorldCom to sell the vast majority of Intermedia's remaining assets. WorldCom had trouble finding a buyer.

WorldCom vice chairman John Sidgmore was blunt: "That is the stupidest deal of all time. We paid over $6 billion for a company that's worth maybe $50 million."

In April 2001 WorldCom disclosed it was paying a $10 million bonus to Ebbers. This seemed odd, as WorldCom's stock price had dropped to below $20 a share. He was also awarded 1.2 million in stock options.

WorldCom also disclosed that Ebbers was forced to sell three million shares to cover a margin call on WorldCom stock. Ebbers had sold about 1.2 million shares of WorldCom stock for a value of $23.5 million. Additionally, WorldCom provided $100 million in loans to Ebbers. He explained the loans were to help cover the margin call so he wouldn't have to sell more WorldCom shares, which could hurt the stock price.

Ebbers also signaled a for-sale sign by saying multiple times that he would support an acquisition of WorldCom for the right price. Verizon and SBC nibbled at the bait but didn't bite.

"Verizon and SBC may have also been scared by WorldCom's accounting practices, according to some telecommunications executives. Some investors and analysts have worried that WorldCom has used aggressive accounting techniques in the past, particularly in reporting revenue," reported the *New York Times* in February 2002. "[More serious concerns], unsubstantiated thus far, [are that] WorldCom may be more aggressive than its peers in recognizing revenue."

An analyst said, "There was some rumbling in the investment community that they may have done some things with the accounting in order to make their numbers."

WorldCom's *Titanic*-like iceberg was looming. The company's stock price sank to single digits. The next few months would be a cascading series of events and discoveries that would reveal one of the largest scandals in the history of business.

Whether the Bandwidth industry knew it, the crash had begun, and WorldCom was to take center stage.

Level 3—Jim Crowe and Kevin O'Hara

The Gold Rush welcomed a new entrant, Level 3, led by a familiar name: Jim Crowe.

I am certain Crowe's sale of MFS to WorldCom was prompted by the fear he would soon disappoint Wall Street. Maybe he believed he was entering a long-term partnership with Ebbers. If so, the WorldCom onboarding must have been a thunderbolt to Crowe, as he listened to Ebbers making it clear to the room full of executives that Crowe was chairman in title only.

It may have been this humiliation that motivated Crowe to dive back in and become bigger and better than Ebbers. He was compelled to finish what he started at MFS. Crowe partnered again with Walter Scott of Kiewit, who did extraordinarily well when he backed MFS. Scott rinsed and repeated, backing Crowe again and bracing for another rewarding strike of lightning.

Crowe teamed up with O'Hara to lead the business. I joined shortly thereafter. Though I think of myself as the fourth person to join, this technically wasn't true, in part because my separation from WorldCom would take multiple weeks to maneuver. In any case I was the first non-corporate, non-Omaha, and non-unemployed executive to join.

I worked out my deal directly with Crowe, as O'Hara's non-solicit agreement with WorldCom prevented him from hiring me. What made my decision easy was the sign-on bonus of $500,000 to be part of Crowe's yet-to-be-named start-up, which had yet to formalize on a business plan, but which had $2.5 billion of assets to invest.[13] I, too, was eager for another bite at the apple.

---

13 To fund Crowe, Kiewit put him in charge of Kiewit Diversified Group, a $2.5 billion collection of noncore businesses, such as an energy provider, coal properties, and a private toll road, that Kiewit accumulated over the years. Crowe's task was to monetize these KDS investments and redeploy the money to build another fiber network provider.

When I joined, dozens of others followed. We handpicked the strongest. We reassembled our band.

WorldCom's attorneys pulled out my MFS employment agreement, preparing to sue me for violating my non-solicit agreement. I imagine they were disappointed to find O'Hara's signature representing MFS, but the line over my name was blank. I hadn't countersigned the non-solicit, thus making it irrelevant.[14]

This was when I learned my original MFS offer letter said that I'd be eligible for pre-IPO stock options, which I never received. Shame on me for not paying more attention. Shame on MFS for not following through. In all the excitement of building MFS, we overlooked some important details.

Crowe turned his attention to how best to brand this new venture. True to his engineering background, he followed his technical instincts. The Open Systems Interconnection model was developed by the International Organization for Standardization in the late 1970s to "provide a common basis for the coordination of standards development for the purpose of systems interconnection." Layer 3 is the network layer, which matched up with Crowe's vision. So we settled on L3 Communications.

Oops. This name was already taken by another business. That's okay. "We will buy them out of their name," we all said with bravado—a hint of arrogance building among us. In fact, L-3 Communications was a multibillion-dollar aerospace spinout of Lockheed Martin. As Mo Green said to Michael Corleone, "I buy you out. You don't buy me out."

We went with plan B: Level 3 Communications!

Crowe viewed a split headquarters as a factor in MFS's organizational dysfunction. This time he insisted we all needed to be together in one city. But what city? Crowe and his corporate lieutenants lived in Omaha. O'Hara and I, along with most of our recruits, lived in Chicago. Crowe was not moving to Chicago, and we were not headed to Omaha.

---

14 O'Hara got sued by WorldCom over numerous noncompete and non-solicit violations, including hiring me. O'Hara prevailed. The court dismissed all charges against him.

To resolve, Crowe commissioned a study to identify the best place to build a fiber and Internet company. Boston, Northern Virginia, Silicon Valley, and Colorado were the finalists.

East Coast Internet talent wouldn't move to the West Coast, and Silicon Valley talent wouldn't move to Boston or DC. Most, however, would consider Colorado. It was a beautiful place to live. A new international airport was under construction. Colorado was the home of the entrepreneurial Cable TV industry. Unlike the coasts, Colorado had limitless space to scale.

We selected Colorado, specifically Interlocken, a new business park close to Denver and even closer to Boulder.

When it came time to announce the birth of our company, Crowe did so with a splash, headlining *USA Today*'s business section on April 1, 1998. Yes, you read that correctly, Level 3 was announced on April Fools' Day, underscoring Crowe's flair for the dramatic.

The headline was "Fathers of Invention: How Level 3 Worked Its Way to the Main Floor." Warren Buffett, Bill Gates, Walter Scott, and Crowe adorned the cover page. The article told of Buffett's 1995 retreat in Dublin for thirty of his billionaire friends. Gates gave a presentation on the Internet that caught Scott's attention. "My gut feeling was, if you weren't part of it, you were going to be left behind," Scott was quoted.

Upon Scott's return to Omaha, he debriefed Crowe and inquired on the implications to MFS, emphasizing, in Crowe's words, "Anytime anyone's told him there's a risk, there's always been an opportunity."

MFS launched Project Silver, with the team visiting twenty to thirty ISPs. This led to MFS's acquisition of UUNET for a stunning $2 billion.

Then came the April Fools' Day announcement: "Today, Level 3 is listed for the first time as a public company on Nasdaq. It begins life with a Market Value of $10 billion and a loud buzz."

Level 3 did not raise money in the IPO. Instead, it transferred its 147 million KDS Class D shares from an over-the-counter bulletin board to Nasdaq. My $500,000 of KDS stock was now worth more than $2 million.

The *Wall Street Journal* and others in the industry were watching to see if Crowe would "make lightning strike twice." One analyst said, "Crowe struck gold last time out. . . . He's trying to recreate that miracle." Another added, "It's going to be a fantastic company, but now the Market Value is out of control."

Crowe was unfazed. "Hopefully, our track record has something to do with it," he countered about Level 3's outsize valuation.

The Gold Rush madness pushed new entrants to make proclamations, one-upping their rivals. Crowe, with no shyness, announced his strategy was to build the first global system of Internet-based local and long-distance networks, challenging aging phone companies in major markets. "Moore's Law has come to communications," Crowe said, referencing the Intel co-founder Gordon Moore's observation that semiconductor performance roughly doubles every eighteen months. "The current phone networks in place date from the turn of the century and were designed to handle primarily voice," Crowe said. IP technology promised far faster transmission and lower operating costs than the legacy networks of AT&T and the Baby Bells.

Crowe had conviction on the competitive landscape: "For as far as the eye can see, the real competition is between the new entrants and the older companies." This proved to be misguided, as competition among new entrants turned out to be the big obstacle.

USA Today fed the drama, reporting Ebbers as a "dangerous enemy" to WorldCom's rivals and suggesting Ebbers remained rankled by Crowe's leaving WorldCom just three weeks after the close of their deal and then enticing me and others to join his new Level 3 venture. "WorldCom would like nothing better than to bury Level 3 in the marketplace," *USA Today* reported.

USA Today included Qwest in its soap opera. Crowe, who was put on Qwest's board by Anschutz, told the newly hired Nacchio he might start a competing business. When Crowe picked Colorado for Level 3's headquarters, Nacchio quibbled, "Jim asked me, 'What is the best place to locate a company?' I said Denver. I didn't know he was going to be a direct competitor. If I'd known that, I would have said Pennsylvania." Nacchio also suggested Crowe "was pursuing a 'copycat strategy.'"

Adding to the drama was *USA Today*'s foreshadowing of Level 3's disruption to the entire telecommunications industry.

> [Level 3's] business plan, if correct in its assumptions, could drive down the cost of long-distance phone calls to almost nothing and hasten the decline of the big phone companies.
>
> Crowe is betting that the dominant networks of the world—the ones that carry phone calls using circuit-switching technology based on a 100-year-old design—are dinosaurs.

The article finished with a consultant's quote that was thrilling for me to read at the time but chilling as I reread it twenty-five years later. "It's a dream team with a dream network and a killer business plan. Just as long as they don't screw it up."

Before the move to Boulder, the majority of our team worked out of an office building in a suburb of Chicago. Our office space was divided in half with tape, reminding us that our rent covered only the west side of the vacant floor.

In an era where office size mattered, O'Hara had the biggest office, and mine was the second largest. All the network responsibilities were assigned to me. With the hoopla behind us, it was time to get to work. What, though, were we supposed to do?

At the time of Level 3's unveiling, the *Wall Street Journal* wrote, "[Level 3's] backers are hoping . . . that Level 3's novel technology will be a hit." I wondered what technology the journalist was referencing.

"Level 3 wants to spend $3 billion developing a national fiber-optic telephone network using Internet-protocol technology to send voice, data, and video over the same lines with lower costs," explained the *Wall Street Journal.* That shed some light on the aspiration, but the question of how lingered.

We needed answers so that we could begin to execute on whatever it was we were going to build.

I had a restless week. The network team was looking toward me for guidance. Toward the end of the week, I decided it was best not to fake it. "Just ask others what they think" was my path to a solution.

So I asked the two members of my senior team. They both said, "I thought you knew and just weren't ready to reveal it to us."

On Friday I approached O'Hara. He was COO. Surely he would know. Nope. He, too, didn't have the answer.

I didn't sleep well that Friday and Saturday night. We had $3 billion in funding, the cover story in *USA Today* announcing our arrival, and no concrete idea of what to do.

On Sunday a lightbulb flashed. I drew pictures, documented call flows, and showed how phone call revenue and costs would pass between customers, local carriers, long-distance carriers, and the new breed of carriers—Voice over Internet Protocol providers—that used Internet technology for phone calls. I mapped how the IP would insert itself into the public switched telephone network and forever change the way traffic flowed. I called it the SS7 to IP Operating System (SIOS). I was excited and anxious to reveal this to my colleagues.

First thing Monday morning, I showed it to my team. As key team members on the MFS co-carrier effort, they had the technical and business background to understand the details. They knew immediately that this was the answer.

I showed it to O'Hara. His reaction was less certain and noncommittal. He decided the two of us should fly to Omaha to see Crowe.

A few days later, Kevin and I sat down with Crowe in his Omaha office. His right hand for corporate marketing was at his side, both still giddy from the *USA Today* article. I walked Crowe through the slides, showing how calls and cash would flow between our IP network and the Bell system. At the end he just chuckled and said something dismissive like, "Keep up the good work. It will lead to something useful down the road."

As we waited for our return flight to board at Omaha's Eppley Airfield, O'Hara asked, "How did you think the meeting went?" "Horrible," I answered while wondering if I should have clung to WorldCom.

"Don't be so hasty. Crowe likes to absorb new information before reacting," said O'Hara. He added, "I suspect we will hear from him soon." Did he know more than he was letting on?

Two days later Crowe called out of the blue. "I was looking at the SIOS materials. I think you are onto something. This is what we need to build."

A few months later, we hired a young engineer at MCI who was working on how to use IP to route phone calls. From his MCI long-distance vantage, he did not understand the local carrier business. I showed him the SIOS material during the interview process, and he said, "Oh yeah, I call that a softswitch!" He joined us, wrote a patent for the softswitch, and led its development and deployment. Managed modem service, which was enabled by the softswitch, became the biggest source of revenue and cash flow during Level 3's first seven years.

———

While we had the patent on softswitch, a company called XCom Technologies, run by Shawn Lewis, had a head start on us. Lewis, to put it mildly, was an interesting character. In his early twenties, he pleaded guilty to writing a series of bad checks and served a year in jail. The charge "was supposed to be expunged," Lewis said.

Lewis's grandfather bought him a build-it-yourself computer kit when he was thirteen. Lewis had an aptitude for computer programming. When he was released from prison, he partnered with a Boston businessman and co-founded XCom Technologies. From a strip club in Boston, Lewis wrote software that terminated dial-up modem traffic to the local telephone companies, generating a steady stream of cash flow for delivering these calls to Baby Bell NYNEX. Lewis was still in his twenties, and yes, he claimed to write the software while hanging out in a strip club in Boston.

Soon after Level 3 went public, Crowe stumbled across XCom, noting that what they built was the first commercially deployed version of what we called the softswitch.[15] Crowe offered XCom 1 percent of Level 3's stock—worth about $165 million. They accepted in a meeting that was said to last about five minutes.

"XCOM holds the rights to technology which provides key components of the bridge necessary to fully interconnect the public-switched telephone network with new networks based on the Internet Protocol, or IP" was the *Wall Street Journal*'s takeaway from Crowe.

"This bridge will allow Level 3 to offer a broad range of communications services, including voice and fax services, to customers at lower cost and over time with new features and enhancements," Crowe explained. "Moreover, customers will be able to enjoy these benefits using their existing telephone equipment and without changing dialing patterns or adding additional devices."

So Crowe was listening to me when I explained my SIOS concept.

---

15 XCom's architecture required a traditional circuit switch. Crowe concluded that Level 3's and XCom's engineers could eliminate the circuit switch from the architecture and thereby create the first true Softswitch. Our teams achieved this vision, and the multimillion-dollar circuit switch was now unnecessary. Level 3 sold the switch to Focal Communications, a start-up led by my MFS and Ameritech colleague Robert "Bob" Taylor.

I was told Lewis would report to me. Soon after the acquisition closed, Lewis visited Boulder. An engineering team that joined us from Internet MCI was charged with rewriting XCom's software to industrialize it, allowing us to scale across all our geographies. We had a productive week of meetings. Knowing Lewis liked to party, a colleague and I hosted Lewis for a night out at Juanita's, the infamous taco and margarita hangout on Pearl Street in Boulder.

"We were drinking a lot. Dan snuck a hundred dollars to the waitress, asking her not to put tequila in our next drink," recalled my colleague. "When we tasted alcohol in the next round, the waitress explained, 'Your friend Shawn trumped you by giving me $200.'" My colleague and I hung with him until the joint closed.

On Friday he said goodbye as he headed to the airport with his girlfriend, who had joined him on the trip. On Monday he didn't call into the staff meeting. We didn't hear from him on Tuesday and Wednesday. On Thursday I called his Boston office and prodded one of his engineers until he spilled the beans. "When Shawn's flight out of Denver was delayed, he decided to fly to Alaska instead of Boston. From there, he and his girlfriend jumped on a cruise."

"Where is he now?" I asked.

"Shawn met a girl on the cruise. His girlfriend got upset. Shawn had a helicopter pick her up from the boat, and he stayed on the cruise. That's the last we heard from him."

Lewis was about 6'1" and hit the gym hard. He liked wearing tight shirts that showed off his muscles. He was a control freak. Lewis kept the master version of his XCom software on a server under his desk in his locked private office at XCom's headquarters in Boston, which was nearby but no longer in the strip club. If the server went down while Lewis was MIA, we were totally screwed.

A week or so later, we heard that Lewis was back in Boston. I had a colleague join me on a surprise visit to Boston on Level 3's private plane. We showed up to the XCom location unannounced and walked into

Lewis's private office. I told Lewis we would be taking the server with us back to Boulder. Lewis stood up, put his chest into mine, and neither of us said a word while my colleague disconnected the server and took it to our black SUV. That was the last time I ever saw him.

Lewis moved to Miami's South Beach. He met a former Miss Hawaiian Tropic and then-current Miami Dolphins Cheerleader. Within weeks of meeting her, he bought her a $30,000 six-carat diamond ring, and she accepted his marriage proposal. He arranged for a $1 million wedding on a private island. As the wedding day approached, they had a spat, and the wedding was canceled. The island bash money was already spent, so Lewis threw a party without her. The festivities included eight hundred oysters, four kilos of beluga caviar, rivers of champagne, dwarfs dressed as jesters, and women in mermaid costumes. O. J. Simpson was among the guests. Deborah Norville covered the story on *Inside Edition* in an episode called "The Million-Dollar Wedding with the Surprise Ending."

A few weeks later, Lewis convinced Miss Hawaii to marry him. During the honeymoon they split up again for good. The terms of the prenuptial agreement were sealed in a court settlement.

Lewis bought multiple clubs in Miami, including the Krave Nightclub, the Living Room, and the Red Lounge. Florida's Division of Alcoholic Beverages and Tobacco and Department of Revenue crashed his clubs after being tipped off about his ten-year-old felony conviction in New Jersey. He sold his clubs soon thereafter.[16]

In the end XCom turned out to be a great acquisition for Level 3. We rewrote the code, used our knowledge of local interconnection and our fiber networks to build the nation's largest managed modem termination network, and raked in millions of dollars a month of high-margin revenue

---

16 Shawn Lewis returned to telecom, applying his software engineering skills to several voice companies. According to LinkedIn, he was chief technology architect for TTEC Holdings in 2023.

that was instrumental in keeping Level 3 out of bankruptcy during the industry's meltdown.

The softswitch and managed modem service were a sideshow to Level 3's main story, which centered on fiber. In April 1998 the *Wall Street Journal* published "Surge in Internet Optimism Fuels Decade's Biggest Junk-Bond Deal," which covered Level 3 completing a "blockbuster $2 billion bond deal." A Yankee Group analyst said Level 3 "has not only a 'Dream Team' but also extremely deep pockets going for it."

The construction of fiber networks was the main thrust of Level 3's business plan. Crowe determined intercity networks were more important than metro this time around, though later we also built metro fiber networks.

These weren't novel thoughts. Qwest, Broadwing, and others were already building a new national fiber network. AboveNet, MFS (now part of WorldCom), Teleport (now part of AT&T), XO, and others were deploying metropolitan fiber networks.

So what would make Level 3 unique? Crowe changed the focus from fiber to conduits, with a vision that a multi-conduit network would allow for future proofing. Fiber technology would improve every few years, Crowe reckoned. To prepare for these continuous technological advancements of fiber, Level 3 would build multiple conduits and keep all but one vacant for new fiber technologies. Over time, according to Crowe's hypothesis, other fiber networks would become economically obsolete by quickly and cost-effectively pulling new generations of fiber through the empty conduits.

"Our goal is not simply to deploy one generation but to build an entirely new model that assumes technology will change quickly, and at times, unpredictably," Crowe said.

The question was, How many conduits were needed? Crowe hired operations research scientists to answer this seemingly all-important question. The Project Max team built wickedly complex optimization models that hinged on entirely unknowable assumptions and, not shockingly, confirmed Crowe's intuition. Six was optimal.

Then Qwest claimed they would be able to put two fiber cables in each of the three smaller tubes (innerducts) that were tucked in its larger tube (duct).

At Crowe's urging, our operations research scientists studied the problem more. The unknowable assumptions were tweaked. Now the answer was conveniently twelve, which also provided extras for leasing or swapping with rivals.

The cost of the network went up materially from these added conduits. In expensive build geographies (e.g., cities, river crossings, and mountain passings), construction methods were more limited. For this and many other reasons, the construction costs were going up and up. The initial estimate for the nationwide build was $3 billion. We spent in the vicinity of $5 billion.

This period, 1997 through 2000, is a thick Gold Rush fog for me, and I was working nonstop. I built up an organization of hundreds of people in the United States and Europe. We hired Kiewit to be our general contractor. We built fiber networks that connected the major cities of the United States and another one connecting those of western Europe. We also built metro networks in most of the major cities.

On top of that, we built a tier-one Internet Backbone that soon eclipsed all others. And we built data center facilities in most of the big cities.

We deployed a vast softswitch network with an interconnection network that made us a local exchange carrier throughout the United States. We used this network to provide managed modem services, becoming the largest network for connecting consumers and businesses to the Internet.

In a sign of confidence, Level 3 executed a 2:1 stock split in July 1998.

The following May, Crowe announced plans to sell four thousand of his shares a day for the next 250 days. Assuming the price remained steady, this sale of one million of his eleven million shares would amount to $85 million of proceeds. Instead, the stock doubled over the next ten months, reaching its peak of $130 a share in March 2000. Between 1999 and 2001, Crowe sold $115 million of stock.

I should have pulled my Bandwidth Boom head out of the fiber trenches we were digging and dumped all my vested stock. Instead, I held, which, as we will see, proved to be a costly mistake.

———

Around the turn of the century, 70 percent of Level 3's organization worked for me. The network organization, which included network planning, construction, engineering, and operations, reported to me. As we looked forward, this would undoubtedly evolve. Sales professionals, marketing and product development, information technology, and corporate would grow faster than the network organization. Still, if unchanged, my scope would represent 50 percent of the company.

In late 1999 Level 3 decided to do a major reorganization to prepare for commercialization. Crowe's vision was to create a matrix organization. Each region—North America, Europe, and Asia—would have its own president. Most of the network organization would report to the regional president. I was assigned to be the group vice president of Global Network Services. I would be "dotted line" responsible for the network.

I don't think it is a self-serving exaggeration to say I contributed as much as anyone, other than, of course, Crowe, during Level 3's first three years. Yet the new organizational plan placed me in an untenable role as an influencer. Though I remained in one of the senior-most positions on the organization chart, this didn't sit well with me.

In hindsight I should have just gone along with it. I should have slowed down, enjoyed the reduced workload, and focused on my family. I should have sold the $65 million of my stock that was already vested. And I should have enjoyed watching others work hard while the remaining $65 million or so vested. Then I should have quit, bought an oceanfront home in South Beach, and acquired the Miami nightclubs that Shawn Lewis was forced to sell.

That, unfortunately, wasn't how I played the cards. I lasted in the role for less than a year and hated every moment. From the dotted line role, I couldn't drive execution. I watched as others less competent than me chipped away at what I had created. I became an unpleasant distraction to the senior team.

I requested to be removed from the role. The logical conclusion would have been for me to leave Level 3. Instead, Crowe created a position for me to take a deep dive into how Level 3 could scale. I hated this as well, but I learned a lot about systems, data, and processes that I applied for the rest of my career. Yet my role as an annoyance to the senior team intensified.

Mid-2001 was upon us. The Bubble was beginning to burst, and Level 3 was in full panic mode. The only good news was my time-out was about to end, and I once again was to become one of the most important executives at Level 3.

Global Crossing—Gary Winnick

Next up is the crazed story of Global Crossing and its investment-banker-turned-Bandwidth giant, Gary Winnick. He sniffed out the Gold Rush and became a leader of the Boom stampede.

"For most of his 56 years, Gary Winnick's life followed a storybook plot: a kid from humble circumstances vaults, like a Horatio Alger character, into the upper tiers of American business and wealth," wrote the *New York Times* in its 2004 article titled "A New Legal Chapter for a 90's Flameout."

Winnick's father died when Winnick was eighteen years old. "Two hours after dinner he . . . went to his Thursday night card game [and had a heart attack]," Winnick told the *Times* reporter. "I got the call. My mother handed me his [diamond-and-ruby wedding] ring that night and I've worn it ever since."

Winnick commuted to a nearby college while working to help with school and family expenses. After graduating with an economics degree, he became a furniture salesman. A few years later, he transitioned to Wall Street, becoming a successful bonds salesman. Then he opened up an investment firm in Beverly Hills, pounced on a big investment idea, founded a company, raised $20 billion in debt and equity, took his company public, saw its stock value rise to $47 billion, and watched his net worth skyrocket to $6 billion.

Alas, the above summary left out some salacious details, including the un-fairytale ending.

The juice behind Global Crossing was junk bonds. Drexel Burnham Lambert was the Wall Street firm that launched Winnick's financial career. Winnick started as a salesman on the bond desk. His associate was none other than Michael Milken, who was on his way to become the king of junk bonds.

Milken was the visionary who pioneered the lending of large sums of money to support high-risk ventures. The high-yield loans became known as "junk bonds" and became a major funding source to support the build-out of Cable TV, mobile, and fiber networks. Milken, who ran Drexel's high-yield bonds business, further rocketed the use of junk bonds to fuel corporate takeovers and to fund the build-out of the cable and cellular empires. *Den of Thieves* and *Barbarians at the Gate: The Fall of RJR Nabisco*, two of my all-time favorite books, chronicle the junk bond era. These junk bonds were risky and, as such, carried much higher yield rates than traditional corporate debt.

When Milken moved his junk bond operation to Beverly Hills, Winnick followed. Before long, Drexel's relentless ambition and perpetual success led to SEC investigations of Milken. In 1989 Milken was accused of insider trading and related crimes, which led to him pleading guilty to securities fraud. He served twenty-two months of a ten-year sentence, paid $600 million in fines, and was barred for life from the securities industry. Milken was pardoned by Pres. Donald Trump in 2020.[17]

Winnick, who prospered at Drexel, received immunity after agreeing to testify against Milken.

Winnick left Drexel in 1985 and launched Pacific Capital Group. In the March 2002 article "The Rise and Fall of Global Dreams," the *New York Times* outlined two foreshadowing transactions: RB Furniture and Ortho Mattresses. In both, Winnick might have profited by selling his interests a few years before the companies filed for bankruptcy.

In 1996 Winnick targeted AT&T. "I knew the company was in turmoil. I thought there might be some crumbs swept off the table that would be good opportunities," Winnick said.

AT&T was laying fiber cable to connect New York to London, a project named Atlantic Crossing. Winnick took over the financial responsibility for AT&T's $750 million project.

In the past undersea fiber cables were constructed by consortiums of traditional telecommunications carriers such as France Telecom and British Telecom. The consortiums, or "clubs" as they were called, would split the cost and divide the capacity among them. The club members would haggle over details such as the specific path for laying the cable and where to land it so it could be attached to land networks, completing the transoceanic connection. The projects took years to complete, as the monopolists were not known for their urgency. Technology evolved slowly, leading to high prices for the constrained capacity.

------

17 Following his release from prison, Milken has dedicated his life to philanthropy for education and medical research through his Milken Family Foundation and the Milken Institute think tank. His contributions to cancer research and educational reform are particularly noteworthy.

Ten previous cables had been built; the most recent was in 1996 and involved seventy-five partners. "It was like doing business with the United Nations," a senior Global Crossing executive said.

Atlantic Crossing consisted of eight strands of fiber, which together would carry more capacity than the ten existing cables. Winnick was off to the races.

In March 1997 Winnick raised $750 million, including $20 million of his own money, mostly in the form of junk bonds. And Global Crossing was born.

In August 1998 Winnick took Global Crossing public, raising $399 million in the IPO at a share price of $9.50. The price jumped to $13.40 on the first day of trading. Another $500 million was raised three months later. When the stock price peaked at $64 a share in 2000, the Market Value of Global Crossing was $47 billion, which valued Winnick's $20 million investment at $6 billion.

"Ambitious, yes; implausible, no," summarized *Forbes*'s assessment in its 1999 feature titled "The $20 Billion Crumb."

In October 1997 Winnick hosted customers at a Marriott on New York's East Side. He told them he would charge $8 million for each unit of Atlantic Crossing capacity, which was 40 percent of the existing market price for capacity across the Atlantic. The rapid growth of the Internet and limited supply of transatlantic Bandwidth made for a seller's market. During the meeting he sold 10 percent of Atlantic Crossing capacity for an aggregate price of about $350 million, already recouping half the cost of the project.

Bolstered, Global Crossing soon announced several more projects: New York to the Virgin Islands; Pacific Coast to Panama to St. Croix; United States to Japan, a Pan-European cable; and even another cable across the Atlantic.

With Global Crossing's success, Winnick lived large. The company had five private planes, including a Boeing 737 and a Gulfstream. Winnick paid $65 million for an estate in Bel Air, the former home of the famed hotelier

Conrad Hilton. He made his office in Beverly Hills, California, a plush Hollywood setting that previously housed the leaders of MCA Studios. "I visited Winnick in his office," recalled CEO headhunter and former *Wall Street Journal* reporter John Keller. "It had a Hollywood Mogul over the top vibe with supermodel assistants and Art Masters hanging on the walls."

Winnick claimed his new best friend was David Rockefeller, after Rockefeller gave him a private tour of the Museum of Modern Art, prompting, in turn, a $5 million donation from Winnick. He gifted a Rolls-Royce to Lodwrick Monroe Cook III, his Global Crossing co-chairman and the former chairman of oil giant Atlantic Richfield. Constantine II of Greece was among Winnick's 240 guests who dined on Aberdeen Angus at Claridge's Hotel in London; US Secretary of Defense William S. Cohen was the featured guest speaker.

All the while Winnick kept the media hype machine humming, boasting to a *Sunday Telegraph* reporter, "Right before I left home, I had a call from the President of the United States to talk about something." He added, "Yesterday, I got a call from Buckingham Palace because one of the people there wanted me." Perhaps this call was to invite him to the one hundredth birthday party of Britain's Queen Mother, which Winnick attended in August 2000.

In April 1998, Winnick plucked Motorola executive Jack Scanlon to be Global Crossing's first CEO. In an interview the executive expressed his excitement at "a chance to build a global telecommunications company." His chance didn't last long. Ten months later he was replaced. No worries. He retained options worth $170 million while continuing as a board member and, later, the head of Asia Global Crossing.

Scanlon's short stint as CEO began a mad game of CEO musical chairs, with five executives sitting in the chair during Global Crossing's first four years.

Bob Annunziata was the second, following his second departure from AT&T. The first time was in 1983, when he resigned to become CEO of Teleport. The second time Annunziata left AT&T was in February 1999, eight months after he sold Teleport to AT&T for $12 billion. Winnick gave him a $10 million signing bonus and stock options worth another $10 million.

Annunziata said he joined Global Crossing for the opportunity to "build a company from start to finish." He lasted longer than the first CEO but just barely—thirteen months. Though his tenure was short, Annunziata left his mark.

On his seventeenth day of joining Global Crossing, Annunziata cut a deal to buy Frontier, the independent phone company headquartered in Rochester, New York, which had acquired fiber strands on Qwest's under-construction network, for $11 billion. In month four he signed an agreement to acquire Rocky Mountain Baby Bell U S West for $32 billion. Both were odd transactions, as each company was a bland, traditional telephone company.

Did Global Crossing know its stock was inflated? Was the end game to ensure it had a viable business underpinning its story-driven stock price by merging with local telephone companies?

Qwest, under CEO Nacchio, was in a similar situation as Global Crossing. They, too, had a soaring Market Value based on a network that was under construction. Nacchio decided to crash Global Crossing's party.

Both Frontier and U S West were public companies. Like most acquisitions involving public companies, the merger agreements allowed a window of time—called a go-shop period—for other bidders to offer a higher price. Qwest offered a price of $35 billion for U S West while also signaling they would also interlope on the Frontier merger.

The Global Crossing-U S West merger agreement included a provision that required U S West, even if another bidder prevailed, to purchase 9.5 percent of Global Crossing for $62.75 per share. This became a lucrative bargaining chip for Winnick.

After initially threatening to match Qwest's $35 billion offer, Global Crossing instead negotiated with Qwest and U S West. Qwest agreed to drop its pursuit of Frontier. Global Crossing agreed to terminate the Global Crossing-U S West merger agreement. U S West and Qwest agreed to purchase 9.5 percent of Global Crossing for $2.4 billion, paying $62.75 per share, which was a 26 percent premium to the market price of $49.75. Qwest also agreed to pay Global Crossing $400 million, which was half of the breakup fee.

On July 19, 1999, Qwest signed the agreement to purchase U S West. As payment, Qwest issued 882 million shares to U S West shareholders when the deal closed in July 2000. Qwest stock had risen to $50 per share, which drove the cost of the acquisition to $44 billion, up from the $35 billion when it was announced.

At first, Nacchio's outmaneuver of Winnick was seen as clever. Qwest's stock, which went public at $22 per share, rose to $50 per share during the year it took to close the U S West acquisition. Qwest, post the combination with U S West, was valued at nearly $100 billion. Yet Nacchio's behavior during the closing process suggested something was amiss.

Especially in light of what was soon to transpire, Winnick was the big winner. He personally pocketed $350 million by selling shares to U S West at the inflated price of $62.75 per share.

Frontier shareholders were the big losers. They received Global Crossing shares valued at $62. Soon after the merger, these shares would become worthless.

Annunziata declared of the Frontier transaction: "Clearly, [the Frontier] fiber optic network in the US puts us in an end-to-end situation globally." Annunziata's "clearly" said something else: he appeared to be stretching to make sense of the acquisition.

Annunziata wasn't finished with his acceleration of Global Crossing's plan. In September 1999 Global Crossing announced a $1.3 billion joint venture with Microsoft and Venture Capitalist Softbank to construct a seven-thousand-kilometer fiber network that was envisioned to eventually connect a dozen countries in Asia. He and his partners called the venture Asia Global Crossing.

"The future of our business and telecommunications is increasingly linked, and our interest in the telecom field is broad," said Microsoft president Steve Ballmer. He added, "Global Crossing and Softbank have complementary capabilities that make them ideal partners for Microsoft." Microsoft and Softbank each committed to purchase $200 million of Global Crossing network capacity.

Later in 1999 Annunziata acquired a UK Bandwidth provider for $1.65 billion in cash, adding a five-thousand-mile fiber network that reached more than two thousand cities and towns in the UK. With Gold Rush bravado, Annunziata explained to a reporter, "If the market is going to be worth $1 trillion by 2005, about a third of the traffic will be in Europe and a third in the US, so we were very clear that in our strategy we wanted to make sure we could have and own our own network here in Europe 100 percent."

Annunziata's final 1999 magic trick was another Asia joint venture, this one in Hong Kong. Hutchison Whampoa, one of the largest public companies in Hong Kong, operated a Hong Kong metropolitan fiber network. To do the deal and own 50 percent of the combined entity, Global Crossing contributed $400 million in stock, $50 million in cash, its 50 percent ownership in the Asia Global Crossing venture, and $350 million worth of capacity on the Global Crossing network. The total value ascribed to the Hutchison Global Crossing joint venture was $1.2 billion. Annunziata declared, "We are executing on our strategy to be the most enviable fiber-optic network worldwide."

Annunziata's transaction spree continued in early 2000 when he acquired a minority ownership position in a private company that did not own a fiber network. The peculiar transaction required Global Crossing to provide stock for a 27 percent stake in a company valued at $3.4 billion. The company provided communications services to the financial sector, including most of the largest investment banks.

"We will now move to the next level beyond providing building-to-building connectivity in major cities—providing desktop-to-desktop connectivity for multinational corporations worldwide," said Annunziata. "Hmm" was my reaction, seeing through the grand folly of this pronouncement.

In Global Crossing's earnings release in the first quarter of 2000, Annunziata framed a positive story: "These outstanding results demonstrate the rapid pace at which we continue to expand." Revenue for the fourth quarter of 1999 was $1.1 billion. However, net losses of $199 million had tripled since the fourth quarter of 1998. The day after the earnings release, Global Crossing announced it was weighing several options to raise more cash to further accelerate capital spending.

Annunziata resigned from Global Crossing in early March. The stock had hit a high of $61 that month. The following month, in April, Global Crossing's stock plunged to $25—less than half its price when Annunziata joined.

———

Around the time of Annunziata's departure, Level 3 CEO Jim Crowe had Kevin O'Hara and me tag along for lunch with Winnick and co-chairman Cook. We met at Dark Horse, Boulder's oldest beer and burger joint. Winnick was dressed in a three-piece suit, something rarely seen in circa 1999 Boulder. Crowe was quite the contrast, dressed as if he were a Stanford engineering professor in his first meeting with a Venture Capital firm. We discussed how to combine Global Crossing with Level 3. I was struck by the craziness of having cheeseburgers and fries with two billionaires discussing the possibility of a gigantic merger. The boy from Glenwood, Illinois, had come a long way.

All was chummy at lunch, with both CEOs expressing confidence in their business models. Reflecting, we all were hiding our growing nervousness about what loomed over the horizon.

Next up in the Global Crossing CEO lineup was Leo Hindery Jr., the former head of AT&T's cable television and Internet unit. Hindery joined Global Crossing a few months before being named CEO in April 2000; he resigned seven months later, taking with him Global Crossing stock valued at $247.5 million.

Two months into his short reign, Hindery sent a memo to Winnick and two other executives, warning of a disastrous outcome.

> Like the resplendently colored salmon going up river to spawn, at the end of our journey our niche is going to die rather than live and prosper. . . . The stock market can be fooled, but not forever, and it is fundamentally insightful and always unforgiving of being misled. . . . Without looking like we are shaking our bootie all over the world, [we must] sell ourselves quickly to whichever of the six possible acquirers offer our shareholders the highest value.

The memo was not disclosed to the public shareholders. Its existence was uncovered years later by the House Energy and Commerce Committee as part of its investigation into Global Crossing's wrongdoings. Winnick sold shares valued at $100 million after the memo was received. Does that smell like insider trading to you?

With his colored salmon analogy in mind, Hindery abruptly changed his swim path from Annunziata's buying spree and began a sell-off of Global Crossing assets.

In July 2000 Global Crossing agreed to sell the local telephone business it acquired from Frontier for $3.6 billion in cash, a small fraction of the $11.2 billion Global Crossing paid to acquire Frontier. I guess Hindery concluded a bird stuffed with cash is worth three plumply overvalued birds stuck in a bush.

Later in 2000, Hindery agreed to merge GlobalCenter with data center operator Exodus Communications. Global Crossing received $6 billion of Exodus stock, which represented 17 percent ownership in Exodus. As part of the transaction, Exodus agreed to purchase 50 percent of its future non-Asia network capacity from Global Crossing for the next ten years.

In actuality both Exodus and GlobalCenter were wildly overvalued. By the time the transaction was completed in January 2001, the 17 percent stake in Exodus was valued at $1.8 billion because of a steep decline in Exodus's stock price. Exodus's financial duress continued, resulting in a bankruptcy filing in late 2001. The shares Global Crossing received and Exodus's ten-year commitment to purchase capacity were basically worthless.

The final event that happened during Hindery's short CEO stint was the IPO of Asia Global Crossing. Winnick had to push hard to execute this IPO. Global Crossing had hoped to raise over $900 million by selling shares at $15 each. However, with market receptiveness soft, the shares were priced at only $7 per share.

Demand was so limited that Winnick had to pressure his bankers—Goldman Sachs and Salomon Smith Barney—to buy fourteen million of the shares. A colleague of Winnick's explained how Winnick strong-armed Goldman and Salomon to buy the shares: "'I don't want to hear any whining from the bankers,' Winnick snapped."

Hindery must have been spooked. He resigned on October 12, 2000, six days after the Asia Global Crossing IPO.

Batting cleanup was Global Crossing's vice chairman, Thomas J. "Tom" Casey, a lawyer who was cohead of the global telecom investment banking group from Merrill Lynch before joining Global Crossing in late 1998. Keeping with the pattern of his predecessors, Casey lasted eleven months as CEO before returning to the board.

Casey was a close confidant of Winnick. In stark contrast to Hindery's dire warnings, Casey offered a positive outlook for Global Crossing, positioning that the decline in the stock price was due to the broader economy: "The fact that we're in a troubled environment is causing people to lose sight of the substantial accomplishments that Global Crossing has made over the last four years."

Casey reaffirmed Global Crossing's strategy and executive team: "Our strategy is not changing. . . . My job as chief executive is to lead a team that is already in place." Casey also reiterated targets for 30 percent growth in cash revenues and 35 to 40 percent growth in profitability.

How, though, was Global Crossing going to deliver on these financial promises? Accounting gimmicks turned out to be the answer. And when this was uncovered, Global Crossing would forever be remembered as "Global Crashing."

Qwest—Joe Nacchio

Nacchio swiped U S West from Global Crossing and, in doing so, saw his rival Winnick put $350 million in his personal bank account. Nacchio whined that Crowe left his board and copycatted Qwest's strategy, then watched Crowe sell $115 million of his $1.1 billion Level 3 stock. Were these the early indications that a mad dash to cash in before the Gold Rush was soon to unravel? If so, when and how would Nacchio get his payday before the Boom exit window was to slam shut?

In March 2000, rumors swirled that Qwest, which had not yet closed the U S West merger, was considering selling itself to Germany's phone company. "[Qwest] could still end up striking a deal with Deutsche Telekom. Such a deal would allow [Nacchio] to walk away with hundreds of millions of dollars and [Anschutz] to expand his bankroll by billions," wrote the *New York Times*.

The pending U S West deal proved to be an insurmountable hurdle to the potential sale. Qwest's interest in the deal raised questions. "It smacks of Monday-morning quarterbacking to point out that the deep strategic rationale underlying Qwest's bid for U S West was not crystal clear when it was announced last summer," the *New York Times* reported. "The conventional wisdom at the time was that rather than being about far-reaching strategic synergies, the deal was more about Qwest's using its high-flying stock to go after an undervalued company, with concrete assets and customers."

On June 18, 2001, Nacchio's crosstown rival, Crowe, announced that Level 3 was lowering its earnings forecast and was laying off almost a quarter of its employees. The next day Nacchio poured salt into Crowe's wounds by reaffirming Qwest's financial guidance. "Despite some softening in the general economy, we believe that demand for communications services is stronger than what others have reported," Nacchio told *Forbes* in its article titled "Is Qwest the Best?"

Forbes praised Nacchio: "Nowadays the Bells are looking a lot stronger than most of the upstarts, and Nacchio is looking like a genius for having a foot in each camp. [A Merrill Lynch analyst] reiterated his buy rating on Qwest, which he hailed as a 'unique hybrid' of next-generation network and traditional local service provider, 'supported by a very strong management team.' And the man heading that team, one Joseph P. Nacchio, appears stronger than ever."

Perhaps *Forbes* and others were focused on the wrong question. Instead of asking *whether* Qwest would achieve its revenue targets, they should have been asking *how*.

Broadwing—Ralph Swett

In the Renegades we learned that Broadwing's founder, Ralph Swett, cashed in his chips by selling his long-distance fiber company to Cincinnati Bell for $3.2 billion. The Gold Rush lured the quiet and long-static phone company into a bold and, in the end, unwise action. It was an uncharacteristic move that defied Cincinnati Bell's century-long evolution from its start as the City and Suburban Telegraph Association in 1873.[18] The company was founded to provide telegraph services in the Cincinnati area. Five years later, in 1878, the local service provider was awarded the Bell telephone franchise in Cincinnati and became the tenth telephone exchange operator in the United States. In 1961 the business was rebranded as Cincinnati Bell, reinforcing the phone company's geographic focus on Cincinnati and the surrounding area.

In early 1999 Cincinnati Bell was granted permission to compete with the Ohio Bell division of Baby Bell Ameritech. Since Cincinnati Bell was not a spin-off of AT&T, Judge Green's consent decree did not apply, and the business was permitted to offer long-distance services. Cincinnati Bell's leaders surveyed the landscape of competitive fiber buildouts and decided to remake their company from a boring telephone company into a national Bandwidth provider, thus their acquisition of Broadwing for $3.2 billion.[19]

"We will be able to offer long distance, local, data, and the full suite of telecommunications services," proclaimed Cincinnati Bell CEO Richard G. Ellenberger.

Cincinnati Bell's shareholders were not enamored. They invested in the Bell company for its dependable, steady stream of dividends. Now the transformation would turn the low-risk Bell company valued on earnings and dividend yield into a high-risk stock valued on revenue growth and profitability multiples.

Ellenberger's solution was for the old shareholders to exit. He raised $400 million from investor Robert M. Bass, one of the billionaire children of the founder of Bass Brothers Enterprises. Ellenberger used half of Bass's investment to buy back stock from existing shareholders. Cincinnati Bell's stock price dropped from $49 to $41 per share.

---

18 In 1903 the name was changed to Cincinnati and Suburban Bell Telephone Company.

19 The $3.2 billion price paid by Cincinnati Bell for Broadwing included $1.1 billion in assumed debt and stock that was valued at $2.1 billion.

Soon after the merger was closed, Cincinnati Bell announced it was changing its name to Broadwing. Beaming with confidence, Ellenberger declared on a conference call, "The broadwing hawk soars at about 1,000 feet above its prey. And when it strikes, it strikes." Dennis Hopper, best known for directing and acting alongside Peter Fonda and Jack Nicholson in Easy Rider, was featured in a series of commercials that touted Broadwing's ability to fly while characterizing AT&T and the Baby Bells as dinosaurs.

In addition to completing the construction of a 16,500-mile network, Broadwing developed a data center and a web hosting business. At its peak, Broadwing, excluding the value of Cincinnati Bell's local telephone companies, was valued at approximately $5 billion.

In April 2001 Broadwing issued a press release. "In a period where established providers are losing market share and emerging providers are struggling, Broadwing continues to shine," said Ellenberger. Before long, the shine would lose its luster.

PSINet—Bill Schrader

With UUNet part of WorldCom and BBN absorbed into GTE, PSINet came to be the only remaining standalone national Internet Backbone. In his Gold Rush celebration, Schrader committed $100 million for the naming rights to the Baltimore Ravens' new stadium. This was a small price to pay for a company whose Enterprise Value soared to $10 billion.

PSINet acquired over seventy companies to chase Schrader's vision to deliver "all media to all the customers on earth." Most of these were small regional ISPs, both in the United States and other parts of the world. For example, in 1998 PSINet exchanged $25 million of its stock for a Canadian Internet provider, which made PSINet Canada's largest ISP.

Schrader must have been feeling pressure to use acquisitions to bulk up revenue and diversify his product offerings. In 1999 he bought Transaction Network Services (TNS) for $720 million in cash and stock. TNS was founded in 1990, went public in 1995, and became a leading global provider for the high-speed, secure electronic fund transfer of credit card and automatic teller machine transactions. TNS's product focus was a departure from PSINet's positioning as an ISP.

PSINet entered the new millennium with a lot of confidence. On January 10, 2000, the company announced its quarterly revenue was expected to nearly double from the fourth quarter of 1998—from $93.9 million to $185 million. The forecast update included the expectation that cash burn would also double—growing to $17 million from $9.9 million. The market responded positively, as PSINet attributed the increase in losses to its expansion initiatives.

"We are now set up for 2000," said PSINet's treasurer. An analyst with Wasserstein Perella put a positive spin on the announcement: "They are investing in their business and finding strong demand across the wholesale end, web-hosting end and e-commerce end." Wasserstein Perella was an underwriter for PSINet's TNS acquisition.

In a show of grand confidence, Schrader launched LeapFrog in February 2000—a $1 billion corporate venture program using a services-for-equity investment model. Instead of collecting cash from its start-up, Dot-Com customers, PSINet would provide services in exchange for equity. What could possibly go wrong?

In late February 2000, less than two months after the market responded positively to PSINet setting expectations of its fourth-quarter revenue and cash burn, PSINet reported its results and the resignation of its CFO. Operating losses jumped to $135 million, which was more than double the $53 million in the fourth quarter of 1998. This time the market's response was not so chummy, driving PSINet's stock price down by 15 percent to $39 per share.

Schrader stated that PSINet remained on track to profitability in 2002 and dismissed the likelihood of being acquired: "We have exactly what the largest telephone companies want—network, hosting, expertise, and customer base—but at this point, we're not for sale."

The following month PSINet announced the $1.9 billion acquisition of Metamor Worldwide, a professional outsourcing services firm that migrated its traditional systems to web and e-commerce applications. Like the TNS acquisition, this purchase was a major departure from PSINet's core business of being an ISP. Proforma, the combination of the three companies, represented $1.1 billion in annual revenue and ten thousand employees.

Schrader did his best to explain the rationale: "What we had before this deal was the best long-haul fiber carrier and also the best [data center] facilities. What we were missing was the skills to . . . build the system it takes to live on the Internet."

In March 2000 the *New York Times* published an article titled "An Iconoclast Goes It Alone; on the Net PSINet's Schrader Is a Force to Reckon With," glowing that "PSINet, with fiber optic lines around the globe, some of the world's most advanced Internet systems and a fast-growing customer base, looms as a highly attractive acquisition candidate—one whose shares are up nearly 300 percent since October, as investors have noted its potential takeover value."

PSINet's COO added fuel to the acquisition speculation. "Some day, they will buy us and we will be sold, and my job is to make it more expensive for them every day that they wait. What they could have bought for $200 million, they will now probably have to pay $20 billion for."

"They" waited longer, and then "they" paid almost nothing for Schrader's PSINet.

Touch America (Montana Power)—Bob Gannon

Robert "Bob" Gannon, the CEO of a power utility in the sparsely populated state of Montana, was not to be left out. Fiber, he concluded, was his path to glory and riches. Gannon risked everything with an all-in Gold Rush bet.

The *Wall Street Journal* published "For Montana Power, a Broadband Dream May Turn Out to Be More of a Nightmare" in August 2001 and brilliantly captured an unfolding disaster.

This is a gamblers' town. More than a century ago, Marcus Daly turned a bet on a copper vein under Butte into the giant Anaconda Co. The gambling game Keno was invented here for bored miners. In 1974, stuntman Evel Knievel, a Butte native, made headlines betting—unsuccessfully—that he could jump the Snake River Canyon on his rocket-powered motorcycle.

Now, Bob Gannon, another Butte native . . . is betting he can pull off his own breathtaking stunt: He plans to turn his staid 89-year-old utility into a high-tech high-flier.

In one dizzying two-year stretch, the Butte-based utility has sold just about all its assets—power plants, transmission lines, dams, gas, coal, and oil fields—pouring much of the $1.6 billion into a coast-to-coast fiber-optic network. By the end of this year, Mr. Gannon boasts, Montana Power will have 26,000 miles of fiber in the ground, a new corporate name, Touch America, and it will be among the nation's largest half-dozen broadband companies.

So far, Mr. Gannon's gamble seems to be following the same trajectory as Mr. Knievel's. The joke around here is that at least Mr. Knievel wore a parachute.

Snake River Canyon, located in the Magic Valley region of southern Idaho, is 500 feet deep and 0.25 miles wide. Knievel was a famous motorcycle stuntman known as much for his bone-breaking crashes as for his record-breaking jumps. His capstone jump over the Snake River was September 8, 1974. A rocket-powered cycle, featuring steam that heated the engine to five hundred degrees, was custom designed for the stunt. A parachute malfunction doomed the jump, leaving Knievel sailing to the bottom of the canyon, landing a few feet from the water on the same side of the canyon from which it had been launched. Knievel, who was strapped to the cycle, said he would have drowned if he had landed in the water. Instead, he had only minor physical injuries. Building upon the *Wall Street Journal* analogy, Gannon landed in the water.

The Montana Power Company was the electric utility company headquartered in Butte that served Montana from 1912 to 2000. In the late 1980s, Montana Power, similar to many other power companies, monetized its pipeline rights-of-way by constructing fiber networks. Instead of leasing land rights to telecommunications carriers, though, Montana Power favored owning the fiber, lighting it up, and providing Bandwidth services.

In June 1999 Montana Power completed the route between Spokane, Washington, and Boise, Idaho. This route connected the utility's Chicago-to-Seattle route with its Seattle-to-Los Angeles route. Denver to Dallas was completed in late 1999. In May 2000 Montana Power announced a partnership with Sierra Pacific Resources to construct a fiber route between Sacramento and Salt Lake City.[20]

This was just the beginning. "Contracts are pending for system expansions from Los Angeles to Dallas through San Diego and Phoenix, from Dallas to Houston through Austin and San Antonio, and from Houston east to Florida through New Orleans," touted Montana Power's president. "These and other builds will increase our network to 18,000 route miles by year end 2000 and 23,000 route miles by year end 2001." In late 1999 Montana Power was one of the three companies that AT&T selected to partner on a nationwide fiber network deployment.[21]

In March 2000 Gannon announced that Montana Power was exiting the power business to fund its transformation into a national fiber company. Wall Street reacted positively, as evidenced by the stock price jumping to an all-time high of $65 a share, implying a Market Value of $6.5

---

20 Montana Power also built fiber rings in Denver, Reno, and Las Vegas.

21 When Qwest merged with U S West in 2000, the combined company was required to sell Qwest's mountain-west, long-distance business. Gannon paid Qwest $200 million and gained 250,000 long-distance customers.

billion. In previous years the stock typically traded between $10 and $14 per share.

Montana Power sold its hydroelectric dams, coal mines, power plants, oil and gas assets, and energy distribution business for a combined $2.1 billion.[22] To capstone this business transformation, Gannon rebranded Montana Power as Touch America. The story of its reboot might best be titled "Oops!"

WilTel 2 (Williams Communications Group)— Howard Janzen

When Williams Companies sold WilTel to WorldCom in 1995, it agreed to not compete with WorldCom for three years. The noncompete expired just as the Boom was exploding. After watching Jim Crowe jump back into Bandwidth Babylon, Williams Companies crafted its reentry, naming it Williams Communications Group and, a few years later, most was carved out and rebranded as WilTel. I will refer to Williams Communications Group and the new WilTel as WilTel 2, to avoid confusion with the original WilTel that was sold to WorldCom. In its deal with WorldCom, Williams kept its video unit Vyvx, which rolled into WilTel 2. Vyvx's leader, Howard E. Janzen, was named the CEO.

WilTel 2's first announcement was a joint venture with Enron and Touch America to construct a $100 million, 1,620-mile route connecting Los Angeles, Salt Lake City, and Portland, Oregon. "We've studied the

---

22 Montana Power sold the company's hydroelectric dams, coal mines, and power plants to Pennsylvania Power and Light in 1998. In the early 2000s, the oil and gas business was sold to a Canadian company. Westmoreland Coal purchased the coal unit. When Montana Power sold its energy distribution business to Northwestern Corp. for $1.1 billion in October 2000, the exit from power was completed. In total over $2.1 billion was raised by selling the assets.

hot fiber routes, and this is definitely one of them," said Janzen. He set a goal of having a 19,000-mile network in place by the end of 1998.[23]

Williams's announced reentry into the fiber business got a thumbs-up from the market, and its stock shot up by 51 percent, hitting a fifty-two-week high of $30. Some were skeptical. The *New York Times* questioned "whether the nation needs another high-speed communications network."

But in early 1999, WilTel 2 speared a whale. The San Antonio–based Baby Bell SBC committed to a $500 million investment in WilTel 2.[24] "We've said all along that we want to be a global telecommunications player," said SBC CEO Edward E. Whitacre Jr., who will come up later in our story. Referencing SBC's pending acquisition of Ameritech, Whitacre added, "The alliance with [WilTel 2] gives us the line to connect all these dots." The alliance made SBC a strategic customer of WilTel 2. SBC grew to be WilTel 2's largest customer, generating $400 million or one-third of WilTel 2's 2002 revenue.

Shares in Williams Companies, which remained the owner of WilTel 2, rose to $36. Plans were underway to spin WilTel 2 into a separate publicly owned company.

In preparation for the IPO, WilTel 2 executives were loaned $13 million to purchase shares of WilTel 2 at management-friendly prices, of which $5.6 million was allocated to Janzen. The expectation was that Janzen would eventually sell the stock that was purchased using these loans, paying capital gains instead of income tax, and retaining a hefty profit after paying back the loans. I suspect the board never thought through what they would do if the value of WilTel 2 cratered before the

---

23 WilTel 2 also announced an 1,800-mile build from Houston to Washington, DC, using its Transcontinental Gas Pipeline right-of-way. WilTel 2 promised Broadwing fiber on this route in exchange for fiber on Broadwing's 4,500-mile fiber network. WilTel 2 also acquired a 350-mile fiber network connecting Jacksonville, Florida, to Miami from MediaOne.

24 Intel invested $200 million in WilTel 2.

loan repayments. When the stock did indeed collapse, the handling of the loans became a thorny problem.

WilTel 2 executed its IPO in September 1999, achieving a Market Value of over $10 billion.[25] Williams Companies continued to own 86 percent of the company. Janzen was pleased with the market's reaction: "Becoming a publicly traded entity . . . will give telecom-focused investors and analysts the opportunity to better value WilTel."

Through construction and fiber exchanges with other network providers, WilTel 2 grew the network to 33,000 miles connecting 125 cities on five continents. The stock price peaked in mid-2000 at over $40 per share, implying a value of $15 billion.

The future looked sunny indeed. In July, WilTel 2 moved into its newly built fifteen-story, high-tech Tulsa headquarters, right next to the home of its former parent. I hope the move-in party was a hell of a bash, as the WilTel 2 team would have no more reasons for celebrations.

AT&T—Robert E. Allen and C. Michael Armstrong

The once all-mighty AT&T wasn't about to be stampeded by the Gold Rush! The company's brass were obsessed with reclaiming AT&T's prior glory as the unquestioned leader of the telecommunications world.

To put this in perspective, while rivals such as MCI had been seeking to kneecap their giant rival and compete, AT&T, with its vast resources, was still able to be the architect of the consent decree that Judge Green approved. AT&T's actions in reaching that divestiture accord implied that the company believed it would remain in the driver's seat post breakup, albeit in a transformed, upscale vehicle. Yet one decade later, AT&T found itself highly vulnerable in the fiber-Internet Boom. What

---

25 $300 million was raised in the IPO, with the stock opening at $23 per share and closing the day at $28. The investments by SBC and Intel made earlier in the year resulted in a 6 percent ownership position.

would AT&T's response be to Qwest, Teleport, MFS, AboveNet, Broadwing, WorldCom, and XO? And could AT&T prepare well for its burgeoning battles with its Baby Bell offspring?

These questions developed under the reign of Robert "Bob" E. Allen, who became CEO in 1988, just a few years after the Bell System breakup. Born in Joplin, Missouri, and raised in Indiana, this tall, soft-spoken Hoosier spent his entire forty-year career at AT&T, starting with AT&T's Indiana Bell upon graduating from tiny Wabash College in 1957.

Allen's biggest accomplishment was leading AT&T post-divestiture and into mobile services with the 1993 acquisition of McCaw Cellular, the nation's largest independent mobile services business, for $16.5 billion. This made AT&T the largest wireless carrier in the United States. Allen's biggest blunder was acquiring National Cash Register (NCR) for $7.5 billion in 1991, only to spin it out for a loss in 1995.

Allen, in search of answers in the wake of the NCR debacle, concluded the first step was to name his successor. He surveyed his executive team. The leading candidate had been his president, Alex Mandl, who, as CFO, had executed the McCaw deal. However, Mandl resigned in April 1996 when he was offered a $20 million sign-on bonus to lead a wireless telecom start-up that was backed by Microsoft and Nippon Telegraph and Telephone (NTT).[26]

Earlier, I mentioned Joe Nacchio's departure from AT&T. Nacchio, who led AT&T's consumer and small business unit, also would have been a leading candidate, but Allen and Nacchio had a difficult history, and Allen didn't want to hand the reins to Nacchio. Nacchio resigned to become Qwest's CEO.

After an extensive external search, Allen picked John R. Walter, the CEO of R.R. Donnelley & Sons, the world's largest printing company.

---

26 Teligent went public in 1997, and the stock price peaked at $97 per share at the height of the Boom. MFS's Buddy Pickle was named COO. However, Teligent's wireless access service didn't gain traction. Pickle departed in 2000, Mandl resigned in April 2001, and Teligent declared bankruptcy in May.

For unclear reasons, Allen thought Walter's twenty-seven-year career printing Yellow Pages was sufficient preparation for running the very complex and increasingly high-tech AT&T.

The *Washington Post* conveyed Wall Street's tepid reaction to Walter in "Heir Is Unknown, Undaunted."

> Allen needed a new president—someone who could succeed him at the helm in a couple of years, pep up the company's anemic earnings, battle an invasion of new competitors and push AT&T into such new markets as local phone service, the Internet and satellite television. So yesterday he and the AT&T board hired an unknown guy who prints phone books.

Noting AT&T's stock dipped on the announcement, a Legg Mason investor expressed the general confusion surrounding the Walter hire: "As best as I can tell, it's very difficult to get a good fix on John Walter's executive abilities."

In any case AT&T named Walter president and Allen's would-be successor in October 1996. AT&T had expectations to promote Walter to co-CEO in early 1998 and, upon Allen's retirement in late 1998, to CEO—with the AT&T board's approval, of course.

"The best laid plans of mice and men often go astray," as Allen and Walter were soon to learn. Soon after Walter joined, Allen lost confidence. Allen confided to a *Wall Street Journal* reporter at the time that Walter had failed to deliver on the very first order Allen had given to his new president. "I told him to fire Nacchio," Allen said. "And he didn't do it." It took several more months for Nacchio to leave and join Qwest.

In July 1997 the *New York Times* reported that AT&T executives "began to have doubts about Walter almost from the moment he came to work last November. These executives said Walter . . . did not immerse himself in the mechanics of the telephone business."

The *New York Times* added that "people familiar with [AT&T] said Walter found himself at loggerheads with Allen on basic strategic issues: whether the company should aggressively enter the local phone business [Walter argued that it should not]; whether it should continue to invest heavily in wireless [he said no], and where AT&T should put its resources [he favored international investments over domestic initiatives]."

The tension between Allen and Walter was exposed when Baby Bell SBC expressed interest in merging with AT&T in the spring of 1997. The *Wall Street Journal* broke the news: "The $50 billion-plus plan [for AT&T] to combine with SBC Communications Corp. could, in one swoop, erase years of unsuccessful strategies under Allen and allow him to retire on high ground." Allen expressed his support for SBC's CEO, Edward Whitacre, to be named the CEO of the combined company.

The merger faced stiff backlash. "The forces of evil unleashed by the new telecom act are now obviously running far ahead of the forces of good. This cannot happen. This can't go forward," said the research director of the Consumer Federation of America. A former Justice Department antitrust chief who lobbied in favor of the telecom law added the proposed merger "creates huge competitive problems. This isn't what was supposed to happen."

In June, Whitacre called Allen and told him SBC no longer wanted to pursue the merger. The *Wall Street Journal* reported that Whitacre blamed Allen, writing that SBC "bailed out because the chances of winning approval for a merger were hurt by Mr. Allen's unusual public campaign in support of the idea of a Bell merger." Specifically upsetting to SBC was Allen's repeated claims that "the Bells weren't opening up their markets as federal law demands."

With the merger now dead, Allen and the AT&T board had to deal with the CEO succession quagmire. AT&T's independent directors concluded Walter was the wrong person for the CEO job.

Prompted by *Wall Street Journal* reporter, John J. Keller, at a hastily called press conference to elaborate on why it had been decided that Walter shouldn't succeed Allen as CEO, an AT&T board member said the board had determined that Walter "lacked the intellectual leadership to lead AT&T." The board member added, "He's a bright guy, but the complexity of the business is far greater than he might have realized." The fallout of these comments certainly provided employment attorneys a mess to clean up.

Reacting to the decision, Walter countered, "I believe I am perfectly qualified to be CEO of AT&T right now." Nonetheless, Walter exercised a clause in his employment contract that enabled him to leave with a large severance payment if he was not elevated to CEO by January 1997. He received a one-time payment of $3.8 million, plus the $22 million enticement he had been promised to leave his Donnelley CEO role—not too shabby consolation for nine months of craziness.

The AT&T board still had to sort through CEO succession. The urgency was apparent to the *Wall Street Journal*, which reported, "In undermining his designated successor, Allen may, in fact, have inadvertently hastened his own departure." To attract a strong replacement, the board needed to offer the CEO title from the start.

The search led them to C. Michael Armstrong, a 1961 graduate from Miami University of Ohio who spent thirty-one years rising up the ranks of IBM, where he pinnacled as the president and chairman of Europe, Middle East, and Asia. In 1992 Armstrong left IBM to become CEO of Hughes Electronics, a satellite communications company owned by General Motors. During his five-year stint at Hughes, Armstrong built DirecTV into the leading satellite television service.

Armstrong's reputation as an aggressive transformational executive with experience in complex industries made him an attractive candidate to AT&T's board, who chose him as Allen's successor in October 1997.

Armstrong's compensation package included stock and stock equivalents valued at $15 million. Interestingly, AT&T guaranteed the minimum value of the equity would be $10 million after five years. With his downside protected, Armstrong was incentivized to swing for the fences.

Though its traditional long-distance phone service business was in steady decline, AT&T was the nation's largest wireless carrier. AT&T had a strong balance sheet, with annual cash flows in the billions of dollars and, resulting from its $3.5 billion sale of its credit card business, was almost debt-free.

Investors reacted positively toward the choice of Armstrong, as evidenced by AT&T's stock price rising 44 percent in Armstrong's first few months, to over $60 a share, resulting in AT&T's value reaching $105 billion. Armstrong was fully armed and, with swagger, ready to make his mark.

Armstrong got to work quickly. He struck the deal to acquire Teleport for $11 billion in January 1998, an event that further ignited the Bandwidth Boom.[27]

"This is a good deal for us, for business-phone customers across America, and a good deal for the shareholders of both companies," explained Armstrong to investors in a conference call. "Teaming with Teleport will speed our entry into the local market."

The *Wall Street Journal* captured the market's positive reaction to the announcement. "From AT&T's perspective, it solves the critical question mark that hung over the company's head for some time: What is their local strategy?" said an analyst at Deutsche Bank. "[Teleport] is by far the most strategic asset in the telecom business today."

A Lazard Frères & Co. analyst added, "It makes the most sense to go after the biggest piece of the puzzle first. Having the marketing clout

---

27 The transaction also required AT&T to complete Teleport's pending purchase of ACC for an additional $1 billion.

of AT&T behind Teleport will make Teleport that much stronger a player. It justifies the premium they're paying."

"This is the best news they've had in a long time," another telecom analyst agreed.

Annunziata, who left AT&T fifteen years earlier to be Teleport's CEO, seemed eager to partner with Armstrong. Annunziata said returning to AT&T was "a great feeling. . . . This is a new AT&T." Annunziata's great feeling must have been a reference to his newly bloated bank account, as less than a year later, he resigned to become CEO of Global Crossing.

———————

Teleport was AT&T's answer to metropolitan fiber. Armstrong's transformation also required an intercity fiber network. Acquiring Qwest, which recently went public and was now valued in excess of $5 billion, was a possibility. Broadwing, which was not yet acquired by Cincinnati Bell, was another potential target, with a more reasonable price tag of $3 billion.

AT&T presumably conducted a build-versus-buy analysis. Instead of buying Qwest or Broadwing, they would fund their own fiber build-out. Their plan was to leverage their existing rights-of-way to allow other fiber companies to deploy a state-of-the-art fiber network. They issued an RFP with a design that matched their existing paths. The RFP allowed respondents to place extra facilities for their own use, enabling AT&T to benefit from more aggressive bids.

I was the leader of Level 3's response to the AT&T RFP. At Crowe's direction we teamed with WilTel 2, which was building out an intercity fiber network, to provide a joint bid. We saw ourselves as the two primary contenders and, by teaming up, would make it difficult for AT&T to leverage us against one another. Since both of our networks were under construction, our response deviated significantly from AT&T's specific routes.

AT&T wasn't happy, both because of our teaming approach and the noncompliant design. Moreover, I suspect AT&T viewed Level 3 and

WilTel 2 as competitors, and as such, it was not in AT&T's best interests to support our business plans. So AT&T decided instead to fund three start-ups—Velocita,[28] CapRock,[29] and Touch America.

"AT&T is responding to unprecedented growth in demand for high-speed Bandwidth and Internet-based services by building a new, state-of-the-art fiber-optic network that will link 30 major metropolitan areas nationwide," explained an AT&T executive.

Next, Armstrong turned his attention to AT&T's biggest competitive gap—how to reach its customers without relying on its Baby Bell off-spring. Armstrong concluded the solution was to acquire Cable TV providers.

Perhaps Armstrong was egged on by Cable TV Cowboy John Malone. Soon after AT&T announced Armstrong as CEO, Malone said, "I've met every AT&T chairman since 1964, and this is the first one that gets it."

When Malone pays a CEO a compliment, CEOs wise to their ways know to grip their wallets.

---

28 See appendix 3, "More Bandwidth Stories," to learn more about Velocita.

29 CapRock Communications was founded in 1981 in Dallas, with a focus on providing remote communications to oil and gas customers. As the fiber build-outs gained momentum in the mid-1990s, CapRock aspired to become "the leading facilities-based telecommunications provider in the Southwest region." They undertook a fiber expansion with the goal of building 6,100 miles of fiber across Texas, Louisiana, Arkansas, Oklahoma, New Mexico, and Arizona. In 1999 AT&T selected CapRock as one of its three partners to build out a national fiber network. CapRock's stock price leaped to a high of $58.50, which implied a Market Value approaching $2 billion.

As the Bandwidth Bubble began to burst, CapRock's stock dropped to $10 per share, which led to McleodUSA purchasing CapRock for $535 million, of which $335 million was in assumed debt. A total of 5,200 miles of fiber were complete at the time of the sale.

McleodUSA soon found itself in bankruptcy, which led it to sell CapRock to Texas investors for significantly less than it paid a year earlier. CapRock, whose ownership changed two additional times in the ensuing twenty years, continues to operate its southeast communications business under the name SpeedCast.

Malone was a major investor in Teleport, so he had already benefited from Armstrong's dealmaking. However, Malone had something else that Armstrong coveted. TCI was the second-largest Cable TV provider with eleven million subscribers.[30] TCI's cable systems passed one out of every three American homes.

Led by Malone, the Cable TV industry made a compelling case that its architecture, which was based on coaxial copper cable, was a better thoroughfare into the home than the twisted pair copper used by phone companies.[31] A hybrid fiber/coax architecture would allow Cable TV companies to compete with the local phone company for voice while also providing a higher bandwidth Internet service than the dial-up access offered by the phone company.

In June 1998 Armstrong agreed to buy TCI for $48 billion.[32] The deal was completed in eight days. As part of the deal, AT&T also agreed to purchase, for $2.8 billion, the AT&T stock that TCI received when AT&T bought Teleport. Malone joined the AT&T board and became AT&T's biggest individual shareholder.

AT&T's stock dropped by 10 percent. Unfettered, Armstrong sought more Cable TV assets.

In March 1999 Comcast, the third-largest Cable TV provider with six million subscribers, announced an agreement to acquire MediaOne, the fourth largest with five million customers, for $48.6 billion. If completed, Comcast would have over eleven million subscribers, slightly surpassing AT&T to become the second-largest Cable TV provider.[33]

---

30 TCI also jointly owned with other CATV operators an additional eleven million subscribers.

31 Hybrid fiber/coax is an architecture where fiber is used between the Cable TV headend and a node in a neighborhood. Coaxial cable then carries the signal from the node to a cluster of homes. The use of the fiber allows more information per home. The fiber also enabled additional services, such as voice and Internet, to be bundled together with the Cable TV service.

32 The $48 billion price consisted of $31.8 billion in AT&T stock and the assumption of $11 billion in TCI debt.

33 Time Warner Cable had 12.6 million subscribers.

"Today, with one fell swoop, we have created the pre-eminent broadband company in the world," said Comcast president Brian Roberts. Armstrong called Roberts and congratulated him on the announcement.

Then Armstrong reflected on the situation. The terms of the deal included a forty-five-day window for MediaOne to accept a superior proposal from another suitor. If this were to happen, Comcast would be owed a $1.5 billion breakup fee.

Armstrong decided to go hostile, penning a letter to MediaOne's CEO and publishing it in the *Wall Street Journal*:

April 22, 1999

Mr. Charles M. Lillis
Chairman and Chief Executive Officer
MediaOne Group,
188 Inverness Drive West
Englewood, Colorado 80112

Dear Chuck:

AT&T is pleased to offer to acquire all of MediaOne Group for cash and AT&T common stock. AT&T will pay $30.85 in cash.

. . . Our offer represents a 17 percent, or $8.6 billion, premium to the value of MediaOne's previously proposed merger with Comcast. . . . This is a 44 percent premium to the trading price of the MediaOne shares prior to the announcement of its merger agreement with Comcast and a 26 percent premium to today's MediaOne closing price.

. . . With the addition of the MediaOne local broadband systems, we will offer customers superior connections for not only traditional video services, but also new competitive voice and data services. In particular, this will enhance our ability to deliver competitive local telephone services to millions of Americans.

Given the clear superiority of our offer to the proposed Comcast merger, we would like to meet with you and your advisors as soon as possible to finalize a definitive agreement between our companies.

Sincerely,
C. Michael Armstrong

The hostile takeover ploy worked. In June 2000 AT&T completed its purchase of MediaOne. AT&T was now the biggest Cable TV provider with sixteen million subscribers. The $5,000 price per subscriber was 67 percent higher than the $3,000 Armstrong paid for TCI subscribers.

AT&T's stock was now down to $31 per share.

Acquiring Teleport, constructing fresh intercity fiber networks, and being the number one Cable TV provider wasn't enough for the ambitious Armstrong. He also sought to cement AT&T Wireless as the top dog in US wireless. In 1999 AT&T acquired Vanguard Cellular Systems for $1.5 billion.

In 2000 AT&T added $3.3 billion of mobile properties—assets that Vodafone and Bell Atlantic were required to divest when they combined to form Verizon Wireless. The result made AT&T the largest wireless carrier in the world.

Armstrong's ambitions were global. Continuing with his "go big or go home" style, AT&T snagged the partnership with BT after MCI was purchased by WorldCom. Both AT&T and BT would merge their international operations into a $10 billion joint venture. AT&T began to sell Concert services in late 1998.[34]

As the century was coming to a close, Armstrong had nearly every item on his transformation checklist covered.

- ✔ Local Services to Enterprise Customers: Acquire Teleport
- ✔ Deploy state-of-the-art intercity fiber network: Partner with Velocita, CapRock, and Touch America to build a fourteen-thousand-mile fiber network connecting thirty major cities
- ✔ Be the largest rival to the Baby Bells in providing broadband and voice services to households: Acquire TCI and Media-One and partner with Time Warner
- ✔ Cement position as largest mobile service provider: Acquire multiple wireless properties
- ✔ Be the leading provider of global communications services to multinational enterprises: Align with BT using Concert as platform

But one big item did not have a check next to it.

- ✗ Maintain a strong balance sheet and high stock price

By the time midnight bells rang in the new century, the *Wall Street Journal* captured the doubts that were mounting on Armstrong's transformation.

> Armstrong's bet is one of the biggest ever in business. . . . He is gambling that he can transform an American icon into an entirely new company. And he is doing it at a time when his firm's core business is being hammered by low-cost competitors.

---

34 Armstrong's international ambitions included Canada. See the Appendix section on Allstream/AT&T Canada to learn how Armstrong's Canuck adventure played out.

But already, some on Wall Street have expressed concern about how AT&T—never known for nimble management—will absorb all of its acquisitions.

Armstrong was not in denial. "There is great risk in our future, and I don't mind. I don't think the sky is falling," he said with words that lacked clarity or conviction. His executive Leo Hindery, who would soon leave AT&T to join Global Crossing, also acknowledging the challenge ahead, added, "Mike is turning this aircraft left when it used to go right."

As it turned out, the sky was falling, and the aircraft was difficult to navigate and losing altitude. The events, as they played out, would lead to the end of an icon.

360networks—Greg Maffei

Ledcor Group was seen as the Kiewit of Canada. Ledcor observed how well Kiewit did on its MFS investment and no doubt took notice when Crowe was backed again by Kiewit to launch Level 3. Armed and dangerous, the Ledcor executives were anxious to wet their fiber beaks.

William "Bill" Lede founded Ledcor Group in 1947 in Vancouver, British Columbia, and became one of North America's largest privately held industrial companies. Lede died in 1980, when a large gravel pile collapsed on him at a job site. His sons, Herb and Dave, took over as chairman and president.

At the time Ron Stevenson, who later became vice chair of Ledcor Group, was executive vice president for Ledcor Industries. Stevenson invited Kevin O'Hara and me to visit Vancouver in 1998. David Lede and Stevenson took us to their private club, where we followed up a round of golf with dinner and plenty of expensive wine. They spent the day and evening picking our brains on MFS, Level 3, and the fiber industry in general.

"What was that all about?" I asked O'Hara on the way home.

"I don't know, but I suspect we will find out," he replied.

Ledcor launched 360networks on May 31,1998, when it acquired the rights to construct fiber alongside Canadian railroad tracks.[35] When 360networks expanded into the United States in 1999, Ledcor doubled down and then some by hiring Greg Maffei, the former CFO of Microsoft, as CEO.[36] They also plucked away our colleague Jimmy Byrd, who had recently been named as Level 3's president of North America.

Maffei took 360networks public in March 2000, just months after his start day. He touted an ambitious plan to construct eighty-nine thousand miles of fiber network across many parts of the world. Four months later the stock hit its peak valuation of $24 per share, a 70 percent increase from the IPO price of $14. 360networks was valued at $15 billion, and on paper Maffei was a billionaire.

Maffei leveraged his Microsoft credentials to attract John Malone of Liberty Media and Kevin Compton of Kleiner Perkins to his board. He enticed Rupert Murdoch of News Corp and Michael Dell of Dell Computers as advisors. A combined $1 billion investment was made before the IPO by Comcast, Liberty Media, Shaw Communications, and Michael Dell.

In 2000, 360networks constructed an $850 million undersea cable in the North Atlantic Ocean. Dubbed Hibernia Networks, the fiber would connect Canada, the United States, Ireland, the United Kingdom, and mainland Europe.

Maffei opened a fancy headquarters in Dublin for Hibernia. "Surrounded by a landscaped moat stocked with koi, the sleek $70 million facility was designed to project 360networks' global ambitions," wrote the *Wall Street Journal* in 2003. Before long, the building was yet one more symbol of the excess of the Bandwidth Boom.

---

35 360networks's original name was Pacific Fiber Link. In 1999 it was renamed Worldwide Fiber Networks. 360networks was adopted when Maffei joined in 2000.

36 In Maffei's last year as Microsoft CFO, he completed ninety investments or acquisitions valued at $10 billion. In July 2000, after Maffei's departure, Microsoft announced a $4 billion write-down of its investment portfolio, which Microsoft attributed to cable and telecom stocks bought by Maffei. Microsoft offered a nonfinance job to Maffei in late 1999, which was perhaps a contributing factor to his resignation.

In early 2001 Maffei told his shareholders, "People around the globe will be able to see and hear one another immediately, with superb [nearly unimaginable] clarity, at the cost of pennies." 360networks assured the market that its business plan was fully funded, and there was strong demand for its network services.

In September Maffei told the *Wall Street Journal*, "We are surviving the telecom nuclear winter." Despite this bravado Maffei must have known he was putting lipstick on a pig. Maffei's survival of the deadly winter would include a bankruptcy filing.

Genuity—Charles Lee

CEO Chuck Lee was swept by the same Bandwidth Boom euphoria that nabbed the CEO of Cincinnati Bell. He, too, was determined to transform his quiet regional phone company, GTE, into a global Bandwidth leader.

We encountered Lee twice earlier in our story: first, in "The Inventors of the Internet Backbones," when GTE acquired BBN, and second, earlier in this section, when Lee took a valiant run at outbidding WorldCom during its hostile takeover of MCI. The *Wall Street Journal* described Lee, who was named GTE CEO in 1992, as a "Country Boy" who learned business leadership early in life.

> [Lee] grew up in rural Wexford, Pa., 25 miles outside of Pittsburgh. He worked as a boy on his grandfather's farm. A neighbor gave young Chuck a steer, which he sold for $600, using the proceeds for a used car, "a dirt-brown Chevrolet" he called the Beer Wagon. After his father had a heart attack at age 35, the younger Lee had to support the family.
>
> . . . in those days when you had a heart attack the doctors didn't let you work anymore; [my dad] died when he was 52," Mr. Lee says. "I was 29."

GTE, short for General Telephone & Electronics, traced its roots to a Wisconsin telephone company founded in 1926. GTE was formed in 1955 and, in 1959, merged with Sylvania Electronics, which produced flash cubes for still cameras, consumer light bulbs, and color TV picture tubes. GTE owned 50 percent of Sprint and other wireless ventures, and it dabbled in Cable TV services.

In the 1960s GTE acquired multiple telephone properties. By the early 1990s, with the acquisition of Continental Telephone, GTE was the largest telephone company that was independent from AT&T.

In 1986 GTE Laboratories announced it had discovered the first efficient, practical way to send two light waves of different frequencies through the same fiber with little deterioration in either signal. Dense wavelength-division multiplexing (DWDM) paved the way to fiber leapfrogging copper in telecommunications networks and becoming the technology breakthrough that spiked the Bandwidth Boom.

GTE sold the businesses that were viewed as noncore, including GTE Sylvania, its 50 percent interest in Sprint, and other wireless holdings. Armed with fresh cash, Lee wanted in on the flamboyant valuations being enjoyed by AboveNet, Broadwing, MFS, Teleport, Qwest, and XO. He set the course to transform GTE from a local telephone company into a national communications powerhouse.

Lee made 1997 a busy year of transformation for GTE. Early in the year, GTE acquired BBN for $616 million in stock. Next, GTE entered into a $485 million contract to buy fiber from Qwest on its under-construction national network. Late in 1997 GTE purchased Genuity, a web hosting provider based in San Francisco that boasted Microsoft and Excite among its biggest customers.

Lee explained, "We are undergoing very dramatic, fundamental change in our industry. We are becoming a nationwide provider of telecommunications services—voice, data, and video."

In late 1997 Lee made his bold run at MCI. GTE was likely fortunate that it lost the MCI takeover battle.

In 2000 the brand Genuity was chosen for the business unit that combined BBN, the Qwest fibers, and Genuity. Paul R. Gudonis, Genuity's CEO, explained, "We researched and considered a number of possibilities for the new name and Genuity was selected because we felt it captures the ingenious, creative spirit, and new focus of our organization and provides us with a name that will sustain us well into the future." If "well into the future" meant only two years, he was spot on.

Genuity's chief strategy officer added,

> It's not about just having great technology, fiber miles, modems or individual services. It's about understanding your customers' business and architecting infrastructure solutions that advance their time-to-market.
>
> Only a company with a fully operational, next generation network, Tier One Internet Backbone and a comprehensive set of e-business services can deliver on this vision.
>
> Genuity is that company.

But this was not to be.

GTE, with its new entity Genuity, caught the attention of Bell Atlantic, the Baby Bell that combined with NYNEX in 1997. The merged Baby Bell had a Market Value of $50 billion. NYNEX's CEO, Ivan G. Seidenberg, was named Bell Atlantic's CEO in June 1998.

Seidenberg immediately set his sights on GTE, with his interest being in both GTE's local phone properties and the Genuity subsidiary. Bell Atlantic agreed to buy the GTE phone business in a stock swap valued at $53 billion. Upon closing, the combined company was rebranded as Verizon, becoming the largest telephone company in the United States, with a Market Value of $125 billion.

Since the Baby Bells remained prohibited from providing long-distance services, Seidenberg had to be creative in structuring Genuity. Verizon would initially own 9.5 percent of Genuity while having a five-year option to increase its ownership to 80 percent. The remaining 90.5 percent of Genuity would be spun out in the form of an IPO.

In September 2000 Genuity unveiled its new platform, named Black Rocket, with the tagline "Do You Want to Change the World?" Genuity CEO Gudonis explained, "We have created an entire new industry category: Network Services Platforms." He defined this platform as the combination of Internet access, security, and web hosting.

Genuity lined up an all-star list of partners, including Cisco, IBM, Capgemini, Ernst & Young, and PricewaterhouseCoopers. Genuity named them ePartners, responsible for developing, building, and deploying solutions in eCommerce, eMarkets, eMedia, eSupply Chain, and eCRM.

"Enterprises are accelerating their moves to do business on the Internet," proclaimed Gudonis. "And now here comes Genuity with a complete solution."

The announcement was bold and brash, and many of us thought it was nonsensical. Genuity spent $20 million on the marketing campaign to promote Black Rocket.

Bolstered by its partnership with its shareholder Verizon, Genuity launched its IPO in June 2000, raising $1.9 billion at $11 per share and representing a $2.1 billion market valuation. On the first day of trading, the stock dropped to $9.40 per share and never traded higher. At the time of its IPO, Genuity had $1.5 billion in debt.

In 2001 Genuity entered Europe through a $114 million acquisition of Integra Telecom, a French web hosting company. The French target's chairman said his company had gotten itself out of a tight spot by selling to Genuity. "The market had concerns about the ability of the company to raise funds," he said. "It's why we have decided . . . to find a big brother."

Soon, Genuity would need a big brother. Would they find one?

XO—Craig McCaw and Dan Akerson

As the 1990s were coming to a close, XO had Gold Rush mojo. In late 1998 an analyst touted, "With XO, McCaw has a winner. In terms of management, there's no better team. They have the McCaw clout, which gives them the money that they need, and they are good executors."

Chicago, Atlanta, San Francisco, New York, and New Jersey were its newest markets, with Miami, Dallas, Denver, San Diego, and Seattle soon to follow.

In a bold Bandwidth Boom maneuver in 1998, XO acquired two conduits and twenty-four fibers across Level 3's intercity build for $700 million up front, plus recurring expenses.[37] XO's lead negotiator was Wayne Perry, who joined McCaw in 1975 and was his long-trusted deal-maker on spectrum auctions and other acquisitions. Perry was viewed as McCaw's most savvy negotiator. I was Level 3's lead negotiator. It was a mismatch in their favor, but sometimes, the underdog holds his own.

Deep into the negotiation, Perry was stuck on the dynamics around future fiber pulls. Though XO had two extra conduits, they didn't want to shoulder the full cost of adding fiber to the spare capacity. Perry pushed, saying, "Perhaps, instead, we could participate in your future fiber deployments. Sharing the costs between us would be good for both parties." We were not keen to give XO this right, fearing future opportunities to differentiate ourselves would be compromised. This remained a hurdle to the very end.

I came up with a concept called tag-along rights, which provided XO an option to take up to a certain percentage of future fiber in a cost-sharing arrangement. Crowe wasn't pleased when I told him of the solution. The tag-along rights served its purpose—it got the deal over the hump, and it turned out to have no value for XO.

Later in 1999 Daniel F. "Dan" Akerson transitioned from being the CEO of McCaw's wireless provider Nextel to becoming CEO of XO. Akerson was previously the CEO of General Instrument (GI), where he succeeded former United States Secretary of Defense Donald Rumsfeld. Earlier, he was COO

---

37 The transaction was completed by Internext LLC, a new joint venture that was equally owned by XO and McCaw's family office, Eagle River. In late 1999 XO purchased Internext's 50 percent ownership for $220 million in XO shares. At that point Internext had invested only $50 million, netting a fourfold return in the flip from Internext to XO.

of MCI. Akerson's transition to XO during the Gold Rush indicated that both McCaw and Akerson viewed XO as its biggest opportunity.

Akerson saw XO as a platform that could combine wireless with fiber as solutions for its customers.[38] An analyst captured Akerson's vision: "'They are one of the few [Internet providers] that offer [metro fiber, intercity fiber, and wireless access], so there is no limitation on the type of customer they can serve, and that allows them to optimize their margins.' . . . For example, XO can operate in metropolitan areas with its fiber-optic network and offer broadband wireless connections to those in less-populated areas."

At that point XO was valued at close to $9 billion.[39] An investor identified XO as a company whose shares would likely appreciate 50 to 100 percent over twelve to eighteen months. In late 1999 famed investor Theodore J. "Teddy" Forstmann of Forstmann Little & Company elbowed his way into the fiber fury by purchasing $850 million of XO shares at a valuation of $7 billion.[40] *Forbes* reported, "At the center of McCaw's new empire is [XO]," and quoted McCaw saying, "It's the best thing I've ever done." Actually, McCaw would learn it turned out to be his worst venture.

---

38 XO paid $695 million to acquire WNP Communications, of which $152.9 million in cash went to the FCC for wireless licenses. Akerson also paid $137.7 million to Nextel to acquire the 50 percent interest in Nextband that was not already owned by XO, which gave XO full ownership of more than forty wireless licenses.

39 The value of XO included over $1.5 billion in indebtedness.

40 Teddy Forstmann led the founding of Forstmann Little in 1978 with a focus on leveraged buyouts (LBOs). Forstmann was featured in the epic LBO takeover book *Barbarians at the Gate: The Fall of RJR Nabisco*. Forstmann Little lost the epic takeover battle to its archrival Kohlberg Kravis Roberts & Co. (KKR).

XO wasn't Forstmann's only Bandwidth Boom investment. In 1999 he invested $1 billion in McLeodUSA and, during the Bust, an additional $100 million. Forstmann Little ended up losing its entire $1.1 billion McLeodUSA investment. See appendix 3, "More Bandwidth Stories," to learn more about McLeodUSA.

Decades later, KKR would be part of a consortium that sought a hostile takeover of Zayo.

Forbes continued, "McCaw's secret weapon at [XO] is its chairman, Daniel Akerson, the former president of MCI. Akerson quit MCI in 1993, and he still seethes that the storied upstart ended up being bought by WorldCom. Now he relishes the chance to counterattack."

Akerson identified AT&T and WorldCom as the targets that he would overtake. "[XO] has $275 million in annual revenue, and Akerson maintains that he can take it to $15 billion in a few years," Forbes reported.

Wall Street was buying this story, as XO's Market Value reached $10 billion. Yet a warning was included in the article: "[McCaw, the] famously private and self-deprecating billionaire, will need lots of luck to make all of it pay off. [XO] has $3 billion in cash, but it could run out in a year, and its main wireless technology is unproven."

In 2000 XO extended its agreement with Level 3 to include Europe, with intentions to piggyback on Level 3's Europe expansion. Though this deal was executed, no money would ever change hands, an XO expansion to Europe would be dead on arrival, and XO would enter a fight for its survival.

AboveNet (MFN)—
John Kluge and Steve Garofalo

The early Gold Rush years of 1998 and 1999 were glorious for MFN, the company that would come to be known as AboveNet. The business signed a $33.2 million contract to provide services to PSINet and formed a joint venture with a UK company to create a fiber route between the United States and Britain while also expanding to the West Coast, focusing on San Francisco Bay and Southern California.

Early in 1999 the deals continued with AboveNet announcing a strategic relationship with Bell Atlantic (later called Verizon) that included a $1.7 billion investment in AboveNet and Dark Fiber leases valued at

$550 million. These leases had a twenty-year life and, as I gleefully learned when we were pursuing AboveNet in 2012, had no renewal rights.

What followed was unforced, monumental, Gold Rush–inspired errors.

In mid-1999 AboveNet announced its $1.76 billion acquisition of a San Jose company that positioned itself as a provider of higher-layer network services that reduced Internet congestion by directing online traffic to Internet Backbones.[41] The acquisition represented a meaningful departure from AboveNet's focus on core Bandwidth Infrastructure. Investors reacted poorly, resulting in an 11 percent decline in AboveNet's stock. Less than a year later, AboveNet made another ill-conceived acquisition, this time a $1.4 billion purchase of a provider of managed services.[42]

In 1999 AboveNet also announced a vast European expansion that targeted eleven markets in seven countries. To fund the capital, Verizon augmented its investment with an additional $975 million, this time in the form of convertible notes.

In actuality the $3 billion combined price of these acquisitions, and the grand expansion across Europe, was a gigantic and untimely strategic blunder just as the industry was becoming a Weeble that was wobbling and about to fall over.

In the early 2000s, the valuations of the fiber companies were ridiculous.

AT&T, Qwest, and WorldCom all topped off in the vicinity of $100 billion. The other heavyweights—Global Crossing and Level 3—each enjoyed valuations of over $50 billion. 360networks, AboveNet, PSINet, WilTel 2, and XO each surpassed $10 billion. Even Broadwing, Genuity, and Touch America attained peak valuations of several billion dollars.

---

41 At the time of this acquisition, the name AboveNet was not yet adopted by MFN. As MFN navigated bankruptcy, it used the name of this acquired company, AboveNet, as its post-bankruptcy name.

42 AboveNet's president explained the logic behind the scope creep of its strategy: "[AboveNet] plans to extend its leadership position in the market by combining fiber-optic infrastructure, world-class data centers, Internet connectivity, and managed services."

Soon, the combined Market Value for all these companies combined would tank to $15 billion. The Bandwidth Gold Rush glow faded to dark.

CHAPTER V

THE BUST OF THE BANDWIDTH BUBBLE

It's only when the tide goes out that you learn who has been swimming naked.

—Warren Buffett

Years before the Bandwidth Bubble had begun to take shape, data had surfaced that indicated something major was amiss. In 1998 an AT&T Labs Internet researcher named Andrew M. Odlyzko published a paper debunking the widely held view that Internet traffic was doubling every three months. Odlyzko explained that the implications of this rapid traffic doubling would imply that by 2001 Internet traffic would have grown by a factor of seventeen million. The amount of Bandwidth, Odlyzko calculated, would require every Internet user on Earth to stream video twenty-four hours a day. Clearly, that wasn't the trajectory.

"[Doubling every three months] was an extremely convenient myth," said Odlyzko. "Every entrepreneur who was getting financing could quote it." In contrast, Odlyzko's analysis suggested Internet usage was doubling only once a year.

Odlyzko's inconvenient truth was either unnoticed or, more likely, ignored.

As the Boom reached its height, the twentieth century was coming to an end. Are you old enough to remember the Y2K hysteria and how computers everywhere could fail or run afoul because of a software glitch as midnight turned to January 1, 2000? Hindsight now shows this proved to be a creepy distraction from the signs that a Bandwidth Bubble was forming.

I was an executive eyewitness to the Y2K obsession forcing our company and customers to be prepared with a solution. Crowe named me the executive on duty in Level 3's network control center on December 31, 1999. The world of business and government had deep fears that infrastructure of all types would malfunction precisely as that day, and that century, would end. Air traffic control systems, railways, shipping, power grids, the stock market, communications networks, and pretty much everything else that depended on computers would break. The fear was the entire world would come to a halt, and the path to returning to normal would be uncertain.

When computer software was written in the 1960s through 1990s, not much thought was given to the date convention. Most software used the last two digits of the year instead of the full year. For example, '93 was used instead of 1993. The Y2K fear was that software wouldn't know what to do when '99 turned to '00. Did '00 mean 2000, or did it mean 1900? Would computers continue to function, or would they malfunction?

Gobs of money were spent on consultants before New Year's Eve. The scope was so vast and much of the software so archaic that no company could know if ticking time bombs were embedded in their software.

So I spent New Year's Eve in our Boulder, Colorado, control center, unsure of what we would do if problems arose. As the clock struck midnight, we watched with high anticipation. Nothing happened on the Level 3 network. US corporations spent an estimated $100 billion responding to this nail-biter, including technical adjustments and consultants who helped fan the flames with media support. When nothing of significance transpired, we were all left wondering whether the money spent had saved us from disaster or if we had been victims of a financial con job.

At around 2:00 a.m., I went to a colleague's house in Boulder to catch the tail end of our company's New Year's Eve celebration. Almost every guest was from Level 3. I recall the atmosphere—partying was hard while anxiety was thick in the room. Y2K was not the source of the anxiety. It was the sense that along with the end of the century, the Bandwidth Boom was kaput.

The third event foreshadowing the collapse was the 9/11 terrorist attack on the World Trade Center's Twin Towers. This moment, the desperate images I saw unfold of the destruction, desperation, and murder of innocent lives will forever be linked to the Bandwidth Boom's inglorious dismount.

The new decade had begun with Level 3 experiencing more than its share of dysfunction. The solution: engage a corporate shrink. In our case our coach was a husband-and-wife team branded as the Woodstone Consulting Company of Steamboat, Colorado. The consultancy focused on "streamlining business relationships, eliminating barriers, and attaining higher performance."

Woodstone recommended retreats, which began with the entire executive team. The idea was to make us a stronger and more cohesive leadership unit. One for all and all for one. I know we needed help, though in hindsight, I don't know that Woodstone ended up as our solution or panacea.

After a series of sessions that involved the senior executive team, each executive hosted sessions for our own teams. Mine was September 10–12, 2001. My team and I arrived on Monday, September 10, and had an afternoon session and a dinner at the Woodstone Ranch.

The next morning was 9/11. I walked into the hotel lobby coffee shop and saw all eyes fixated on the TV.

"What happened?" I asked my colleague.

As he tried to articulate an answer, the second 767 crashed into the World Trade Center South Tower. I recall the TV newsperson trying to imagine how this could be an accident. "Perhaps the smoke from the first tower caused the second plane to lose its course," the news anchor said in a tone that made it clear the news reporter knew that explanation was nonsensical.

Still, we left and drove to the Woodstone Ranch, where we watched more TV, as everyone everywhere did, numb from the horror of watching the world's two tallest buildings collapse floor by floor in a massive cloud of smoke and fire. Watching this, you just knew there could be thousands of people trapped inside the Twin Towers, which together claimed to house upward of seventy-five thousand to one hundred thousand office workers on any given day. Now these people might have been pulverized by the implosion. We felt shock in our stomachs, called our families to check on their well-being, and headed to our homes immediately.

The stock market was gut-punched too. Both NYSE and Nasdaq did not open on Tuesday September 11 and remained closed until the following Monday. The markets fell by 15 percent by the end of that first trading week.

The effect of this event, and the resulting disruption to the financial markets, perhaps masked the turmoil that was already in play for the fiber gladiators.

"Unfortunately, we see a slowdown in business as a result of the shock of Sept. 11, which hopefully is a short-term reaction to the attacks," an industry executive speculated, being correct about the slowdown but wrong about the cause.

With abundant clarity, the Boom times were over.

In late 2001 the *New York Times* foreshadowed the historic nature of the Bandwidth disaster.

[R]ecent telecom failures and those still likely to occur . . . represent one of the most spectacular investment debacles ever. Bigger than the South Sea bubble. Bigger than tulipmania.

Bigger than the dot-bomb. The flameout of the telecommunications sector, when it is over, will wind up costing investors hundreds of billions of dollars.

The *New York Times* followed this coverage in early 2002 with "The Fiber Optic Fantasy Slips Away."

Only a few years ago, the dream of striking it rich by transmitting Internet data and telephone calls across continents and under oceans, through endless ribbons of fiber optic cable, captivated one company after another. But rarely in economic history have so many people with so much money got it so wrong.

Instead of a stampede of customers to fill up these fiber optic highways, the industry found itself with too many vacant lanes—way too many.

In March 2002 the *New York Times* remained relentless, publishing "Telecom, Tangled in Its Own Web."

Thanks to a star-quality cast, the Enron wreck has been riveting theater. Greedy executives concocting transactions to inflate company earnings, grasping Wall Street bankers eager to assist, pliant accountants and analysts looking the other way—Broadway's finest could not have come up with a better script.

Yet while all eyes remain on Enron, a tragedy of identical plot but with far more damaging implications has been playing out on another stage. Unlike Enron's saga, this drama is not about a single, rogue company operating to enrich its executives. [Instead,] this tale is about an entire industry—telecommunications—that rose to a value of $2 trillion based on dubious promises by Wall Street and company executives of an explosive growth in demand for telecommunications services. When that demand failed to materialize, the companies were left with mountains of debt and little revenue.

Since the telecom sector peaked in the spring of 2000, some $1.4 trillion in investor wealth has evaporated. Almost 400,000 jobs in the telecommunications sector have vanished as well . . . many companies in the industry are still operating on the edge.

Some analysts were on to the folly: "All the companies in the emerging telecom area are candidates for restructuring given the over leveraged balance sheets."

He nailed it. Most of the multibillion-dollar fiber companies ended up declaring bankruptcy, and those that didn't barely escaped the fate. Not one sailed through unscathed.

WorldCom—Bernie Ebbers

When we last left the WorldCom story, the company was in a 2001 free fall. Ebbers was forced to back out of WorldCom's $115 billion acquisition of Sprint and followed up with the acquisition of Intermedia—a deal his vice chairman, Sidgmore, called the "stupidest deal of all time." The drop in WorldCom's stock drove Ebbers into deep personal debt. Moreover, rumors were swirling that accounting misbehavior was taking place inside WorldCom. The Bandwidth Boom was over; the wide breadth and vast depth of the Bust, which was on the cusp of being revealed, was beyond anyone's imagination.

In March 2002 the SEC opened an investigation into WorldCom. The *New York Times* reported,

The government appears to be looking into many of the business practices that helped WorldCom emerge from obscurity

the last 15 years by acquiring dozens of companies, most notably MCI Communications.

[The SEC inquiry includes] a 24-point bill of particulars,
asking for a trove of documents from as far back as January
1999 . . . [and indicating the SEC] appears to be taking an
interest in WorldCom's acquisitive streak . . . [requiring] all
WorldCom's documents concerning its accounting for
goodwill—the accounting term for the difference of the purchase price in an acquisition and the assessed value of the
acquired company's assets. Writing off goodwill in a one-time
charge can be perfectly acceptable, even required, under
accounting rules. But such a charge could also enable a company improperly to avoid a long-term drag on its earnings.

An analyst helping to sound the alarm added, "I've been following this
industry for 12 years, and I have never seen an SEC request like this. Nothing
good can come of it. WorldCom was always an entrepreneurial company. If
you're being entrepreneurial, you're shooting from the hip, you're taking risks,
and maybe you take a few too many risks with your accounting."

Forbes, seeing a financial and human disaster for the ages unfolding,
summoned none other than the legendary journalist and author of *The Right
Stuff* to call 911 in its article titled "The Rise and Fall of Bernie Ebbers."

Memo to Tom Wolfe: It's time to start writing up the epic story
of the Great Telecom Debacle. The fall of Bernard Ebbers
provides the climax for Act III, in which the hero pays for his
hubris and is subjected to the ritualized humiliation of a company press release thanking him for his services.

"Tom Wolfe, The Bear has now arrived. Glad you missed this memo.
I'll take it from here," is my unsent 2024 memo to Wolfe.

The news of the SEC investigation drove WorldCom's stock price to below $5 a share, pushing its Market Value below $7 billion. Just two and a half years prior, WorldCom was valued at over $100 billion. Now WorldCom's debt of $37 billion was looking larger than the entire value of the company.

WorldCom also came clean on the full extent of the personal loans provided to Ebbers. At WorldCom's peak value, Ebbers had assets valued at $1.3 billion. However, he sold only a tiny fraction of his World-Com shares. "I was one of the founders of the company. I was extremely proud of WorldCom, and I did not want to send a message to the shareholders that when the stock was struggling, that the CEO didn't believe in it," Ebbers explained.

Ebbers took out $400 million of bank loans, collateralized by his boat and yacht company, his timber operation, a sprawling 164,000-acre ranch in Canada, his luxury yacht, *Aquasition*, and his ten million shares of WorldCom stock. With WorldCom's stock price trending toward $0, Ebbers's assets were worth far less than the loan. He was bankrupt.

Ebbers claimed his board discouraged him from selling his stock and instead suggested that the company loan him the money to pay back the bank. Years later, a congressional panel asked John Sidgmore and Bert Roberts to defend the loans to Ebbers. They each meekly responded, "[If I] had to vote for it again, I wouldn't do it."

On the last Friday in April 2002, the board probed Ebbers, trying to understand the steep decline in the stock price and the ramifications of the SEC probe. Ebbers offered to resign. The board deliberated over the weekend, and Ebbers turned in his resignation on Monday.

Ebbers's departure was respectful. The board agreed to pay him $1.5 million a year for as long as he lived. Ebbers would retain the title of chairman emeritus and would be entitled to use the WorldCom plane for thirty hours a year.

Forbes continued,

> Everyone knew WorldCom and its chief executive were in trouble, but the sudden fact of his demise was shocking nonetheless.
>
> WorldCom was the emblematic telecom firm of the go-go 1990s, and Ebbers was the visionary who stitched it together like Frankenstein's monster from pieces of the many firms he acquired over the past two decades. His departure today will serve future business historians as [the event that marked] the formal end of the telecom frenzy.

In May the *Economist* jumped in with "Yesterday's Man."

> In his cowboy boots and Stetson hat, Bernie Ebbers made a splendid cover for any magazine that wanted to be abreast of the telecoms revolution of the late 1990s. His background, too, had just the sort of mix of hard grind, modest intellectual prowess, and sporting vim that America likes in its hero-bosses. Born in Canada, he worked as a milkman by day and a bouncer by night. He dropped out of college twice, eventually graduating with a degree in physical education from Mississippi College . . .
>
> With this perfect pedigree, Mr. Ebbers coupled a talent for deft deal-making and cosmic quantities of chutzpah . . . he gobbled up more than 75 companies to become a brand-new telecoms giant.
>
> The press, for whom a takeover battle makes far more exciting copy than the humdrum daily business of minding the shop, happily collaborates. And the smoke and mirrors of

acquisition accounting offers plenty of scope for creative rear-
rangement of the balance sheet.

No wonder that, at the peak of the telecoms frenzy, Mr.
Ebbers was the subject of as much admiring prose . . . *Forbes*
listed him as the 174th richest American. *Time Magazine*
included him in its cyber-elite. George Gilder, an Internet
guru, [declared] "It's all over. Bernard Ebbers of WorldCom
has won."

The same month, *Time* magazine piled on with "The Rise and Fall
of Bernie Ebbers."

A transplanted Canadian, Ebbers, 60, teaches Sunday school
at his Baptist church each week, serves meals to the homeless
at Frank's Famous Biscuits in downtown Jackson, Miss., lives
modestly in a prefab home, and wears jeans and boots to the
office.

Right now, though, it's hard to find a friend of Bernie outside Mississippi.

WorldCom, like much of the U.S. telecom industry, looks
as broken as a coin phone in a bus station. And so, sadly, does
the executive image of Ebbers.

"I feel like crying," Ebbers [said] after his resignation last
week. "But I am 1,000 percent convinced in my heart that this
is a temporary thing."

The thing Ebbers was referring to was anything but temporary.

What *Forbes*, the *Economist*, and *Time* magazine hadn't acknowledged in its condescending recap of media coverage of the WorldCom/
Ebbers downdraft and scandal was its own failure to scoop other media
in shining a light on Ebbers and the implausibility of WorldCom's
numbers.

WorldCom's board, seeking to recover, named Vice Chairman John Sidgmore to replace Ebbers.[1] We first met Sidgmore in The Inventors when he was hired to be CEO of Internet pioneer UUNET. Sidgmore sold to MFS, only to see Crowe sell MFS to WorldCom less than a year later, a transaction that made Sidgmore an extremely wealthy man and World-Com's Vice Chairman. I teased that Sidgmore would come to regret selling UUNET. Here's why.

When he accepted the CEO position, Sidgmore did not yet know what he was up against, so his first move was to declare a bankruptcy filing would not be necessary. "In the reasonable, foreseeable future, we don't see any scenario under which we are going to run out of cash or go into bankruptcy. This is a real company that makes real money." Boy, was he wrong!

Sidgmore, at first, felt comforted by having WorldCom CFO Scott D. Sullivan by his side. Both analysts and WorldCom executives viewed Sullivan as the brains behind WorldCom's ascent, as captured by the *Wall Street Journal*.

Many Wall Street professionals thought Scott Sullivan was one of the best CFOs around. He was seen as the key to World-Com's financial credibility, known for having instant answers

---

1 Sidgmore claimed to have quit WorldCom in 1999, after Ebbers embarrassed him by pulling out of a deal to acquire Nextel on the day it was to be executed. Sidgmore told Keller, the former *Wall Street Journal* reporter turned CEO headhunter, that World-Com's board later pleaded with him to come back. Eventually he would replace Ebbers as CEO.

"They gave me $12 million to come back, but that wasn't why I agreed," Sidgmore told Keller at the time. "I'm already very wealthy and didn't need their twelve million bucks. I love UUNET. I didn't want UUNET and its employees to get hurt." Sidgmore told Keller he had no sooner returned when two accountants showed up at his office one day and told him they had discovered a massive fraud of $4 billion. "I felt like someone had punched me in the stomach," Sidgmore told Keller. "And as the next couple of months went by and more accounting fraud was uncovered by our accountants, I had to keep upping the number making public statements and the fraud was reaching into the billions. I had no choice in the end but to put the company into bankruptcy."

to even the most obscure financial questions—and being willing to have substantive discussions about them.

...Sullivan...acted as a sort of straight man and temper-
ing influence for Ebbers....Mr. Ebbers had the charisma and
the vision; Sullivan had the answers. Ebbers did the talking,
but often what he said was "We'll have to ask Scott."

Sullivan had offers from six of the top eight accounting firms when he
graduated from college in 1983. KPMG was his choice, and he did well
during his four years with the firm. He left to join a company acquired by
Ebbers in 1992. Within two years Sullivan was named CFO and was
already emerging as the loyal hand of Ebbers. The wonder boy was only in
his early thirties.

The *New York Times* referred to Sullivan as the "whiz kid," quoting
an analyst, "[Sullivan] was the most credible C.F.O. in the entire industry.
He knew the minutiae of the numbers and he could go all the way up to
the big picture."

———————

I had a different view of Sullivan, stemming from an encounter I had
with him in 1997. Within days of completing the MFS acquisition, most
of the MFS executive team parted ways. My expertise in MFS's network
business made me uniquely valuable, so they retained me. I was assigned
to run network planning for WorldCom's local networks, reporting to
former MFS executive Ron Beaumont, who was named president of the
network organization.

Soon thereafter WorldCom discovered a fatal flaw with its acquisition
of MFS. To justify the high purchase price, WorldCom assumed it would
achieve major cost savings by routing its long-distance traffic through

MFS's local network.[2] Though the network technology allowed for this, the regulatory laws did not. It was illegal.

This was unwelcome news for Ebbers and Sullivan as a gigantic portion of WorldCom's promised synergies went belly-up. WorldCom, in its Boom times rush to buy MFS, discovered this restriction after the acquisition was consummated.

Ebbers and Sullivan could have come clean with Wall Street investors by telling them they made a costly and regrettable error. They knew this would cause WorldCom's stock price to crater and their reputations as Wall Street's Batman and Robin Dynamic Duo would be shattered. So they searched for a "better" option.

The resourceful Sullivan conceived of a creative accounting solution. His plan was to treat the planned savings as a write-off associated with the acquisition.[3] The logic was that this "savings" would have been

---

2 Long-distance telephone companies paid local telephone companies to originate and terminate phone calls. The per-minute termination rate was artificially high because of historical reasons tied to the breakup of AT&T. These high access rates were the mechanism for using long-distance revenue to subsidize phone companies so that local phone service would be affordable to most Americans.

In 1997 access rates were set at approximately 2.5 cents a minute, and this expense category was WorldCom's biggest cost. If they could cut this in half, their stock would soar, and they had a plan to do just that: acquire MFS!

MFS was leading the charge in bringing competition to local phone service. Local competition required local phone companies to exchange phone calls with one another. If MFS's customer needed to call Verizon's customer, MFS would hand off a call to Verizon, and Verizon, in turn, would deliver it to the customer. The reverse of this was also true, as Verizon's customers would need to call MFS as well.

Although the technology of exchanging local calls was identical to exchanging long-distance calls, the price was far lower—about 0.5 cents a minute instead of 2.5 cents a minute.

WorldCom's plan was to gain synergies through the MFS acquisition by routing its long-distance traffic through MFS's local trunks. This would save 50 to 75 percent of the fees it was paying the local telephone companies to terminate its long-distance traffic. If successful, WorldCom would save tens of millions of dollars per month.

3 When a company is acquired, the price paid for the company is reflected in the acquirer's balance sheet using the accounting standard referred to as purchase price allocation. WorldCom used this process to modestly inflate the cost of the acquisitions so as to be able to hide from the income statement expenses that WorldCom would incur after the acquisition was completed.

available to WorldCom, but they simply couldn't realize the savings immediately. So Sullivan decided to treat the cost savings as an intangible asset, which they would write-off over a twenty-year period. In doing so WorldCom would realize the savings on the income statement, thereby inflating near-term profits. Wall Street would be deceived into thinking that the promised synergy targets were being met.

Note the cash to pay the access charges would be unchanged. Only the accounting treatment would alter. As long as Wall Street didn't dig too much, WorldCom's near-death experience would be avoided.

To implement this revised plan, WorldCom needed an expert to collaborate with its accounting firm, the famed Arthur Andersen, to provide context for the write-off. Sullivan asked me to be this expert. I explained the scheme wasn't permitted by regulators, and to me, the accounting treatment seemed suspect. Sullivan viewed these as technicalities. He explained that WorldCom was the only account assigned to the Andersen partner, who, not surprisingly, lived in WorldCom's hometown of Jackson.

"You're overthinking this," Sullivan tried to assure me. "Andersen just needs some color."

I overthought it some more. After a sleepless night or two, I decided to tell Sullivan no. I didn't realize it at the time, but my alternative might have involved sleepless nights in a jail cell.

"No problem" was Sullivan's reaction. "We will get someone else to do it."[4]

I knew that I compromised my career at WorldCom. That was okay. I didn't want to move to Jackson, and I didn't want to work for people I didn't trust. I resigned, leaving $500,000 of unvested stock options

---

4 Bill LaPerch recalled a similar event. "When MCI was acquired by WorldCom, I was running operations for MCI. I had a $860 million budget, which included paying network access expenses. When the budget was redone by the WorldCom geniuses, my budget was reduced to $200 million, with the reduction the result of WorldCom deciding to capitalize access cost. I was just a dumb ops guy, so I was not involved with the accounting aspect here, but something smelled rotten in Denmark. I left WorldCom three months after the MCI merger was consummated."

behind. I watched as WorldCom's stock price doubled, wondering if I made the right decision.

In 2002, a few weeks after Sidgmore began his turnaround with Sullivan at his side, the massive fraud of WorldCom was revealed. I felt vindicated and lucky that I didn't get sucked in.

In the weeks before Sidgmore took the helm of WorldCom, a small band of WorldCom accountants conducted an internal, covert, and unsanctioned investigation. The leader of this effort would be credited with unmasking the biggest corporate accounting fraud in business history. Within a year her contributions would be applauded on the world stage.

Every year since 1927, *Time* magazine names its Person of the Year. The first winner was Charles Lindbergh, who flew the first solo transatlantic flight. Martin Luther King Jr. was recognized in 1963, and Ayatollah Khomeini in 1979. Taylor Swift was honored in 2023.

In 2002 a WorldCom employee was named *Time* magazine's Person of the Year. The recipient wasn't Bernie Ebbers or Scott Sullivan. Instead, WorldCom's internal audit director, Cynthia Cooper, was recognized for her relentless pursuit of Sullivan's cooking of the WorldCom books.[5]

Cynthia Cooper launched WorldCom's internal audit team in 1994.[6,7] To convince her bosses of the need for an internal audit, she invited Ebbers

----

5 Cooper was one of three whistleblowers named by *Time* magazine as their 2002 Persons of the Year. Sherron Watkins of Enron and Coleen Rowley of the FBI were the other two.

6 The internal audit function provides independent assurance that a company's risk management, governance, and internal controls are operating effectively.

7 In 1991 Cynthia Cooper had a two-year-old daughter and a failing marriage, which prompted her to move to her childhood home of Clinton, Mississippi. She needed a job to pay the bills, so she jumped at the opportunity to be a $12-an-hour contract employee for WorldCom. In 1993 she married Lance Cooper, a high school classmate who was smitten with her.

and Sullivan to a meeting. Her plan was to show how internal audits could improve profitability.

When Ebbers didn't show up on time to the meeting, Cooper was determined to wait for him. He showed up thirty minutes later wearing a sweatsuit and smoking a cigar. "What in the hell is the purpose of this meeting?" Ebbers asked Cooper. Unflustered, Cooper gave her presentation. She claimed her newly created internal audit group could find millions of dollars in wasteful operations with the use of internal controls. Ebbers was the last person to leave the meeting.

Cooper led internal audit for the next eight years with Sullivan as her boss's boss.

In March 2002 the head of WorldCom's wireless business met with Cooper. He explained that he had set aside $400 million of reserves from the third quarter of 2001, which he intended to tap into to bolster weaker quarters. The $400 million was booked as a reserve to account for the likelihood that some customers would not pay their bills. In practice, the manipulation of reserves was commonly used to smooth out performance. It served as a safety net to be used on rainy days.

The executive told Cooper that he discovered the $400 million reserve was being depleted by Sullivan's office to boost WorldCom's income. Cooper brought the matter to the Audit Committee of WorldCom's board. On March 6 the Audit Committee discussed the topic with Cooper and Sullivan, leading Sullivan to back off from using the reserve.

The following day Sullivan tracked down Cooper by first calling her house and receiving Cooper's mobile phone number from her husband. Cooper took Sullivan's call from a salon and, with highlights and tinfoil in her hair, listened as Sullivan lambasted her for pursuing the issue.

The combination of Sullivan's scolding and the SEC inquiry drove Cooper and her audit team to launch a quiet but intense search for wrongdoings.

In May, Cooper's audit team received an email from a Texas-based WorldCom accountant who tracked capital accounts. Attached to his email was a local Texas newspaper article pertaining to a former employee who had been fired after he raised questions about a minor accounting matter involving capital expenditures. "This is worth looking into from an audit perspective" was his message to Cooper.

Cooper's team continued their discreet probe into WorldCom's accounting system.[8] By early June they found $2 billion of questionable accounting entries. Many involved expenses being booked as capital expenditures to erroneously inflate profitability.

On June 11 Sullivan called Cooper. He asked her to swing by his office. As the meeting was ending, Cooper casually asked one of her auditors to share with Sullivan what he was working on. This was preorchestrated, designed to gauge Sullivan's reaction to the audit they were performing. Sullivan urged them to delay the audit until after the third quarter, suggesting he was already intending on cleaning up some problems. Cooper said the audit would continue.

Alarmed by Sullivan's response, Cooper reached out to WorldCom's Audit Committee chair, Max E. Bobbitt, the former President of Alltel, who redirected Cooper to KPMG, the recently hired public accounting firm that was supporting the SEC probe. KPMG urged Cooper to validate her findings.

---

8 Capital expenditures had already surfaced as a focal area of Cooper's probe. Her team discovered $2 billion of capital expenditures that were booked in the first three quarters of 2001 that lacked authorization. When Cooper's team probed, they were told it was for "prepaid capacity." When they asked what this meant, they were redirected to David Myers, WorldCom's controller.

Cooper followed up by asking the Texas accountant to explain "prepaid capacity." He was cryptic while citing the source of the entries as being Buford Yates, WorldCom's director of general accounting.

On May 28 an accountant on Cooper's team stumbled across an accounting entry for $500 million in computer expenses with no supporting documentation.

The audit team was reaching a conclusion that operating costs were being masked as capital expenditures. By doing so near-term profits were being vastly inflated. The capital expenditures would be spread across many years, and as they were depreciated, they would affect net income but not the important measure of profitability.

Cooper's team included Myers in an email in which they asked for explanations of prepaid capacity. Myers replied that his staff did not have time to support internal audit inquiries.

On June 17 Cooper asked the director of management reporting for documentation to support the capital entries. She confirmed she made the entries without any support for them. Next, Cooper stopped by WorldCom's director of general accounting. He was unfamiliar and pointed to WorldCom's controller, who acknowledged there wasn't support for the entries.

On June 20 Cooper and her team met with WorldCom's Audit Committee. Sullivan defended the entries but asked for the weekend to prepare a full justification. On June 24 the Audit Committee told Sullivan he had until the end of the day to resign.[9]

When Sullivan chose not to resign, he was fired. He was only forty years old.

On June 24 WorldCom disclosed that it had inflated its earnings by $3.8 billion over the previous three quarters. The SEC immediately filed a civil fraud lawsuit, and trading of WorldCom's stock was halted.

Sidgmore released a statement: "Our senior management team is shocked by these discoveries . . . I have committed to driving fundamental change at WorldCom, and this matter will not deter the management team from fulfilling our plans."

Arthur Andersen, the world's most prestigious accounting firm, also issued a statement suggesting they, too, were misled: "It is of great concern that important information about line costs was withheld from Andersen auditors by the chief financial officer of WorldCom. . . . Our work for WorldCom complied with SEC and professional standards at all times."

At first, WorldCom's collapse was tied primarily to its failed acquisition of Sprint. The *Washington Post* wrote,

---

9 David Meyers, WorldCom's controller, was also given a day to resign. Unlike Sullivan, he did.

The MCI deal was the culmination of a relentless growth-through-acquisition strategy that finally hit a wall two years ago, when the government blocked Ebbers' boldest acquisition ever, WorldCom's $129 billion purchase of Sprint Corp. It forced him to do something he had never done before: make money by running what he already owned rather than buying something new.

It proved beyond him. And that is how WorldCom, a global company sprung from the unlikely soil of small-town Mississippi, apparently came face-to-face with an unpalatable choice. It could admit to Wall Street that it was losing money, staggering under the debts that came with its many purchases and its miscalculations about long-distance telephone service and the Internet. Or it could cook the books and keep showing profit.

WorldCom used the acquisitions to distort future profits by creating capital reserves to offset expenses. This allowed them to show a predictable increase in operating profits in the quarters that followed the acquisitions. So long as the acquisition gravy train continued to run, the fictional story of synergy achievement would persist.

"The kiboshing of the Sprint merger was, for all intents and purposes, the end of WorldCom," an accounting expert said. "When you have companies that have to make acquisitions to survive, once the music stops, the dance is over."

Further investigation revealed that the scope was far greater and the period during which the fraud was being committed was far longer. The number kept growing until finally $11 billion was restated over a period that began in 1997. This coincided with my turning down Sullivan's request to help with a suspicious write-off, which also prompted me to leave WorldCom.

Congress launched an investigation, suggesting Ebbers would be implicated for his Gold Rush antics. "Ebbers was aware that hundreds of millions of dollars had been moved from expenses to capital accounts" was Sullivan's account, according to a spokesman for the House Energy and Commerce Committee, as told to the *New York Times*. "This is the first evidence we have seen showing that the muddy little footprints may lead back to Bernie Ebbers' doorstep."

On July 8, 2002, the Congressional hearing featured Ebbers, Sullivan, Salomon analyst Jack Grubman, and an Arthur Andersen partner.

In Ebbers's opening statement, he declared his innocence while exercising his Fifth Amendment rights.

> When all of the activities at WorldCom are fully aired and when I get the opportunity, and I'm very much looking forward to it, to explain my actions in a setting that will not compromise my ability to defend myself . . . I believe that no one will conclude that I engaged in any criminal or fraudulent conduct during my tenure at WorldCom. Until that time, however, I must respectfully decline to answer questions . . . on the basis of my Fifth Amendment privilege.

Sullivan was brief and to the point when pleading the Fifth: "I have no prepared statement. Based upon the advice of counsel I respectfully will not answer questions based on my Fifth Amendment right."

The Andersen partner cited his willingness to cooperate with the investigation while denying wrongdoing.

> Let me state clearly, and without any qualifications, that prior to June 21st, 2002, when Andersen was first contacted about this matter, neither I, nor any-to my knowledge, nor any of my team members had any inkling that the [fraudulent accounting entries] had been made.

The Andersen partner blamed management.

The fundamental premise of financial reporting is that the financial statements of a company . . . are the responsibility of the company's management, not its outside auditors.

. . . In performing our work we relied on the integrity and professionalism of WorldCom senior management. . . .

If the reports are true that Sullivan and others . . . improperly [misstated] the company's actual performance, I'm deeply troubled.

Grubman started by expressing his joy for being asked to testify to the committee. He, too, claimed he was misled by management.

I appreciate your invitation to appear before this committee today and I voluntarily did so. I want to commend you and everybody on this committee for acting quickly to try to find out what went wrong here.

. . . WorldCom is a company that I believed in wholeheartedly for a long time. It fit my long held, honestly held investment thesis that the newer, more nimbler companies would create value. It evolved into a company that had an unparalleled array of global network assets and a huge customer base after its merger with MCI.

As far as the topic of this hearing today, the fraud at WorldCom, of course, it influenced our analysis. . . . If the public statements are fraudulent, I and other analysts have flawed information to go on. If we had a truer picture of WorldCom's financial results earlier, no doubt our opinion would have changed.

He acknowledged the inherent conflict of interests within investment banking firms while defending his integrity.

As far as investment banking conflicts with analysts . . . in full-service firms that research is a product used by the bank. If a research analyst has stature and credibility with investors and happens to have a favorable view of a company, that will help get banking business. If, on the other hand, you have an unfavorable view, that will hurt.

In all instances, the life blood of an analyst's reputation and credibility is integrity and honestly held research and opinions with investors and that is something that I have always practiced.

Grubman estimated that his firm received $80 million in fees from WorldCom but denied any connection between these payments and his compensation.

Grubman was asked about spinning, a practice where investment banks allocated lucrative IPO shares to CEOs as a reward for their business. Grubman answered,

I'm trying to think if I can answer that specifically yes or no. I just don't recall, because that's not something that I would be involved with, so I can't recall. I'm not saying no, I'm not saying yes. I just can't recall. . . . My company is a big company, so therefore I cannot say definitively one way or the other if what you're saying is true or not.

A panel member barked, "For an analyst, brilliant as you are, you have a terrible recollection."

––––––––––––––––

Back at WorldCom, the company filed for Chapter 11 bankruptcy protection on July 21, 2002. Eleven days later, the US government began moving on the company's executives, with the FBI making its first arrests of CFO Sullivan and his controller David Meyers. Each was charged with securities fraud and conspiracy. Sullivan was hit with seven counts of fraud charges and was released on a $10 million bond. Charges of bank fraud and making false statements were added in April 2003.

In March 2004 Sullivan changed his plea from not guilty to guilty. In the hearing he admitted to falsifying WorldCom's financials even though he knew "it was wrong" and told the judge, "I deeply regret my actions." Sullivan implicated "management at the highest level." He agreed to direct the $13 million in proceeds from the sale of his Boca Raton house to a restitution fund administered by the government.

On the same day as Sullivan's March 2004 hearing, the government filed an indictment against Ebbers, charging him with securities fraud, conspiracy to commit securities fraud, and making false filings with the SEC. Sullivan would be the star witness. The trial was set for early 2005.

In June 2004, as part of his cooperation deal with federal prosecutors, Sullivan pleaded guilty to one count of conspiracy to commit securities fraud. He faced five years in prison and a $5,000 fine, with sentencing delayed pending the outcome of his cooperation.

On January 25, 2005, opening statements were made in Ebbers's New York City trial to a twelve-person jury: "When he said hit the numbers, it was a command to commit fraud," Assistant US Attorney David Anders told jurors. "Bernie Ebbers lied not once, not twice, but again and again and again. He made sure that the people who worked for him, lied for him."

Ebbers's personal motives were tied to the $400 million personal loans he owed to his company. If the stock price collapsed, Ebbers would be unable to pay these loans. He'd be bankrupt.

"He lied in an effort to inflate WorldCom's stock price and to avoid personal financial ruin," said Anders.

So he conspired, claimed Anders, with several other WorldCom executives to conceal WorldCom's deteriorating financial condition. "Ebbers knew that WorldCom's books had been cooked," Anders said. "He knew because he told people who worked for him to do it."

Reid H. Weingarten, Ebbers's attorney, punched back, attacking the credibility of Sullivan, pointing out he was "very capable of deception, that is, he's a liar." Weingarten pointed out to jurors that Ebbers hadn't sold his WorldCom shares, which countered the claim that Ebbers knew the stock was inflated. "There are a zillion documents in this case and there ain't one smoking gun," Weingarten told the jurors.

WorldCom's controller, Meyers, indicted and under arrest, agreed to cooperate with the government in its fraud case. Meyers was among the first witnesses called by the prosecution.

In February 2005 Meyers testified in the US District Court in Manhattan that, starting when he became controller in 1997, he and Sullivan had agreed to make various adjustments to WorldCom's books. Some of these adjustments were appropriate, and some "could be considered gray," Meyers testified.

But Meyers's most damning testimony pertained to false entries that were made at Sullivan's direction in the third quarter of 2000. Meyers testified that Ebbers approached him in the hallway following the release of third-quarter results, telling him, "I'm sorry that you were asked to do what you were asked to do. You should not have been put in that position." Meyers added, "[Ebbers] gave me his word that we would never have to do that again." This, along with other Meyers's testimony, rebutted Ebbers's defense that Sullivan alone was to blame.

Meyers also added credence that the $400 million in loans to Ebbers were Ebbers's motive for committing fraud. Meyers recounted a 2001 meeting when Ebbers pushed finance managers to cut costs and turn around the business: "[Ebbers] said, 'I'm sure all of you guys are aware of my personal situation.' Bernie told them if WorldCom's stock price dropped below a certain point, banks would start calling in his loans, 'and everything I've worked for since I've joined WorldCom would basically be wiped out.'"

In cross-examination Meyers acknowledged that he and Sullivan had hidden their transgressions from Ebbers "to some degree." When asked why he agreed to cooperate with the prosecution, Meyers said, "I hope not to serve any prison time."

Next on the stand were WorldCom's director of management reporting, Betty L. Vinson, and another accountant, Troy Normand, on Meyers's team. They told similar stories. "Scott said that Bernie didn't want to lower the third-quarter expectations so that we needed to make the entry," testified Vinson. Both Vinson and Normand pushed back, the latter telling Sullivan that he didn't want to follow directions to hide certain operating expenses and that he "was scared, and I didn't want to put myself in a position of going to jail for him or the company." Both threatened to quit, with Vinson's draft resignation being presented as evidence. "The actions proposed regarding quarter-close entries have necessitated this action," Vinson wrote. "I feel that upper management has forced my decision surrounding my resignation."

Vinson said Ebbers received a budget report in the first quarter of 2001 that showed both WorldCom's true line costs and the lower number that was reported to the public, which again indicated that Ebbers was aware of the fraud.

On February 7 Sullivan, the star witness for the prosecution, took the stand. He testified, "[Ebbers] was very hands-on, very detailed. . . . he has a good grasp of accounting concepts." Sullivan also described his tight relationship with Ebbers, saying that he was in Ebbers's office seven to ten times every day, that the two executives ate lunch together most days, and that they traveled together worldwide.

Sullivan recounted when he urged Ebbers to cut the growth estimates given to Wall Street. Ebbers insisted they wait until the fourth quarter. Sullivan recalled Ebbers saying, "Scott, we can't issue an earnings warning to the marketplace until we tell them what we're going to do about it. . . . We have to stick with our guidance for the third quarter. We have to hit our numbers." Sullivan said he explained to Ebbers that the only way to meet Wall Street targets was to lower the company's reserves while booking revenue that was not collectible, emphasizing there was no business justification for making these adjustments. "I told Bernie, 'This isn't right. This is the only way we can get the numbers to market expectations,'" Sullivan said. He added Ebbers replied, "'We have to hit our numbers.'" Sullivan said he took Ebbers's words to mean that Sullivan's team should make the accounting adjustments.

This adjustment, Sullivan recalled, was what led to Vinson's threat to resign. Sullivan said he sent a handwritten note to Ebbers that read, "In the future, it will be up to the operations of our company to hit our earnings targets and not up to the accounting department." Sullivan claimed he met with Ebbers to discuss the note, and Ebbers told him, "We shouldn't be making these adjustments, we need to improve operations." The note was not produced as evidence.

WorldCom lowered guidance in November 2000. However, the situation continued to deteriorate, making it clear by January 2001 that the company would fall short of its revised forecast. When told of this, Sullivan said Ebbers replied, "'We can't lower our guidance [again]. We just announced new guidance on Nov. 1. Get to work on it.'"

Sullivan described how he first depleted tax reserves to artificially reduce expenses. Next, he reclassified line costs from an expense to a capital expenditure. In total, $771 million in expenses were removed from the first quarter of 2001 results. Sullivan claimed that he explained this to Ebbers at a steakhouse in Washington in March 2001.

The prosecution played a voicemail that Sullivan left for Ebbers in late June 2001. In the voicemail Sullivan said revenue growth "keeps getting worse and worse," and reported financials "already have accounting fluff in it." Sullivan also flagged low margin equipment sales, which were used to bolster revenue but added to the challenge of achieving ongoing growth. "We're going to dig ourselves into a huge hole because it disguises what's going on," Sullivan said. "It's the worst kind of revenue that you can have."

The prosecution introduced as evidence an early 2002 videotape of an Ebbers interview with CNBC. "We're ahead of where we thought we would be," Ebbers told CNBC viewers. "We have been a sound financial company. We have been very conservative in our accounting." He also said the dividend the company was paying on the MCI tracking stock wasn't in any danger.

Also put into evidence was a long handwritten note from Sullivan to Ebbers that preceded the interview. The memo expressed concerns about WorldCom's financial well-being.

Sullivan also discussed Ebbers's $400 million loan, claiming Ebbers told him the banks were demanding more collateral for loans secured by Ebbers's WorldCom shares. Sullivan recounted that Ebbers said in a meeting with the accountants, "[I]f our stock goes down below $12, I'm wiped out."

Sullivan also told of Verizon's interest in potentially acquiring World-Com in August 2001. Verizon CEO Ivan Seidenberg expressed interest in completing the acquisition by Labor Day, Sullivan testified, adding that he voiced concern to Ebbers that Verizon would discover, during its due diligence examination of WorldCom's finances, that the company was manipulating its accounting statements. According to Sullivan, Ebbers responded, "You're right. This probably isn't a good time to be talking to Verizon anyhow, because our stock price isn't where it should be."

Sullivan claimed Ebbers blocked information about accounting adjustments from being shared with WorldCom's board. He cited a specific example in which Ron Beaumont, WorldCom's COO, presented details on these adjustments in the June 2001 board meeting. After the meeting, one of the directors asked Sullivan, "You're not really going to do that, are you?" Sullivan shared this with Ebbers.

"Bernie told us to keep the presentations at a higher level, that our jobs were on the line," Sullivan told the courtroom. "And he told Ron Beaumont not to get into the 'close-the-gap' [revenue adjustments] items."

Under cross-examination, Weingarten emphasized that the prosecution's star witness, Sullivan, had a track record of lying. "If you believe something is in your interest, you are willing and able to lie to accomplish it, isn't that right?" Weingarten pressed Sullivan. More specifically,

Weingarten raised the June 2002 board meeting in which Sullivan denied wrongdoing. "You looked at those 12 (board members) and lied your head off, didn't you?"

Sullivan's response to both those questions and others was to acknowledge he lied.

Weingarten also probed whether Sullivan had witnesses to his claims that he kept Ebbers informed of the accounting misrepresentations. "It was Bernie and I," Sullivan replied. "Those conversations were always Bernie and I. There was never anyone else in these conversations."

When pressed by Weingarten about committing the fraud, Sullivan acknowledged, "I knew it was wrong, I knew it was against the law, but I thought we were going to get through it. I have no excuse for what I did."

Soon after Sullivan's testimony, the prosecutors rested their case.

The first witness called by the defense was whistleblower Cooper. She recounted Sullivan's role in the cover-up, including how he aggressively sought to suppress her discoveries. She testified that when Sullivan blocked her from sharing her findings with the Audit Committee, she appealed to Ebbers. "Tell the audit committee what you need to tell them" was Ebbers's response. Her testimony seemed consistent with the defense's claim that Sullivan, not Ebbers, was to blame.

Weingarten also used Cooper to describe the complexity of the fraud. "It was a spider web of transactions, and it was not easy to trace," she testified.

WorldCom chairman Bert Roberts testified that he spoke with Sullivan in the summer of 2002. "I asked Scott [Sullivan] if Bernie [Ebbers] knew, and Scott's answer to me was that Bernie did not know of the journal entries."

On February 28 Ebbers took the stand. As expected, he claimed Sullivan kept him in the dark. "He's never told me he made an entry that wasn't right," Ebbers testified. "If he had, we wouldn't be here today."

Ebbers was asked about the dinner with Sullivan in Washington. He acknowledged having dinner in Washington around that time but didn't recall who was with him. "I don't recall a conversation about capitalizing line costs," he testified.

Ebbers denied apologizing to Myers for cooking the books, testifying, "I didn't have anything to apologize for."

Ebbers painted a picture of his interactions with Sullivan that was a stark contrast to Sullivan's account. Ebbers said he wasn't close to Sullivan and rarely met with him alone. "You talked to Mr. Sullivan several times a day?" Anders asked. "Oh no," responded Ebbers, "I don't ever remember speaking to him three to four times a day." Ebbers's characterization was they spoke three to four times a week.

Ebbers likened his CEO job to his experience as a basketball coach. "I believe that a coach's job is to get the best players that he can, or she can, and get them to play together, sometimes massaging egos." This philosophy implied that Sullivan was largely autonomous in his management of accounting.

Ebbers explained his reaction when he was told by WorldCom's attorney of the accounting problem: "I was shocked. I couldn't believe it. Never thought anything like that had gone on, never—I mean, I put those people in place, I trusted them, and I just had no earthly idea that anything like that would have occurred."

Ebbers testified that he didn't take accounting courses during college. "The closest thing I ever had to an accounting course is a preliminary course in economics," he told Weingarten. Under cross-examination, the prosecution was able to establish that Ebbers was aware of accounting write-offs.

ANDERS: You are familiar with what a write-off is, correct?

EBBERS: Yes.

ANDERS: For example, you took a $4 billion write-off for research and development after the MCI deal, right?

EBBERS: I don't recall the number, but there was a write-off.

ANDERS: Do you recall that initially you and others had wanted to write off $7 billion, correct?

EBBERS: There was initially a larger number. I don't recall the relationship with the 7 to the 4.

ANDERS: Do you recall there was a large number and then you were forced to bring it down to a smaller number, right?

EBBERS: Yes.

ANDERS: That's an example of a write-off that you were aware of, right?

EBBERS: Yes, yes.

Anders was also able to corner Ebbers into admitting he signed SEC quarterly financial filings (10-Ks) that contained false statements.

ANDERS: You would agree that you now know that there were false statements in the 10-Ks you signed, right?

EBBERS: I don't know that there were false statements in there.

ANDERS: We have talked about these line cost transfers, right?

EBBERS: Right.

ANDERS: They are not reflected in those 10-Ks, right?

EBBERS: I don't believe they are.

ANDERS: So you would agree that the line cost numbers are wrong in the 10-K, right?

EBBERS: If it's reported in there that way, yes.

ANDERS: As we talked about, the line cost number flows all the way down the income statement, right?

EBBERS: Yes.

ANDERS: To net income, right?

EBBERS: Right.

ANDERS: Earnings per share?

EBBERS: Right.

ANDERS: So if the line cost transfers aren't in there, then the 10-K
　　is wrong, right?

EBBERS: Yes.

Anders resurfaced several instances in which Ebbers received reports showing that monthly line costs fluctuated wildly. Ebbers conceded that the swings should have been a red flag but added, "I just didn't see it."

"WorldCom reduced its line costs through adjustments by over $2 billion and you had no idea?" Anders asked.

"I did not," answered Ebbers.

The prosecution delivered their closing comments on March 2. "Money, power, and pressure corrupted Bernard J. Ebbers and motivated him to commit fraud," Anders told the jurors. "He lied and concealed a fraud of more than $5 billion. WorldCom had truly become WorldCon. Ebbers . . . was the leader of WorldCom and the leader of that con."

The prosecutor told the jury, "He lied right to your face. The 'Aw Shucks' defense insults your intelligence."

Weingarten countered Ebbers's crucial mistake was delegating financial responsibility to the wrong man, Sullivan. "Ignorance of Sullivan's wrongdoing isn't criminal." He claimed Sullivan lied and Meyers exaggerated in both cases to reduce their sentences. Weingarten mocked Sullivan's claims that he was in constant communication with Ebbers but never with others present. The prosecutor's case, Weingarten claimed, was "Ninety-eight percent Sullivan, two percent Myers."

Weingarten also emphasized that Ebbers bought stock after he left WorldCom. "With his remaining available cash, he buys more WorldCom stock," Weingarten told jurors. "The only conclusion you can come to is that [Ebbers] didn't have a clue the books were cooked."

Jury deliberations began on March 5. In the first vote, ten jurors favored conviction, and two voted for acquittal. As they discussed, two of the for-conviction votes shifted to supporting acquittal, resulting in a tighter eight for conviction to four against. On the eighth day of deliberations, one of the four holdouts changed their vote. The other three followed twenty minutes later.

On March 15 Ebbers was found guilty of all nine counts against him, including conspiracy and securities fraud. He faced life in prison.

A bus driver from Manhattan and one of the four holdouts focused on the judge's instruction that "in order to find the defendant guilty, it would suffice to determine that he turned a blind eye to the fraud, even if he did not partake in it directly."

The bus driver's conclusion was, "A blind eye is not going to get him an excuse." He added, "He was the man who was in charge. It's just kind of hard to sit there and think he didn't know what was going on."

A second juror said, "I think Ebbers pretty much hung himself. How could he be up that high in a company that he started, and then he says I didn't know anything?"

In July Ebbers was sentenced to twenty-five years in prison. He also forfeited most of his remaining wealth. On September 26, 2006, Ebbers surrendered himself to the Federal Bureau of Prisons in Oakdale, Louisiana. After spending thirteen years behind bars, he was released in late 2019 for health reasons and died in February 2020. He lived to be seventy-eight.

Sullivan, who faced 165 years behind bars, received a sentence of five years. The reduced sentence was because, as Anders advocated, Sullivan was a "model cooperator." Sullivan also forfeited his assets to settle a class action lawsuit. He spent four years in prison and one year in home confinement in a new home in Boca Raton.

Meyers and Yates each received a prison sentence of one year and one day. Vinson was sentenced to five months in prison and five months of house arrest. Normand received three years of probation.

Sidgmore, Beaumont, and Roberts were never charged with any misconduct or wrongdoing over WorldCom's collapse. Sidgmore died from kidney failure in December 2003, at the young age of fifty-two, only a few months after the FBI arrested Sullivan. The fear of what the FBI might pursue against Sidgmore must have weighed on him. Beaumont left WorldCom when the fraud was uncovered but continued to be an executive in telecom for the next twenty years.

In April 2004 WorldCom emerged from bankruptcy, changing its name to MCI and relocating its corporate headquarters to Ashburn, Virginia.[10] The $41 billion debt was reduced to $5.5 billion.[11] Revenue was roughly $20 billion.[12]

In 2005 Verizon acquired WorldCom for $6.6 billion, prevailing against a hostile counter bid in a battle that involved a WorldCom rival.[13] Ironically, the other bidder was Qwest.

Global Crossing—Gary Winnick and John Legere

When we last visited Global Grossing, the Gold Rush was rapidly losing its luster. Chairman Winnick had named his fourth CEO in two years.

---

10 Hewlett-Packard executive Michael D. Capellas became chairman and CEO on December 16, 2002. When WorldCom emerged from bankruptcy, Capellas said, "World-Com's turnaround is a tribute to the human spirit and the amazing will of our 50,000 dedicated employees. We are emerging with a new board and management team, a sound financial position, unmatched global assets, a strong customer base and industry-leading service quality. It really is a great day for the company. We come out of bankruptcy with virtually all of our core assets intact. But it's been a marathon with hurdles."

The bankruptcy judge required that the former SEC chairperson Richard Breeden remain WorldCom's court-appointed monitor for an additional two years.

11 Citigroup settled with WorldCom investors for $2.65 billion on May 10, 2004, providing funds for WorldCom to settle with bond holders.

12 MCI shares continued to trade under the symbol MCIAV, though its aspiration was to eventually relist on the Nasdaq market. The stock price was $18 a share.

13 E. Stanley Kroenke purchased the Douglas Lake Ranch in British Columbia once owned by Ebbers for $68.5 million. Sullivan's ten-bedroom, twelve-bath Boca Raton mansion sold for $9.7 million in February 2005.

Thomas J. "Tom" Casey, his new CEO, presented an outlook that was starkly different from that of his predecessor, Leo Hindery. Whereas Hindery advised Winnick to sell as soon as possible, warning Wall Street was about to wake up and realize it was being fooled, Casey told Wall Street that Global Crossing would meet its aggressive 30 percent growth targets. What did Hindery know, and why was Casey painting a different portrait?

As the story played out, I, along with my Level 3 colleagues, was presented a clue to the answer just as the rush for gold was giving way to the panic of the meltdown. The clue came during a visit from my parents on Saturday, September 30, 2000—the last day of the third quarter.

My family was anticipating a fun family weekend. We were all gathered at Boondocks, an amusement park near Boulder, when my mobile phone rang.

"Are you available to do some work over the weekend?" my boss, Level 3 COO Kevin O'Hara, asked. "It's important."

O'Hara explained that Level 3 CFO Sureel Choksi received a call from Global Crossing. "They want to sell $40 million of fiber assets to Level 3, and in exchange, they proposed that we buy $40 million of fiber assets from them." O'Hara said that they are trying to track down Crowe to get his reaction.

> ME: "What fiber assets does Global Crossing want to sell to us?"
> O'HARA: "They say they will take whatever we want to give them, so long as the assets are valued at $40 million."
> ME: "What fiber assets does Global Crossing want us to buy?"
> O'HARA: "They say they will give pretty much whatever $40 million worth of assets we want to buy. They need the transaction signed by midnight today (Saturday), with the details sorted through tomorrow (Sunday)."

Perfect! That sounded like a fine opportunity. Global Crossing needed to pump up their quarterly financial results. We could cherry pick $40 million of valuable assets and, in return, give them $40 million of assets that were of little value to either of us. We would record the transaction in an overtly conservative manner in our accounting statements with full transparency provided to our auditors. "Who cares about the accounting benefit? Let's create some value out of their desperation," I thought and sent texts to my team to get them mobilized for a Saturday and Sunday of work.

Late that evening, O'Hara called back. "False alarm. Crowe and Walter Scott do not want to proceed forward."

O'Hara and Choks tracked down Crowe after his daylong hike in Jackson Hole, Wyoming. Crowe called Scott while O'Hara and Choks remained on the phone. Scott said that no matter how we handled the transaction on our end, it wouldn't "pass the *Wall Street Journal* sniff test." Even if we did the right deal for the right reason with conservative accounting, the headline will be that Global Crossing and Level 3 did a business deal for deceptive accounting reasons.

O'Hara and Choks passed the "not interested" message to Global Crossing. Anticipating an escalation, Crowe instructed O'Hara not to give his mobile number. "An hour or so later, Winnick called me directly, making a forceful attempt to put him in touch with Jim, which I declined," recalled O'Hara as he reflected on the story in 2024.

Roy Olofson, Global Crossing's vice president of finance, also knew what Casey was masking.

In May 2001 Olofson expressed concerns to his boss, Dan Cohrs, Global Crossing's CFO on capacity swap transactions. He used the term "round tripping" to describe the deals, noting the practice enabled both parties involved in the swap to pump up their financials.

Global Crossing was the counterparty to Qwest, and each used the swap to help the other inflate its revenue and profitability. In one deal Global Crossing exchanged $100 million of capacity with Qwest.[14,15]

These reciprocal swap deals were structured such that both Global Crossing and Qwest were able to book $100 million of revenue in the quarter of the transaction (or shortly thereafter). Each party's balance sheet would reflect a $100 million operating lease to reflect the $100 million obligation owed to the other party for use of the swapped asset. By treating the $100 million as an operating lease, the associated expense would be amortized over the next twenty years. As such, the $100 million in manufactured revenue had minimal offsetting expenses. And profitability in the quarter of the transaction was inflated by $100 million.

To an outside investor, the revenue and profitability results would look impressive, while the change in the balance sheet due to the operating lease would go unnoticed because of the lease's small size relative to other costs being spent to build out the networks. These transactions would help both parties report strong revenue and profitability growth to their investors. However, no net cash would change hands.

In early August 2001, CFO Cohrs sent an email to his CEO Casey, citing a news story that questioned the use of these reciprocal transactions

---

14 Others were involved in the accounting scam as well. 360networks bought $150 million in capacity from Global Crossing, while Global Crossing returned the favor by buying $200 million in return. Both these deals were signed just before the end of the first quarter of 2001.

15 Another abuser of these accounting scams was Enron. As Larissa Herda, tw telecom's CEO, recounted on a entrepreneurship panel at CU Boulder, "Enron came in and met with my CFO, and they presented a scheme." Enron's scheme allowed telecom companies to accelerate revenue from long-term Dark Fiber sales to four years despite the lease term being twenty years. "My CFO said to Enron, 'So after four years you've already accounted for all the revenue, so after the four years you're going to have this kind of revenue drop, what do all these other companies say about that?' The Enron guy said to my CFO, 'Well they don't plan to be around in four years.' My CFO was aghast and told me this story, and I said we are not selling another circuit to Enron, because something is wrong."

by Qwest. "The bad news is that this is raising visibility on the swap issue," Cohrs wrote.

Later in August Cohrs's deputy, Olofson, formalized his concerns in a five-page letter to James Gorton, Global Crossing's general counsel. Gorton waited five months before sharing it with Global Crossing's Audit Committee. Global Crossing's auditing firm was Arthur Andersen. And Andersen learned of this on the same day Global Crossing filed for Chapter 11. Again, none of this was disclosed to shareholders.

To what extent were Winnick and other Global Crossing executives dumping their stock to in-the-dark investors?

Olofson also expressed concern about Global Crossing's executive vice president of finance, Joseph Perrone, who was Andersen's lead audit partner—the outside auditor—on the Global Crossing account before he joined the company in May 2000.

A subsequent congressional investigation also uncovered an August 2000 email from a Global Crossing executive characterizing the value of a capacity swap as "the plug number designed to meet" Wall Street expectations.

Winnick and Casey were directly engaged in these swap transactions. In June 2001 Winnick sent an email to Casey about an Enron swap. "I spoke with [Enron COO] Jeff Skilling and there are three people vying for the business—we're one of them. They too are looking to do something here by quarter end," Winnick wrote.

In April 2005 the SEC completed its investigation and reached a settlement agreement with Casey and two others. The SEC cited,

> [T]he failure of [Global Crossing] and its senior management to ensure that the company provided complete and accurate disclosure to investors concerning certain significant transactions entered into by the company in the first half of 2001. Because of [Global Crossing]'s inadequate disclosure, investors were not given the opportunity to fairly judge the quality of [Global Crossing]'s financial results. . . .

These significant transactions involved [Global Crossing]'s sales of telecommunications capacity to other carriers that were linked to, and in some cases dependent on, its purchase of capacity from the same carrier. [Global Crossing] referred to these transactions as "reciprocal transactions." In early 2001, [Global Crossing] was increasingly reliant on the reciprocal transactions as a substantial source of [Global Crossing]'s announced "pro forma" results, i.e., results of operations that were prepared on a basis defined by [Global Crossing], and that were not in accordance with generally accepted accounting principles ("GAAP"). . . . Without these reciprocal transactions, [Global Crossing] would not have met [financial targets].

The SEC concluded,

[Global Crossing] violated Section 13(a) of the Exchange Act and Rules 12b-20 and 13a-13 thereunder, which require issuers . . . to file accurate and not misleading quarterly reports. . . . [Global Crossing] failed to adequately disclose material events, trends, and uncertainties relating to its past and future financial condition and results of operations. . . . As a result, investors did not have the ability to judge the quality of [Global Crossing]'s financial results.

A director in the SEC's Los Angeles office said, "Global Crossing's senior executives failed to discharge one of their most important responsibilities—to communicate with investors in a clear and straightforward manner about the company's business and financial condition."

Despite these substantive findings, the settlement was favorable to those implicated. Casey and the two others each agreed to pay a $100,000 civil fine. Winnick was not cited for wrongdoing or fined. None faced criminal charges.

Global Crossing paid nothing. John Legere, who became CEO during the investigation, expressed relief: "We're happy to have reached a settlement with the SEC and that we can put these issues solidly behind us without a finding of fraud or a financial penalty against the company."

Legere succeeded Casey in October 2001. With Casey lasting only one year in the revolving CEO chair, Legere became the fifth CEO in Global Crossing's four years of existence.

"Bankruptcy is not a possibility at all," Legere proclaimed in an interview shortly after he was named CEO. Bankruptcy was declared on January 22, 2002, just three months after Legere's declaration.

Meanwhile, as CEO Legere took responsibility for saving the business, Winnick was assiduously working on saving Winnick. He had already embarked on a plan to navigate the company through a rapid restructuring in which he would remain at the helm. The complex plan, which was hatched with his new board member and former US ambassador to Singapore, Steven J. Green, almost sort of worked.

With Global Crossing sailing toward bankruptcy in the fall of 2001, the plan Winnick and Green cooked up was for them to invest $25 million to gain a 13.4 percent stake in the Singapore investment firm K1 Ventures. Green would be named as chairman. Both K1 and Singapore Technologies Telemedia (ST Telemedia) were controlled by Temasek Holdings, an investment fund owned by the government of Singapore.[16]

Now with K1 aligning Winnick's financial interests with ST Telemedia, Winnick negotiated with ST Telemedia to purchase a controlling interest in Global Crossing for $750 million.[17] This transaction—called a prepackaged (pre-pack) bankruptcy—would be implemented as part of a

---

16 Singtel and Singapore Airlines were among Temasek's other holdings.

17 Hutchison Whampoa (Hutchison) was originally a partner in the pre-pack. They pulled out because of regulatory resistance from the US government.

bankruptcy filing that would result in most debt and all equity being wiped out. The K1 investment and the pre-pack transaction were completed before Global Crossing disclosed the high risk of an impending bankruptcy.

Though Winnick informally disclosed the K1 transaction to his Global Crossing board, he didn't disclose K1's relationship to Temasek. Nor did he explain that he was collaborating with another Temasek-controlled entity on a pre-pack bankruptcy buyout. When investigated, Winnick and Temasek denied that the K1 transaction would result in Winnick benefiting from the bankruptcy in ways not available to others. Right . . .

ST Telemedia eventually agreed to buy 61.5 percent of a reorganized Global Crossing for $250 million. This implied the value of Global Crossing as it emerged from bankruptcy to be $600 million, a gigantic haircut relative to the $50 billion valuation a couple of years prior.[18]

Winnick, however, finally met comeuppance. He was forced to resign on December 30, 2002, while bankruptcy was far from being resolved. Winnick would spend the next few years defending himself from questionable accounting, disclosure, and stock selling practices.

In April 2004 Global Crossing emerged from bankruptcy with a cleaned-up balance sheet. Legere continued as CEO.

Amazingly, despite Global Crossing's seismic collapse, Winnick was able to extract $734 million for himself. However, he faced civil and criminal legal challenges. He was investigated but not charged with insider trading. He settled a lawsuit related to the case for $55 million without admitting guilt.[19]

---

18 The $600 million consisted of $400 million in public stock, valued at $10 a share, and $200 million in debt.

19 A total of $324 million was paid to settle a lawsuit with shareholders and former employees. In addition to Winnick's $55 million, an insurance company paid $195 million, and the law firm Simpson Thacher & Bartlett paid $19.5 million.

A former Global Crossing employee and plaintiff characterized the settlement as not even a symbolic victory: "This is not justice. What this man did to the company left whole families devastated. . . . I really feel he should be in a cell with much stronger bars than Martha [Stewart] might get."[20]

CNN Money published an article titled "The Emperor of Greed" in 2002, which provided an unflattering summary of Winnick and his inner circle that included Tom Casey and Steven Green.

> Winnick and his cronies are arguably the biggest group of greedheads in an era of fabled excess. Not only did Winnick sell off stock at huge profits while investors who jumped in later watched their stakes burn to nothing, but he treated Global Crossing from the get-go as his personal cash cow, earning exorbitant fees from consulting and real estate deals between Global Crossing and his own private investment company.
>
> In all, Winnick cashed in $735 million of stock over four years—including $135 million Global Crossing issued to his private company—while receiving $10 million in salary and bonuses and other payments to the holding company.

Winnick died at the age of seventy-six in his lavish Bel Air home in late 2023. The cause of death was withheld. Perhaps it was remorse.

---

20 The contrast between the treatment of Stewart versus Winnick is impossible to rationalize. In 2001 Stewart avoided a loss of $51,000 when she sold stock worth $228,000 in ImClone Systems a day before ImClone disclosed a negative development. Stewart's broker learned of the news and shared the inside tip with Stewart, which prompted her to direct her broker to sell. She was convicted of felony charges related to obstruction of justice. Stewart spent five months in prison and five months under house arrest and paid a $30,000 fine.

Notwithstanding the almost constant drama surrounding Winnick, Legere ran Global Crossing successfully for several more years. In 2011 Level 3 purchased Global Crossing for $3.0 billion.[21] The implied value per Global Crossing share was $23, a 2.3-fold increase compared with the $10 per share value when Global Crossing emerged from bankruptcy. This outcome was a positive one for Legere, and it proved to be a steppingstone for him to become the extraordinarily successful CEO of T-Mobile.

Qwest—Joe Nacchio and Dick Notebaert

In early 2001, Level 3 needed revenue momentum. Glenn Russo, who led the Intercity Dark Fiber Solutions business unit for my product organization, stumbled across a commercial opportunity that seemed too implausible to be true. "Qwest referred one of its customers to us to buy a nationwide Dark Fiber network," Russo told me.

"Why wouldn't the customer just use Qwest's fibers?" I asked. Russo didn't know. On the other end of the transaction was Stephanie Copeland, a longtime colleague at both MFS and Level 3 who had recently joined Qwest. Russo pressed Copeland for more information. Citing confidentiality, the only information she offered was that this was a real opportunity for Level 3.

"Are they looking for a swap?" I asked Russo, assuming Qwest must be looking for a quid pro quo.

"No, it's a straight-up purchase," he answered. The contract value would be multiple tens of millions with a significant portion paid up front.

The sale closed, leaving us perplexed as to how Calpoint and Qwest were benefiting from the Dark Fiber provided by Level 3.

Our questions were resolved when, in October 2001, the Wall Street Journal published "Qwest Chief Defends Calpoint Contract as Deal Fuels

---

21 The $3 billion price consisted of $1.1 billion in assumed debt, and the remainder was in Level 3 stock.

Worries about Revenue." Calpoint was somehow going to create high-capacity voice and data services for Qwest to offer large enterprise clients. To enable this, Qwest sold Calpoint equipment valued at $300 million and Qwest, in turn, would subsequently lease capacity from Calpoint for approximately $700 million over five years.

The transaction made sense only through an unscrupulous accounting lens. If Qwest used its own fiber, Qwest's accountants would require that the equipment sold to Calpoint to be amortizing over the twenty-year life of the contract. If, instead, the equipment was installed on Level 3's Dark Fiber, the accountants would allow the sale to be reflected as upfront revenue. In a sign of the times, Nacchio, who was directly involved in the structuring of the transaction, chose to forgo the cash flow benefits of the fiber sale in order to pump up Qwest's quarterly financials.

Qwest's stock was already down significantly from its high of $57 per share. On the Calpoint news, it dropped another 15 percent to a fifty-two-week low of $16.50 a share.

Nacchio tried to rationalize: "The deal, 'brings more engineering talent, more arms and legs' while helping Qwest avoid having to alter capital spending and engineering priorities in order to add the new services. There's been a huge overreaction. I never would have imagined we'd have such a furor." More accurately, he never imagined he'd be caught with his hand in the cookie jar.

The *Wall Street Journal* carefully surfaced concerns of accounting manipulation.

> Some investors feared that Qwest was using the equipment sale to boost revenue and meet growth targets, while artificially lowering its capital spending by letting Calpoint spend the money to put the capacity into service.
>
> "The market is telling you that there's a management-credibility issue here," said [an investor]. . . . "We don't believe

there's accounting shenanigans here. They're not breaking any rules."

"Everything the company is doing is completely above-board and within standard accounting rules," said Nacchio.

This wasn't the first time Nacchio played it loose with accounting manipulation. Several years before, U S West, which had, earlier in 1999, entered the merger agreement with Qwest, was purchasing $60 million worth of new computers. Qwest suggested it would be better for U S West to buy the computers through Qwest, allowing Qwest to book the $60 million as revenue. Qwest would treat the computer purchase as capital, thereby boosting both revenue and profit. U S West's lame duck CEO balked, saying, "[N]o way we will engage in activity like this." This was the tip of another cook-the-books iceberg.

The pressure continued to build. By the end of 2001, Qwest's stock dropped to $12. The third-quarter revenue was flat, and the company posted an unexpected third-quarter loss. The *Wall Street Journal* summarized Nacchio's challenge: "Probably the most important task facing Qwest is reassuring investors and analysts who say their confidence has been undermined by executive turnover, confusion about complex contracts, and fear that the company's growth rate was inflated by sales that tended to be one-time in nature." The turnover included Qwest's CFO and several senior executives.

In February 2002 the Qwest merger with U S West was finally completed. Nacchio came in hot with his new employees. He sarcastically joked that the new name of the company would combine the first two letters of Qwest's name with the last three letters of U S West, which formed Qwest. He used the expression "US Worst" to describe the culture of U S West and was quoted as using the word "clowns" to describe U S West employees. When he shut down a remodel of the lobby of a U S West New York City headquarters building, he hung a sign saying, "Excuse our appearance. We're entrepreneurs. This building was built in a different era, and we save cash by not remodeling."

Nacchio wasn't frugal about his travel. Though he was said to have been critical of U S West's CEO using a $5 million plane for business travel, Nacchio replaced U S West's Falcon 20 with a $20 million Falcon 50EX. A bigger plane was needed, Nacchio explained, for his frequent trips between his New York home and Qwest's Denver headquarters.

At the same time, Qwest's situation was becoming dire. In February Qwest drew down a $4 billion line of bank credit to cover its short-term cash needs.

At an all-hands meeting soon after the merger closed, an employee asked Nacchio why the board hadn't fired him yet. Nacchio responded that the board "thinks I'm doing a pretty good job."

The stock was down to $7.49 per share.

In April 2002, one month after the SEC started investigating WorldCom, the shit hit the fan. Qwest announced a $20 to $30 billion charge to reflect a decline in the assets that it had acquired. Moreover, the SEC disclosed that it was conducting a broad inquiry into how Qwest booked revenue for 2000 and 2001. Specifically, the focus of the probe was whether Qwest inflated its revenues by engaging in improper swaps of fiber capacity. The Calpoint transaction was part of the investigation.

In 2001 Qwest reported Bandwidth capacity revenues of $1.01 billion, up from $468 million in 2000. In the same year, Qwest reported purchases of $1.08 billion from other carriers, an eerily similar number to the revenue. Qwest's Bandwidth capacity revenue expectations for 2002 were only $212 million, a whopping 80 percent drop from 2001.

The accounting for these capacity transactions obfuscated revenue and profitability growth, which, if not properly disclosed, would inflate Qwest's performance and mislead investors. The SEC's investigators were drilling down on transactions in which Qwest sold capacity to a customer while simultaneously agreeing to buy a similar amount of capacity from the same customer.

The level of accounting deceit uncovered in these investigations was truly breathtaking. WorldCom's fraud was the intentional misclassification of expenses as capital to inflate profitability. Qwest's fraud involved inflating revenue through swap transactions while masking the offsetting costs as capital. Like WorldCom, the magnitude of the fraud was mammoth.[22]

Qwest booked the revenue from these sales up front. Yet the purchase was booked as a capital expense, which meant the cost would be expensed over many years. The pop in revenue was also a direct boost of profit. As it did with WorldCom and Enron, Arthur Andersen advised Qwest on how to carefully structure the transactions to rationalize this accounting treatment.

The obvious question was, how much of this swap activity was designed to inflate the stock price? Were the swaps of meaningful economic value to the parties involved, or was it just a ruse to mislead investors?

Qwest's annual meeting was in early June 2002. The once-swaggering, defiant Nacchio and his board appeared to be running for cover from shareholders, as this meeting would be held in Dublin, Ohio, the first time Qwest held its meeting outside Denver and its fourteen-state home territory since going public in 1997. Once again, an employee went at the CEO, who once called his new employees from a merger "clowns."

"I would like for Phil [Anschutz] and the rest of the board to terminate you, Joe," the employee said. "Investors, creditors, regulators, employees, and pensioners—we have lost confidence in you."

Later in June Anschutz and Qwest's board agreed it was time for Nacchio to get the hook and find a new CEO to course-correct Qwest. Anschutz flew to New York and personally delivered the news to Nacchio

---

22 The SEC filing disclosed that $684 million of the $1.01 billion in 2001 revenue was associated with swap transactions. Similarly, $863 million of the $1.08 billion purchases were also linked to swaps.

that he was being replaced, effective immediately. Nacchio received $10.5 million in severance pay and a $1.2 million partial 2001 bonus.

The following month, in July 2002, the US attorney in Denver began a criminal investigation of Qwest and its executives. Trading in Qwest shares was briefly suspended at the opening of the New York Stock Exchange. When the stock was allowed to trade, its price plummeted and closed at a price of $1.77, down eighty-three cents.

Four Qwest executives were indicted on twelve counts in February 2003.[23] The charges stemmed from a $100 million 2001 contract with the Arizona School Facilities Board, where Qwest was accused of improperly booking $34 million in revenue to meet second-quarter financial forecasts and then lying to auditors and regulators by claiming the Arizona school had approved the accelerated equipment order.

Two were acquitted.[24] The jury was deadlocked on the other two.[25] The US attorney used the threat of a second trial to reach plea bargains with the deadlocked pair to gain their cooperation in the investigation of Nacchio.[26]

The difficulty in securing convictions for revenue fraud led the US attorney's office to take a different approach to Nacchio. Prosecutors focused on Nacchio's stock sales, which occurred while he masked the

---

23 Three were members of Qwest's global business unit: Grant Graham, the CFO; Thomas Hall, a senior vice president; and John Walker, a vice president. The fourth was assistant controller Bryan K. Treadway.

24 Treadway and Walker were acquitted.

25 The jury was deadlocked on eight of the eleven charges for Graham and on all the charges for Hall.

26 Graham and Hall each agreed to cooperate with the Qwest investigation of Nacchio. Hall pleaded guilty to a misdemeanor count of falsifying documents and was sentenced to a year of probation and fined $5,000. Graham pleaded guilty to a felony count of accessory after the fact to wire fraud with reckless indifference. Graham was sentenced to time served, which referred to the three and a half years on pretrial supervision. No fine or restitution was imposed.

company's true financial outlook. In December 2005 Nacchio was charged with forty-two counts of criminal insider trading regarding his sale of $100.8 million worth of Qwest stock in 2001. If convicted, he would face up to ten years in prison on each count.

Nacchio's trial took place in March and April 2007 and ran for sixteen days over a span of three and a half weeks. Twelve jurors heard the case.

Qwest's former CFO testified she told Nacchio that the 2001 revenue targets were as much as $1 billion too high, yet Nacchio reassured analysts about Qwest's financial momentum.

A Qwest sales executive testified he warned Nacchio in late 2000 that his unit, which was responsible for a significant portion of Qwest's sales, would have a hard time reaching its 2001 revenue target. When that target was raised, the sales executive sent a one-word email to Nacchio and others: "Bullshit."

Qwest's president, Afshin Mohebbi, Nacchio's number two for several years, testified he warned Nacchio repeatedly in writing, in late 2000 and early 2001, about the growing gap between financial targets and the revenue projections coming from Qwest's business units. Mohebbi called the 2001 targets a huge stretch. Despite the warnings, Nacchio did not change guidance.

Even a Goldman Sachs analyst weighed in, saying he told Nacchio that Qwest had a big credibility issue because it wasn't fully disclosing one-timers. He characterized some of the one-time deals as bogus swaps between telecom companies, resulting in the booking of phantom revenue.

Nacchio did not testify. Anschutz testified for the defense, attesting to Nacchio's character.

In the closing arguments, federal prosecutor Colleen Conry claimed Nacchio knowingly and unlawfully concealed Qwest's growing financial challenges while simultaneously cashing in on his stock options. "This is a case about choices," the prosecutor said. He "made a choice. He chose not to tell. And he also made a choice to sell."

Nacchio's defense attorney offered the following in his closing arguments. "Maybe his business projections didn't come true. Maybe the same optimism and drive that helped Nacchio build Qwest led him to miscalculate . . . but that is not a crime."

The jury deliberated for six days. The judge read the verdict one at a time for each of the forty-two counts. "Not guilty" was read for the first count, which was repeated twenty-two times. Nacchio's son, Michael, began sobbing. The attorneys on both sides knew it was premature to react.

On count twenty-three, the judge read "guilty," which was the first of twenty "guilty" verdicts. Nacchio was released on a $2 million bond. He locked arms with his wife and son as he walked out of the courtroom.

"Convicted felon Joe Nacchio has a very nice ring to it," said the United States attorney for Colorado. "I couldn't be happier that after five and half years, justice has finally been served." He characterized the proceeding as the largest insider trading case in US history.

In July Nacchio was sentenced to six years in prison. He was also ordered to pay $19 million in fines and forfeit $52 million in ill-gotten gains for illegally selling company stock. The judge denied his request for bail pending the appeal and, in 2010, turned down the appeal for another trial. The US Supreme Court refused to hear his case.

On April 14, 2009, Nacchio and three of his buddies took a three-hour ride to Minersville, Pennsylvania. They listened to a playlist that Nacchio prepared for this special occasion. Nacchio belted out, "You can check out anytime you like, but you can never leave," as the final song—"Hotel California" by the Eagles—played. He then stepped out of the car and entered the Schuylkill prison facility.[27]

---

27 Nacchio shared the story in a 2013 interview with the *Wall Street Journal* for its article titled "Former Qwest CEO Joseph Nacchio: Tales from a White-Collar Prison Sentence."

Nacchio spent the next five years in prisons, specifically at two low-security facilities in Pennsylvania. The detainment camps had no bars or walls, and inmates were allowed to use email daily.

Nacchio worked out most days, leaving in better shape than when he entered. His prison jobs were laundry and tailoring. His inmate friends had nicknames such as Spoonie, Juice, and Spider.

He remained defiant.

> I trust Spoonie and Juice with my back. I wouldn't trust the guys who worked for me at Qwest.
>
> I can't wait for the first person to come up to me and say something to me [about the conviction]. I'm going to look them in the eye and say, "You must be confusing me with someone who gives a fuck about your opinion."
>
> I never broke the law, and I never will.

Nacchio insisted his conviction was government retaliation for rebuffing requests in 2001 from the National Security Agency (NSA) to access his customers' phone records.

Nacchio doubled down in a 2014 interview with Denver's CBS affiliate, which published "Defiant Joe Nacchio Lashes Out against Those Who Sent Him to Prison." He continued to attribute his conviction to his rebuff of the NSA. "I refused to allow Qwest to participate or personally participate in what I believe was an illegal surveillance activity."

When asked why he wouldn't apologize to all the people who were harmed by Qwest's fraud, Nacchio lashed out, "There's nothing to apologize [for]. . . . The people say to me, 'Why don't you apologize?' What am I apologizing for? Not breaking the law?"

He also had parting words for Anschutz, whom he felt abandoned him. "The only thing Phil really worries about is his money. For example, he's never even contacted me. So he's probably too busy counting."

In December 2005 the *Denver Post* published an article titled "A Look at Joe Nacchio," presenting a nuanced perspective of the man. Focusing on his upbringing in "Brooklyn's gritty neighborhood of Red Hook," they described his "journey from near Red Hook to [his multimillion-dollar, nine-thousand-square-foot mansion] in nearby Mendham, New Jersey, as a classic American success story." The *Denver Post* covered his ascent at AT&T and then reflected on his unraveling.

> Nacchio's big leap at AT&T came after the company sent him to the Massachusetts Institute of Technology's Sloan Fellows Program, where he earned a master of science degree in management in 1986.
>
> In his early 40s, he was the youngest head of a major AT&T division. He was well-liked and respected, but some feared him. "He can be very cutting in a meeting where he's in charge. He doesn't suffer fools easily," said [a colleague].
>
> . . . Nacchio's boss for about six years during the 1980s, put it more delicately. "Because he had strong feelings about things, he could get more emotionally involved than others. I counseled him to keep that in check, which he essentially did."
>
> . . . Contrary to AT&T's risk-averse culture, Nacchio thrived on taking calculated risks. Nacchio turned around the flagging consumer long-distance unit, a $20 billion enterprise, in the early 1990s.

The *Denver Post* went on to explain that Nacchio had fired the long-time ad firm that worked with AT&T and ordered up an aggressive new ad campaign from another ad firm that "put its archrival MCI on the run." What's more, AT&T numbers were leaked to the press. This, plus Nacchio's short time in the new role and the public airing of the ad partner's firing, angered AT&T's CEO Bob Allen, according to the former corporate communications chief of AT&T, who told the *Post*. "That convinced Bob that Joe wasn't ready to run the company," said the former AT&T executive. "Of course, Bob thought that of practically everyone who worked for him."

This description of Nacchio reminds me of a younger me. If I had been in Nacchio's shoes in the early 2000s, I believe I would have stayed decisively on the right side of the line. However, I know firsthand the intensity of the public company CEO seat, including the self-confidence of being able to navigate tough situations and the passion for not wanting to disappoint investors.

Maybe the difference in Nacchio's outcome and mine is that I had a chance to learn from the mistakes of him and the others. Or maybe it was just a stroke of luck—bad luck for him and good for me.

In the aftermath of the Nacchio debacle, Qwest made a powerful CEO change, handing the reins to Richard "Dick" Notebaert, the hard-nosed former CEO of my first employer, Ameritech. Notebaert brought with him his former no-nonsense CFO at Ameritech, the much accomplished Oren G. Shaffer. The appointment of Notebaert and Shaffer put both outsiders and those inside Qwest on notice that the nonsense days were over. Anschutz, who still owned 18 percent of Qwest, resigned as chairman while remaining on the board.

Notebaert insisted they would not end up in bankruptcy: "We will not be the next shoe to drop. We believe we're doing the right thing by discussing this issue in a very transparent way."

With the help of KPMG, which replaced Arthur Andersen, Qwest cleansed its financial statements. When the restatement was completed in October 2003, Qwest disclosed that it understated losses in 2000 and 2001 by $2.5 billion.

With most of the fallout in the rearview mirror, Notebaert turned his attention to the future of Qwest. He decided to take a hostile run at WorldCom. In February 2005 WorldCom decided to sell to Verizon for $6.6 billion instead of selling to Qwest for $7.3 billion. Both bids included stock as a significant part of the payment. WorldCom concluded that, despite the lower headline number, Verizon's offer was of higher value.

The Verizon-WorldCom merger required regulatory and shareholder approval. During the go-shop period, Notebaert increased his offer to "best and final offer" of $9.7 billion. *Forbes* covered the offer in its article titled "Qwest's Last Stand." Their prediction was that the offer, despite its boldness, wouldn't change the outcome.

> In many ways, the offer is a blockbuster. It tenders $16 per share in cash, includes new financing details and adds assurances that Qwest will close the deal even if WorldCom's business declines. . . . It's nearly everything WorldCom's board asked for when they . . . gave Notebaert a pie-in-the-sky example of what it would take to change their mind.
>
> But it still isn't enough to seal a deal, and chances are good that Qwest is going home empty-handed.

WorldCom accepted a sweetened offer from Verizon, concluding that Verizon's new bid, which was still 13 percent lower than Qwest's, was superior. WorldCom's board understandably had little confidence in the future performance of Qwest's stock.

On May 3 Notebaert withdrew Qwest's bid and gave up. Though Notebaert remained CEO for another two years, the lost battle for WorldCom was indeed Qwest's last stand. In February 2006 Anschutz, who still owned 17 percent of Qwest, announced he would not seek reelection to the board. In June 2007 Notebaert resigned.

Ed Mueller replaced Notebaert as CEO, which was fitting, as he also succeeded Notebaert at Ameritech. "When Mueller took the helm at Qwest in August 2007, he was charged with leading a financially stable but unwanted company" was how the *Denver Post* characterized the challenge.

Mueller dabbled with potential acquisitions and considered divesting the legacy Qwest fiber business. In the end he realized he was dealt a losing hand.

Three years after Notebaert's departure, CenturyLink agreed to acquire Qwest, which enjoyed a Gold Rush value of $100 billion, for $22.4 billion.[28]

WilTel 2—Howard Janzen and Jeff Storey

When we last left CEO Howard Janzen in mid-2000, he was feeling spunky about WilTel 2. The value hit $15 billion, with Baby Bell SBC as its anchor customer and part owner, and its shiny new Tulsa headquarters was complete. Soon thereafter, the Bubble burst.

By early 2002 Janzen and his WilTel 2 board knew a restructuring was unavoidable, disclosing the likelihood of a Chapter 11 filing. WilTel 2 did not disclose that in January the board awarded the company's five top executives $13 million in retention bonuses for the purposes of repaying the loans Janzen and others were given in 1999 to buy the now worthless WilTel 2 stock. The "heads we win, tails you lose" outcome was not well received by shareholders when this was later revealed.

The woes of WilTel 2 spilled over to its parent. The Williams Companies delayed its fourth-quarter 2001 earnings release, citing the need to study its debt exposure to WilTel 2. In April 2002 Williams decided a divorce was necessary and distributed its WilTel 2 shares to the shareholders of the Williams Companies. Also in April, Williams paid $753 million to WilTel 2 for network capacity it was leasing from WilTel 2. In return WilTel 2 assumed $753 million in unsecured debt owed to the Williams Companies.[29]

In the state of Oklahoma, Oklahomans hear a lot of "whoas" from their cowboys. The Williams Companies heard it loudly as its spin-off of

---

28 The $22.4 million purchase price consisted of $10.6 billion CenturyLink stock and $11.8 billion in assumed debt.

29 The Williams Companies also agreed to take over interest payments on $1.4 billion in WilTel 2 notes that the Williams Companies had previously guaranteed.

The Williams Companies, which was simultaneously hit by the Enron-related collapse in energy trading, was forced to sell pipelines, refineries, and natural gas production leases in an unfavorable market. Williams's stock dropped 89 percent in 2002.

WilTel 2 came under scrutiny. The state of Oklahoma opened an investigation into the collapse of WilTel 2, which came only a few months before WilTel 2's bankruptcy filing. A committee representing unsecured creditors questioned whether the terms of the spin-off were a fraudulent transfer: "More broadly, did the Williams Companies engage in misconduct by spinning off a subsidiary that was over-leveraged and had no reasonable chance to survive as a standalone entity given its capital structure?"

In April WilTel 2 filed for bankruptcy protection. "I'm not happy . . . but I do understand what caused it: the biggest meltdown ever in our industry," Janzen said. Skirting his own participation in WilTel 2's failure, Janzen blamed "a debt-laden balance sheet created by Williams . . . that left WilTel 2 ill-equipped to survive the industry's troubles." It's always your daddy's fault.

"The goal is to come out of it as quickly as possible," said a WilTel 2 spokesperson. "Our customers won't be impacted, our employees won't be impacted, our vendors won't be impacted. It's just our holding company." The reassurances proved to be hollow, as customers, employees, and vendors were all severely affected.

Level 3, navigating through its own challenges, took a run at acquiring WilTel 2 for $1.1 billion. Level 3's offer was conditioned on WilTel 2 having $450 million of cash on its balance sheet, which reduced the effective purchase price to $675 million.

WilTel 2 said no and instead partnered with Leucadia National, a low-profile publicly traded investment firm that focused on special situations. Leucadia agreed to invest $330 million for a 45 percent ownership stake, of which $150 million was distributed to the holders of bank debt and $180 million was paid to Williams Companies to sever its relationship. WilTel 2 also repurchased its Tulsa headquarters from the Williams Companies for a $150 million obligation. Bond holders would own the remaining 55 percent. Shareholders, including SBC, received nada.

WilTel 2 emerged from bankruptcy in October, rebadging itself as WilTel and debt-free except for its headquarters liability. Jeffrey K. Storey, a WilTel 2 operations executive since 1999, received a battlefield promotion to replace Janzen as CEO.[30]

When Leucadia invested the $330 million, it agreed not to make a tender offer to buy the remaining shares until after October 14, 2004. Despite this agreement, it tendered an offer in May 2003, with a simple explanation that its "offer was in the best interests of the company and its shareholders."

When WilTel's independent directors requested time to seek other offers, Leucadia withdrew its offer, adding that a sale to a third party would not be realistic because Leucadia would not sell its shares. In August the independent directors agreed to support Leucadia's offer to buy the remaining shares for $423 million.[31]

In 2005 WilTel reached a settlement with SBC, releasing SBC from its obligations to purchase services from WilTel. The agreement required SBC to make cash payments to WilTel totaling $236 million. Two years later Leucadia struck a deal to merge WilTel into Level 3 for $696 million.[32]

---

30 Jeff Storey spent his earlier years at Cox Communications and SBC.

31 The $423 million equated to $15.50 per share, a 24 percent gain over the first trading day closing price of $12.51 per share.

32 Leucadia received $386 million in cash, plus Level 3 shares that were valued at $310.5 million based on Level 3's stock price of $2.70. Leucadia realized an additional $100 million based on cash held by WilTel at the time of closing and also retained the benefit of payments owned from SBC per the $236 million settlement. Leucadia reported a gain of $180 million on its $753 million investment in WilTel.

Leucadia also retained $4.9 billion of net operating loss carryforwards, which might have added another $500 million or more to the value of this transaction.

Ian Cumming was the CEO of Leucadia and led the WilTel investment. He passed away in 2018 at the age of seventy-seven. His obituary quoted Crain's New York Business: "While Mr. Cumming's track record compares very nicely with Mr. Buffett's, he's as unknown as the Oracle of Omaha is famous." The obituary also included a quote from Cumming and his partner Steinberg's final annual letter to Leucadia investors in 2012: "We have always preferred to make money, rather than headlines." Well, Cumming made money on his investment in WilTel.

Broadwing

Broadwing, the new brand of Cincinnati Bell's business, was in a pickle. The Bandwidth Boom lost its momentum, and flames of the Bust were burning hot. Even after accounting for all the cash flow generated by its legacy business, Cincinnati Bell's burden of funding its newly acquired Broadwing resulted in an overall cash burn of $39 million in 2002. Broadwing shares traded at below $5, which was 10 percent of the share price at the time of the legacy Broadwing acquisition. Cincinnati Bell's leaders determined a harsh pivot was necessary and decided to unload the legacy Broadwing business to anyone that would take it.

An entity named C III Capital Partners offered $129 million cash, plus the assumption of $375 million of associated operating liabilities. For this $504 million, C III acquired the competitive fiber business that Cincinnati Bell bought from Broadwing for $3.2 billion just three years prior. Cincinnati Bell gave C III the Broadwing name and wished them luck.

The legacy telephone business was retained and branded, once again, as Cincinnati Bell, and their $3.2 billion nightmare was over.

PSINet—Bill Schrader

The Bandwidth Boom was a glorious time for PSINet and its CEO, Bill Schrader. The company's name was on a new NFL stadium, and Schrader was confident a buyer would emerge with $20 billion to buy the business. Even as the Bubble began to burst, Schrader remained an optimist. At PSINet's holiday party in December 2001, he told employees, "AT&T is going to go down, but PSINet will survive."

PSINet's board was alarmed by the $3 billion in debt and Schrader's relentless focus on expansion. Schrader "was more interested in building the infrastructure than in achieving profitability with what he already had," shared one of PSINet's directors.

Despite Schrader's positive outlook, PSINet was going downhill— and fast. Cash burn increased dramatically as revenue fell short of expectations. PSINet's stock price was down 95 percent from its peak. Goldman Sachs was retained as an adviser with the hope of raising capital by selling noncore businesses, but there were no diving saves in the PSINet portfolio.

"The risk of bankruptcy can't be ruled out," reported an analyst.

The company warned, "Even if [we are] successful [in our restructuring], it is likely that the common stock of the company will have no value."

In April 2001 PSINet reported fourth-quarter 2000 losses of $3.2 billion and disclosed that it was in default on lease obligations. PSINet's board asked for and received Schrader's resignation. Schrader lost most of his $1 billion in PSINet stock by refusing to sell. "When the stock was at $30 he'd say 'I'm a seller at $75.' When it went to $60, he said 'I'm a seller at $100,'" an insider of PSINet revealed.

In May the company reported it would miss the deadline to file its quarterly financial report and would be unable to pay its quarterly dividend of $14.4 million due to preferred shareholders. Later in May, PSINet, its cash blown and unable to pay its bills, filed for bankruptcy.

In February 2002 the assets of PSINet, including customer portfolio, fiber and systems backbone, and intellectual property, were purchased for a paltry $10 million.[33] Unlike every other company covered so far in this book, PSINet's assets do not belong to any of our four consolidators—AT&T, Lumen, Verizon, or Zayo. Instead, another clever entrepreneur, David Schaeffer, the founder and CEO of Cogent Communications, purchased the PSINet assets. Schaeffer was socially quirky and, as I learned as I got to know him, had an IQ of a genius. He bought several other fiber properties at distressed prices and assembled them into a public company worth multiple billion. Nice job, Dave!

---

33 The $10 million comprised $7 million for the business and $3 million already paid by the buyer for the right to conduct its due diligence.

Genuity

In 2001 all seemed to be going well for Genuity. Verizon, which owned 9.5 percent, was expected to buy the rest of Genuity as soon as Verizon received a green light from regulators. Genuity launched its new bundled and branded Black Rocket offering, boasting big name partners such as Cisco and IBM. Genuity expanded into Europe by acquiring a French fiber provider. Then the good times rapidly turned bad.

Within weeks of the French transaction, Genuity announced it was lowering its 2001 revenue outlook, citing a slowdown in corporate purchases. Genuity reduced its workforce by 12 percent but could not have imagined the implosion right around its corner.

In early 2002 Genuity's relationship with Verizon seemed solid. Genuity was announced as Verizon's preferred provider of Internet Backbone services. "Over the past two years, Genuity and Verizon have uncovered great opportunities to . . . jointly market and sell Genuity's innovative services," declared Verizon CEO, Ivan Seidenberg.

Soon thereafter, Genuity added another billion dollars to its debt capacity, despite the objection of the Verizon representative on Genuity's board. In July 2002, with three years remaining on the five-year term, Verizon surrendered its option to buy Genuity. In announcing its decision, Verizon sought to reassure the public markets that it would not absorb Genuity's $3 billion in debt and $4 billion in 2001 losses. The option forfeiture triggered a covenant on Genuity's debt facility, which set the course for bankruptcy.

Trying to fathom Verizon's reluctance, one analyst speculated, "It sounds like they're anticipating that the business is going to continue eating cash and they don't want to throw good money after bad."

Genuity claimed to not have had advanced warning of Verizon's decision. *Forbes* evidently was tipped off. On the Monday before Verizon's disclosure, *Forbes*, in an article titled "Genuity Is Left on the Launching Pad," poured salt on Genuity's wounds.

Genuity introduced itself to a puzzled public two years ago with an ad campaign that featured a mysterious-looking black rocket and a very 1990s tagline: "Do you want to change the world?" But the '90s, alas, had just ended, and the world had already changed—and not for the better, if you were a tech investor.

Now Genuity and its rocket appear headed for the auction block.

The Verizon announcement ravaged Genuity's stock, which had already fallen to $2.59 a share. It plunged further to 29 cents. With bankruptcy looming, Genuity searched for a white knight, but time was not on the Verizon castoff's side.

Just a day before the Verizon announcement, Genuity drew down $723 million of its outstanding line of credit. "This is absolutely coincidental," said a Genuity spokesperson. "We had absolutely no knowledge of what Verizon was going to do." Genuity's behavior suggested otherwise. The credit was drawn down with unusual urgency, with Genuity demanding the banks complete the transfers by noon.

Genuity, in default on its $2 billion line of credit and its $1.15 billion loan from Verizon, filed for bankruptcy in November 2002.

Genuity found a big brother, but this one was not nearly as loving as Genuity was to the French company. Level 3 agreed to buy Genuity's assets for $242 million. The price was subject to adjustments in Level 3's favor. By the time the acquisition was completed in June 2003, the purchase price dropped to $60 million in cash, plus $77 million in cash prepayments to vendors.

Genuity, inclusive of BBN, was now owned by Level 3 for a purchase price of less than 5 percent of Genuity's peak valuation. The ugly yet desperately needed industry process of consolidation was officially underway.

Touch America—Bob Gannon

In 2000 Montana Power CEO Gannon announced the company would exit the power industry and instead focus on Bandwidth under the brand Touch America. His timing couldn't have been worse. First, Gannon exited the power business in a down cycle that reversed course soon after the exit. "Electricity prices in Montana doubled, then redoubled, and doubled again," reported CBS News. PPL, which bought Montana Power's electric generating plants, sold a third of the power to out-of-state utilities at a huge markup and posted record profits.

In 2001 Bandwidth boinked. By April, Touch America stock dropped to $8 a share, down from $65 the previous spring of 2000. As it dropped, Touch America bought back 6.6 million shares for $204.8 million. These shares soon became worthless. In June 2003 Touch America, then with twenty-one thousand miles of fiber network, declared bankruptcy.

That February CBS News had already aired its segment: "Who Killed Montana Power?"

> For nearly 90 years, the Montana Power Company exemplified the very best of American capitalism. It provided cheap, reliable electricity for the people of Montana, excellent benefits for thousands of employees and generous, reliable dividends for its stockholders.
>
> Everyone was happy, except for the corporate officers and their Wall Street investment banking firm who decided there was more money to be made in the more glamorous and profitable world of telecommunications.
>
> The result exemplified the worst of American capitalism.

Goldman Sachs advised Montana Power for seventeen years. In the years leading up to the power divestiture, Goldman bankers made over one hundred visits to Butte.

CBS News questioned whether Montana Power's directors, the majority of whom were Montanans, "fully appreciated the risks and challenges of the fiber-optic business." A public service commission member questioned the competency of Touch America's leadership: "[Montana Power's

Board was] intoxicated by their stock price and Wall Street was telling them, 'Hey, you're on a roll.' These were utility guys. What did they know about the telecommunications business?"

The board considered multiple options. One was to continue to treat the fiber business as a division within Montana Power. The second was to take Touch America public and spin out the shares to Montana Power's shareholders. At the urging of Goldman Sachs, the board chose the third option—to sell the power assets and transform into Touch America.

Matt Darnell, a Goldman Sachs's managing director, downplayed his firm's role: "There were numerous options explored but we certainly supported the thoughtful and deliberative approach the board took." He added, "The telecommunications industry has taken a turn no one expected."

Gannon defended himself and his board: "There are a lot of suggestions in watering holes around this state that Goldman was flying around the West looking for a bunch of guys to dupe and that we were a bunch of rubes from Montana who got taken in. That's the stupidest thing I have ever heard." To me, this doesn't sound far-fetched—it rings as an accurate portrayal of the situation. The shoe fits. Gannon was CEO and needed to wear it.

"There's a lot of second-guessing taking place now," Gannon said. "But we were the ones sitting in the chair, and we made a decision that was absolutely correct and justified." I ponder what an incorrect and unjustified decision would have looked like to Gannon and his board. Of all the Boom-and-Bust sagas, Gannon's Touch America story may be the saddest for the apparent incompetence of its strategic due diligence and execution.

Shareholders agreed and filed a breach of fiduciary duty lawsuit against the company, board, senior executives, and Goldman Sachs. Six of the state's largest firms collaborated on the suit. "We're trying to salvage the pension funds of thousands of Montanans who are wrapped up in this company," explained a former state Supreme Court justice.

The former judge continued, "[Gannon] was tired of what he thought was a stodgy utility stock. He owned an awful lot of shares. And I think he wanted to be the Bill Gates of Montana." The judge also made clear his view that Goldman Sachs, whose fees were $20 million, shared much of the blame for the Montana Power-Touch America debacle: "The evidence is going to show that the weeks would go by and then there would be memos in which Goldman Sachs would just keep pushing, 'This has to be done now. No better time than now. The market for this can only get worse.' [Goldman Sachs] were definitely the driver. They were pushing it all the time," the former judge noted.

At the height of the Bandwidth Boom, Gannon was praised for his bold makeover and cheered at Touch America's shareholder meeting. One year later, with the Bandwidth Bust in full force, Gannon was a pariah, scorned for selling out Montana Power shareholders. The fury intensified when the shareholders learned that $5.4 million in change-of-control payments were made to the executive team, $2.2 million of which went to Gannon. In 2001 the company's headquarters were picketed. Gannon was heckled—and had raspberries thrown at him—by locals as he walked in downtown Butte. He was refused service in downtown Butte restaurants. A banner was hung a few blocks from Touch America's headquarters that labeled Gannon a "rat bastard" and a "weasel."[34]

"I get pretty emotional about Gannon and his bunch because I think they were motivated by their own greed," said a retiree. "They saw a golden parachute at the end of this."

"You can see a company going out of business," said a woman who lost her job the day before bankruptcy papers were filed. "But when three or four people get rich off it, that's hard to handle," said a colleague who was laid off the same day. He added, "At least I know now how companies should not be run, how customers should not be treated."

---

34 Gannon claimed that some or all the assertions in this paragraph did not happen.

The irony of the situation hung in the air. Gannon was raised in Butte, less than a mile from Montana Power's headquarters. He had a thirty-year career at Montana Power. His father worked for the company. This was not the ending that Gannon had envisioned as his retirement celebration.

In late 2003 Touch America sold its remaining fiber assets to 360networks for a negligible $28 million, bringing an end to both Touch America and Montana Power. Touch America and 360networks sorted through claims with AT&T, Verizon, Qwest, and Sierra Pacific. Each got certain assets to resolve claims.

"It hasn't worked out the way we had hoped and believed was achievable," Gannon explained. "There are reasons for that. That is a hard reality." Indeed it was, and his words were an understatement.

Eight years later I acquired 360networks, which made Zayo the proud owner of most of Touch America's assets. Separately, I also bought a spare conduit on the route that Touch America jointly built with Sierra Pacific.

360networks—Greg Maffei

"When Maffei bets, he bets big. He believed in [360], so he bet big," proclaimed the former head of worldwide sales at Microsoft. Maffei's wager on 360networks was gigantic. He left Microsoft well into the Boom cycle. In hindsight he was late to the party, as rivals Broadwing, Genuity, Level 3, Qwest, and others had a multiyear head start.

When we last spent time with Maffei, he was maintaining a brave front, telling investors and employees that 360networks would survive the troubled times ahead of it.

360networks's stock price began its steep decline immediately following its September 2000 peak of $24 per share. By the end of 2000, the stock price was at $10. By March 31 the price dropped to below $5. In June the stock traded for less than 10 cents. Maffei's $1 billion of paper net worth burned to ashes.

In early June 2001, 360networks disclosed it wouldn't make a scheduled $11 million interest payment on its $2.8 billion debt. Later in June, just eighteen months after Maffei's arrival, 360networks filed for bankruptcy.

Maffei reflected, "It's so easy with 20-20 hindsight to say this was too aggressive, that was too aggressive. The fundamental premise is that the business plans of all of the emerging telecoms were too aggressive, and without access to capital, they weren't able to complete them."

He lamented, "I had great timing on leaving Microsoft. I had less good timing on entering telecom. If I'd known it was going to be this volatile, I wouldn't have come." The word "understatement" again comes to mind.

When he filed for bankruptcy protection, Maffei expressed optimism for the future of a restructured 360networks. "Almost all of our competitors are in financial distress," Maffei said in 2002. "We're more conservatively financed than most of them, and we'll be the first back into the market." Maffei, making good on his word, led 360networks rapidly through the bankruptcy process. He sold off or shut down many of the company's partially completed projects.[35]

In November 2002 Maffei led 360networks out of bankruptcy. He collaborated with Wilbur Ross, founder of WL Ross & Co., to structure a reorganization plan. Ross invested $30 million and received 10 percent of the new equity and 13 percent of the new bank debt.

Ross explained their logic: "After studying optical fiber network opportunities for more than a year, we have concluded that 360networks presents the most attractive combination of balance sheet strength, network map, technology, proximity to cash flow breakeven, and management talent. We see it as a logical base for the imminent industry consolidation."

---

35 During bankruptcy, Hibernia Atlantic, which was the name of 360networks's nearly complete €900 million undersea cable to Europe, was put in receivership. Using his Columbia Ventures Private Equity firm, Kenneth Peterson Jr. of Washington State purchased Hibernia, including 360networks's Dublin headquarters, for $18 million in 2003. The North Atlantic fiber began carrying traffic in 2005. In 2017 Peterson achieved a great exit when GTT purchased Hibernia Atlantic for $607 million. GTT, which completed dozens of acquisitions, went bankrupt in 2021.

360networks emerged with about $215 million in debt and $50 million in cash.

"Instead of being the most risky player out there, we are arguably the least-risky guy to buy from," said Maffei, who remained CEO after the restructuring.[36]

Maffei wasted no time putting his clean balance sheet to work. One week after emerging from Chapter 11, 360networks announced the $260 million acquisition of a Canadian fiber provider, which added two hundred eighty thousand miles across Canada, as well as seventeen markets the fiber business was serving.[37]

"We are very excited about combining these two successfully restructured companies," Maffei said in a release. "Together we will be Canada's largest and most complete full-service competitive telecommunications company."

Two years later 360networks sold most of its Canadian assets to Bell Canada for $201 million. According to Wilbur Ross, 360networks realized a big gain.

In April 2003 Maffei acquired the US Bandwidth business of Dynegy, a Texas power utility, which consisted of sixteen thousand route miles serving sixty-five wholesale customers in forty-four US cities.[38] "[Their fiber] business will complement our US network footprint," said Maffei. "The volatility in the telecommunications industry continues to create consolidation opportunities for those with the right assets and financial backing."

In 2003, 360networks purchased Touch America's fiber business, which was once valued at over $1 billion, for $28 million. "Touch

---

36 Jimmy Byrd, who briefly ran Level 3 North America, remained COO.

37 After accounting for cash on the acquired company's balance sheet, the net purchase price was far lower.

38 Dynegy, a Houston, Texas, energy company, followed Enron's footsteps off a steep cliff. See the Enron Broadband & Dynegy Global Communications section of appendix 3, "More Bandwidth Stories."

America's business enables us to be a leading provider . . . in the western US," explained Maffei to *Photonics Spectra.*

A portion of the power company and Touch America assets were sold or returned to AT&T, Level 3, and Qwest. By 2004, 360networks eliminated all its debt and resolved claims associated with its acquisitions.

In July 2005 Maffei became president of Oracle and, less than a year later, became president and CEO of John Malone's Liberty Media. After these transitions, Maffei retained his position as chairman of 360networks while parsing the day-to-day management to SVP Rick Coma and CFO Chris Mueller. They would run the company for six years. 360networks steadily grew in value in the decade that followed bankruptcy. Kudos to Maffei, Coma, and Mueller for achieving a respectable outcome.

AT&T—C. Michael Armstrong and Ed Whitacre

In his first three years as the CEO, Armstrong transformed AT&T and, in doing so, shaped the Bandwidth Boom, becoming the largest Cable TV provider with plans to layer voice and Internet services alongside interactive video. AT&T Wireless established itself as the biggest mobile carrier. AT&T's enterprise offerings were bolstered by its acquisition of Teleport and partnerships to construct new intercity fiber routes. AT&T was positioned to be a global leader considering it also had its partnership with BT. Yet as the Bandwidth Bust loomed, all was far from going as AT&T had planned.

The cable systems that AT&T purchased from TCI were in dire need of upgrades, causing AT&T to spend several billion dollars a year on the overhauls.

Though Cable TV revenue was growing at 10 percent, profitability had declined to 16 percent, which was far below the industry norm of 35 percent. No doubt the gap also exposed the difficulty of a large corporation with a monopoly history competing in an industry dominated by scrappy operators.

Meanwhile, AT&T's Concert venture with BT was losing $250 million a quarter. AT&T and BT agreed to dissolve Concert in October 2001, causing AT&T to take a $3.5 billion write-off.

All three of AT&T's intercity joint build partners—Velocita; McLeodUSA, which acquired CapRock; and Touch America—became bankruptcy hot messes.

By October 2000 AT&T's stock traded at $23 a share, less than half of the share price shortly after Armstrong was named CEO. AT&T's Market Value declined by $70 billion in less than a year. Debt, which was essentially zero when Armstrong assumed the reins, ballooned to $62 billion!

In October 2000 the *Wall Street Journal* published an article titled "Armstrong's Vision of Cable Empire at AT&T Unravels on the Ground." "The sobering conclusion: It may be time to give up on his ambitious plan to reinvent one of the great American business icons by integrating cable, wireless, and long-distance. Those businesses, [Armstrong] says, 'aren't necessary for each other.'" Though potentially correct, he destroyed tens of billions before having this epiphany.

AT&T's board approved a plan—code named "Project Grand Slam"—to split AT&T into four businesses: AT&T Wireless, AT&T Broadband, AT&T Consumer, and AT&T Business.[39] "We have three really strong growth businesses and we have a huge declining business that shrouds all of these businesses," Armstrong said. "[This is] going to provide a foundation and a path to let these businesses realize their value in the marketplaces." Armstrong aspired to become the CEO of AT&T Broadband.

The *Wall Street Journal* reported on the skepticism surrounding the plan: "Some industry observers question whether the breakup is the right move for AT&T, saying the company has execution problems that can't be solved with tracking stocks and spinoffs."

---

39 By June 2001 AT&T Wireless stock was $17, reflecting a market valuation of only $19.4 billion.

An analyst added, "It's hard to escape the feeling that a corporate funeral took place today. . . . It's really the end of an icon and no matter how they try to put a positive spin on it, it's the death of a corporate giant."

During the Bandwidth Boom, Armstrong went hostile on the Roberts' family in the AT&T-Comcast battle over MediaOne. With AT&T now vulnerable, Comcast decided to exact some vengeance by launching an unsolicited offer to acquire AT&T Broadband for $58 billion.[40] This was half the price AT&T paid for TCI and MediaOne. The hostile bid, if successful, would make Comcast the largest cable company in the world.

"If Comcast gets away with this one we may see the end of AT&T as we know it," an investment fund speculated.

Though AT&T's board initially rejected the offer, claiming the $58 billion "did not reflect the full value of AT&T Broadband," this was just delaying the inevitable. AT&T and Comcast agreed on a price of $69 billion in December 2001.

In July, with AT&T's stock price at $10 a share, David W. Dorman, AT&T president and the former CEO of AT&T-BT's Concert joint venture, replaced Armstrong as AT&T's CEO. Dorman did his best to show optimism: "We've got a lot of wind at our back. . . . We believe we've got some very clear opportunities right in front of us [such as the local phone market]." Perhaps the Concert blowup left a blur in his eyes.

AT&T entered 2003 as a fatally wounded animal with no fight remaining in it. After one more year of trying to regain momentum, AT&T sold itself for $20 per share to one of its seven babies, SBC Communications.[41]

---

40 The $58 billion purchase price consisted of $44.5 billion in stock and $13.5 billion in assumed debt.

41 SBC Communications' original name was Southwest Bell Communications.

Edward E. Whitacre Jr. began his career as an engineer in the Bell System and was named CEO of SBC, the Baby Bell headquartered in Dallas, in 1995, just as the Renegades of the early fiber build-outs were gaining momentum. During the Boom chaos of the turn of the century, Whitacre and his team, regarded by many in the industry as the sharpest guns among the Bell industry executives, rolled up three of SBC's siblings: Pacific Telesis Group in 1997, Ameritech in 1999, and BellSouth in 2006. In 1998 the Texas Bell also purchased Connecticut-based independent telephone company Southern New England Telephone, landing SBC close to Verizon's New York ranch. Things were getting really interesting and aggressive in the land of the Bell Heads.

Then in 2005 Whitacre, whom his executives always referred to as "Mr. Whitacre," made the move for which he will always be remembered: SBC acquired its former parent, the once almighty AT&T, for the bargain price of $16 billion. Soon thereafter, in early 2006, SBC changed its enigmatic, vestigial branding to the saved, reborn, and still enviably iconic AT&T.

"The AT&T name has a proud and storied heritage, as well as unparalleled recognition around the globe among both businesses and consumers," Whitacre explained. "No name is better-suited [as] the brand that will lead the industry in delivering the next generation of communications and entertainment services."

In 2007 Whitacre retired.[42] Instead of gobbling up other fiber properties, AT&T focused its attention on other lines of business, especially mobile and video service. Today, AT&T's Bandwidth Infrastructure assets are the least robust of the four consolidated US platforms.

---

42 Ed Whitacre's retirement didn't last long. US auto giant General Motors filed for Chapter 11 bankruptcy on June 1, 2009. Eight days later the White House announced that Whitacre would serve as chairman and six months later added the title of CEO. Whitacre, who had to learn the auto business fast, was credited with successfully leading the turnaround of General Motors. Whitacre was then succeeded by telecom veteran Dan Akerson.

XO—Dan Akerson
and Teddy Forstmann

XO was on a tear during the Gold Rush. Rockstar Dan Akerson took over as CEO. Gold Glover Teddy Forstmann pumped in $850 million of fresh capital. The $700 million of fibers purchased on Level 3's network were soon to be ready for delivery. Then the sledgehammer came crashing down on the Bandwidth industry.

XO, like all telecom companies, was hit hard. As XO was drifting toward bankruptcy, its ability to fulfill its joint build obligations was in question. At Level 3 Crowe directed me to urgently rework the fiber deal with XO to ensure our payments wouldn't be tied up in the anticipated XO bankruptcy. XO would get less fiber and conduit; we would get fiber on XO's metro networks, and XO's overall price for the joint build would be lowered. It took one long and tense month to complete the restructuring with both XO and Level 3 in a desperate state to find a path through the financial crisis.

Crowe hounded me every one of the last ten days to get it done. I told him XO's terms wouldn't work for us. Crowe didn't want the details. He just wanted it finished. I didn't appreciate the gravity of the situation, as Level 3 was in deeper financial straits than I was privy to.

Toward the end of the negotiations, I was on the phone with XO's COO. We were going back and forth on one of the details when someone shouted to him, "Tell Caruso I'll escalate to Crowe if he doesn't concede this point." I asked, "Who's that talking?" After a moment of silence, a response came: "I'm sorry. Akerson briefly jumped in my office and was listening in. He just left." I'm sure he was listening the whole time, and I doubt he had left.

I'm glad I held out because the details mattered, and we found a middle ground. In the end XO received eighteen fibers on our network, plus the tag-along rights, which were never used, that gave XO the ability to purchase additional fiber associated with Level 3's future capacity overbuilds.[43]

After we executed this deal, I sent an email to Crowe, inviting him to a meal with the two other key Level 3 executives who supported me on the restructuring.[44] Crowe jumped on the invite, and we enjoyed a steak and martini dinner together. This was perhaps my warmest moment with Crowe, as we all let down our guards and relaxed, putting aside for one evening our high-anxiety day jobs.

Since I did the original and the reworked deal, I knew the opportunities and vulnerabilities on both sides. As I left Level 3, XO was near the top of my acquisition wish list of fiber properties.

After XO reported losses of $2.1 billion in November 2001, Nasdaq suspended trading. XO's shares traded over the counter at 8 cents a share. The $10 billion Market Value posted in late 2000 evaporated in full.

Forstmann, with the backing of Akerson, had a plan to protect its $1.5 billion XO investment through a prepackaged bankruptcy. However, the legendary Carl Icahn was working on a different scheme. Forstmann versus Icahn—which heavyweight would win? With the Bear roaming in the background, perhaps both would lose.

AboveNet—Bill LaPerch

In 1999 I joined Level 3 CEO, Jim Crowe, on a trip to New York to meet with Howard Finkelstein and Silvia Kessel, who were representing Metromedia's Kluge. Kessel was short, spoke with a pronounced German

---

43 Level 3 also released XO from its obligations to purchase pan-European fiber in Europe. XO was required to put the $700 million payment associated with the U.S. joint build in escrow, assuring Level 3 of payment even if XO declared bankruptcy.

44 John Scarano was the SVP who was my right hand during the negotiation. John Ryan was one of Level 3's senior attorneys at the time and later was promoted to be Level 3's general counsel.

accent, and proudly displayed an ashtray filled with ashes. Her voice was coarse, which I surmised was the result of chain-smoking cigarettes.

After pleasantries, both parties went silent, waiting for the other to initiate the business conversation. Finally, Jack Grubman, the Salomon banker who brokered the meeting, spoke. As he did it became clear that Grubman told Finkelstein and Kessel that Crowe wanted to meet with them while telling Crowe that Finkelstein requested the meeting. From my seat either one could have been true, or perhaps both fished for the conversation. Crowe and Finkelstein danced around a merger, with Grubman egging it on. Both executives expressed confidence in their business prospects. Neither wanted to show weakness or fear to the other.

Nothing came of this meeting, which was quite fortunate for me because if AboveNet had been acquired by Level 3, Zayo's 2012 acquisition of AboveNet could not have happened. And I wouldn't be sitting at my brand spanking new beachside home at the El Dorado Golf and Beach Club in Cabo writing this book.

In 2001 AboveNet faced an existential problem. Nearly every distressed fiber provider was a customer of AboveNet, and these businesses made up the overwhelming majority of AboveNet's impressive $2 billion revenue backlog. Backed by these customer contracts, AboveNet was well on the way to building the most valuable metro networks in the world.

Then the wobbly status of these customers abruptly destroyed AboveNet's source of funding.

In January 2001 AboveNet announced that it had obtained a firm commitment for a $350 million credit facility from Citicorp that would fund AboveNet to achieve cash flow sustainability. Six months later AboveNet revealed that Citicorp's "commitment" was conditioned on AboveNet securing commitments from other lenders in the amount of $287 million. In August a class action lawsuit was filed against CEO Steve Garofalo, who was once lauded for birthing AboveNet by creatively leveraging the available network of sewers below NYC. Now he was being sued for misleading AboveNet's public investors.

For all these reasons, AboveNet's stock was trading below $1 per share.

In late 2001 AboveNet was able to close on a $611 million debt facility that included Verizon and Citigroup. The loan bought AboveNet a short reprieve.

Another bomb dropped in April 2002 when AboveNet's auditor, KPMG, found the company's internal systems and controls so flawed that they couldn't be reviewed in accordance with normal accounting procedures. AboveNet issued a statement: "[KPMG] identified issues relating to a number of adjustments that it believes will be necessary . . . and is working to determine the nature and amount of such adjustments."

Furthermore, KPMG stated that it couldn't "review the quarterly data that is expected to be included in the Annual Report in accordance with professional standards because AboveNet's internal control structure and policies and procedures for the preparation of interim financial information did not provide an adequate basis for them to complete such a review."

As if April wasn't already bad enough, Verizon and Genuity terminated their fiber-optic agreements with AboveNet.

AboveNet filed for bankruptcy in May 2002.

In September 2003 AboveNet emerged from bankruptcy.[45] Craig McCaw, the original backer of XO and an AboveNet shareholder since 1997, partnered with board chair Kluge to each invest $25 million, which gave them a significant ownership stake in AboveNet.

The restructured company emerged with $276.6 million in revenue, $80 million in cash, $83 million in debt, and an estimated Enterprise Value of between $230 million and $330 million—a far cry from the $10 billion value AboveNet enjoyed during the Boom.

The accounting flaws identified by KPMG proved to be insurmountable to reconcile. In fact, it was 2008 before AboveNet filed a quarterly (10-Q) or annual report (10-K) with the SEC. The 2008 filing covered

---

45 During bankruptcy, AboveNet sold valuable real estate assets in the extraordinarily important data center market of Northern Virginia for a combined value of $63 million, which represented steep discounts to the properties' true values.

2006. When this report was filed, Rob Powell of *Telecom Ramblings* blogged.

> [T]his day marks the end of an era, the final curtain has come down on the telecom nuclear winter that began in 2001 in which hundreds of telecom companies went under—from WorldCom on down.
>
> All those assets filtered through bankruptcy court into new ownership, occasionally more than once, but one case always stood out. . . . AboveNet used to be known as Metromedia Fiber Networks and the depth of its collapse is legendary, involving a 5 year SEC investigation.
>
> While AboveNet emerged from bankruptcy in 2003, they have struggled for 5 years now to find out where the money went back in 2000–2002. They have not succeeded and have admitted they likely never will, nor have they managed to produce a 10-Q or a 10-K since then, despite being a public company. In fact, AboveNet didn't have any quarterly or annual reports on file this century which they hadn't disavowed.
>
> All the other rubble from the crash (MCI, GX, WilTel, Genuity, etc.) has long been cleaned away or refurbished, but AboveNet was still lying in a coma. Until today.
>
> . . . So maybe they're back.

Emerging from bankruptcy with a vacant CEO seat, the board needed to find a replacement. William G. "Bill" LaPerch, AboveNet's operations executive, received a battlefield promotion to CEO and became AboveNet's fifth CEO in three years. Unlike his predecessors, LaPerch would have a long tenure as CEO.

LaPerch was a longtime telecom operator, making his way up through NYNEX and MCI as a highly regarded operations and engineering executive. He served our country as a captain in the army after graduating from West Point. He earned his MBA from Columbia University while at MCI. In 2000 WorldCom acquired MCI, and LaPerch joined AboveNet to lead operations.

"Before AboveNet, MCI was the highlight of my career," LaPerch shared with me in 2024 as he reflected on his career journey. "To say I was devastated when WorldCom acquired MCI would be a big understatement. Within a month I knew WorldCom was not for me." Fifteen members of LaPerch's MCI team followed him from WorldCom to AboveNet.

LaPerch recalled his first board meeting, which was held at Rockefeller Center soon after he joined. "Everybody had an assigned seat, with mine between two elderly gentlemen. I introduced myself to the guy on my left. He responded, 'Hi. I'm John Kluge.' I turned to the unfamiliar man to my right, who introduced himself as David Rockefeller. Red-faced and giggling to myself, I thought I had finally made it to the big time, sitting between two of the richest people in the world."

During his first year as CEO, LaPerch sold multiple data center assets that, if held, could have been worth many hundreds of millions. "We're paying pennies on the dollar for the San Jose building's assets. If we had to go out today and build a building from scratch like this, we would have to deploy $80 million in capital," an Equinix executive joyfully remarked. LaPerch used the proceeds to expand AboveNet's fiber network with a focus on connecting AboveNet fiber to as many data centers as possible.

LaPerch's ten-year execution of the Bandwidth Infrastructure strategy was extraordinarily successful. He captured his focus in his 2006 and 2010 interviews.

> We come to work every day prepared to execute on our organic growth strategy.
> We go into the top metro markets, and we do two things simultaneously. One, we connect it to the rest of our markets using our wide area network. Two, based on the experience

we have in our existing markets, we know that a significant amount of business is going to be focused around the data center infrastructure. So, we establish a fiber ring that hits the key carrier hotels and data centers.

We don't go out and resell someone's service.

We are . . . not interested in being a cloud service provider, a CDN provider, or a professional services company.

In early 2009 AboveNet was relisted on the NYSE. In August 2009, as its stock approached $80 per share, AboveNet executed a stock split. This conveyed leadership's confidence that their business, which was now worth more than $1 billion, would continue its growth in value.

And grow it did. By 2012 AboveNet's revenue reached $540 million, which was double from when it emerged from bankruptcy. The company achieved a very respectable profitability margin of 45 percent, and cash flow was positive. The stock price continued to rise, again approaching $80 per share, reflecting a 200 percent gain since the 2009 split.

Despite posting exceptional results with perhaps the industry's best collection of fiber assets, the $1.8 billion valuation placed on AboveNet by the public markets was low relative to its financial performance. Private Equity took notice. In 2011 a Private Equity firm bought the stakes of some of AboveNet's investors.

LaPerch, no doubt encouraged by bankers and investors, saw the opportunity to take the company private in a management buyout. In doing so they put AboveNet in play and on a collision course with my Zayo journey. The outcome would end up shocking the industry.

Level 3—Jim Crowe

As the first year of the new decade was upon us, Level 3 realized it was overextended. The decision was made to pull out of Asia, which reduced capital obligations by several hundred million.[46] Level 3 took a 2001 write-off of $500 million.

In January 2001 Level 3 posted results for the fourth quarter. "It was another solid quarter and an exceptionally strong year for us," declared Crowe.

However, Crowe's positive spin was not aligned with the duress that was hitting the Bandwidth industry. By June 2001 Level 3's stock price cratered to $5, down 94 percent from its $130 peak in March 2000. The Market Value of Level 3 was $2 billion.[47]

Like with AboveNet at the time, Level 3's revenue was largely associated with customers now in financial freefall. This made the revenue base highly suspect, especially the upfront payments that would be owed for customer contracts for yet-to-be-delivered fiber.

In June the *Wall Street Journal* published an article titled "How the Fiber Barons Plunged the U.S. into a Telecom Glut." Capturing the veiled panic of the times, the *Wall Street Journal* reported, "Now, Level 3 has hit a wall even Crowe may have trouble overcoming. . . . Some acquaintances have noticed that Crowe seems more subdued these days. But in public the executive remains the picture of confidence."

A saltwater fish tank that was the feature of Crowe's office suite was emptied to save money. I don't know what happened to the fish.

Crowe explained how the landscape had shifted, stating only the strong would survive. "A year ago we were in a hothouse environment where every plant, regardless of its strength, prospered. Now, we are outside in the cold world. It is a better environment, as painful as it is."

---

46 The Asian undersea assets, which cost roughly $660 million to deploy, were sold to an Asian carrier for $80 million. This released Level 3 from the obligation to complete construction of the undersea network. A year later Level 3 completed its retreat from Asia by also selling—or, more accurately, handing over—its Japan, Hong Kong, Korea, and Taiwan fiber networks to an Asian company named Reach. This reduced Level 3's capital obligations by $300 million and led to a $500 million write-off.

47 Level 3's $2 billion Market Value implied a value of $8.4 billion, including debt.

Crowe spun the industry turmoil as being advantageous to Level 3's strategy. "The shake-out that is occurring is good for Level 3 in the long term, although it is awfully hard to convince someone who is sitting in a dentist's chair being drilled that this is a good thing," said Crowe. "It hurts."

Level 3's plan was to use its stockpile of cash to outlast its rivals. Level 3 reduced its workforce by 20 percent. "[Crowe] stresses that the company still has $4 billion in cash and says it will emerge from the shakeout," wrote the *Wall Street Journal.*

In September 2001 Level 3 announced a plan to buy back debt at a discount of 43 percent to 73 percent of its face value. Over the next six months, Level 3 bought back $2.1 billion at a steep discount, lowering the debt load to $6 billion.

At the end of 2001, Level 3 had $2.1 billion in liquidity, which it said was sufficient to get to cash flow breakeven. The markets weren't buying it. "Level 3's bonds are trading at about 50 cents on the dollar," a high-yield telecom research analyst stated. "That's basically the market telling the company that it has to restructure."

In January 2002 Level 3 disclosed that if its current rates of sales, cancellations, and disconnects continue, the company may soon violate a "revenue-based financial covenant." If so, the lenders could force Level 3 into bankruptcy.

The loan covenants specified that Level 3 must have $2.3 billion in annual revenue to stay in compliance. Crowe tasked his team to find an acquisition that met three requirements. First, revenue needed to be at least $1.7 billion to satisfy the debt covenants. Second, the acquisition price needed to be as low as possible, as Level 3 had little cash or debt capacity to spend on acquisitions, and Level 3's tiny stock price prevented it from being usable as acquisition currency. Third, the acquired company must be classified as a communications services business, as Level 3's debt covenants counted only revenue of communications businesses.

Crowe needed a diamond in the rough, and the bankers found one, by the name of Howard Diamond.

Diamond was CEO of Corporate Software, the largest reseller of Microsoft Office. His firm would book as revenue the entire cost of the software and return 97 percent of it back to Microsoft.

Corporate Software had a whopping $1.2 billion of revenue but a miniscule profitability of only $18 million. This profile meant Level 3 could acquire a whole lot of revenue for a tiny purchase price. Most importantly, the business was classified as a communications services business.

Level 3 purchased Corporate Software in February for $139 million and, two months later, purchased Software Spectrum, a rival to Corporate Software, for $122 million.

An analyst was blunt in his assessment: "This is creative financial engineering. . . . [T]hey're buying their way out of the revenue covenant. It's not immediately clear to me how a business-applications software reseller that's been around well over 15 years is directly integral to Level 3's communications business."

Crowe did his best to provide a strategic angle. "This transaction affords a number of distinct advantages to Level 3. . . . Level 3 already provides a variety of information technology services. . . . This acquisition will enable [us] to attain scale and to leverage [our] customer base, worldwide presence and IT professional relationships." None of this was true, of course, as the purpose was solely to avoid bankruptcy.

Howard, a true diamond in the rough, wasn't the only gem Crowe had up his sleeve.[48] The other was the famed Oracle of Omaha, Warren Buffett. Level 3 chairman Walter Scott, who also served as a board member of Buffett's Berkshire Hathaway, convinced Buffett to invest $100 million alongside $400 million from Legg Mason and Southeastern Asset

---

48 Crowe had a backup plan. The renegotiation of the XO-Level 3 Joint Build was structured in a manner that would allow Level 3 to reclaim the conduit and six of the twenty-four fibers without any refund. This would allow Level 3 to accelerate the recognition of over $700 million in revenue and satisfy the debt covenants.

Management. *Forbes*'s headline was "When Buffett Buys, a Bottom Must Be Near."

The fresh investment from extraordinary investors positioned Level 3 to shift from defense to offense. Buffett said, "[L]iquid resources and strong financial backing are scarce and valuable assets in today's telecommunications world. Level 3 has both." He added Level 3 "is well equipped to seize important opportunities that are likely to develop in the communications industry."

The endorsement of Legg Mason was every bit as important. Legg Mason Funds Management beat the Standard & Poor's 500-stock index for eleven consecutive years, the longest run for a stock-fund manager at that time. Chairman Bill Miller expressed his confidence: "Level 3 is emerging as one of the ultimate leaders, survivors, and consolidators in the industry." Level 3 stock promptly shot up 66 percent the day of the announcement.

Crowe was ready to go on the offensive. "The company is now able, with this additional capital, to better take advantage of acquisitions and consolidation opportunities in the communications industry today."

Crowe leaned on James Bond for credibility. Actor Sean Connery was featured in television ads touting Level 3's glorious vision of a Bandwidth-enabled world.

Though bankruptcy in the near term was avoided and 007 was endorsing the company, the stakeholders were not believing the message. Level 3's bonds traded at 35 cents on the dollar. The stock remained below $2 per share. These implied that both debt and equity investors believed a bankruptcy was very likely.

Much work would be required to get the company healthy and worth more than its debt.

Jack Grubman—Sarbanes–Oxley

This book would be irresponsibly incomplete without the enthralling story of Salomon Smith Barney analyst Jack Grubman.

The job of a Wall Street analyst is to provide research and analysis on public companies, which typically include recommendations for investors to either buy or sell the stock of the covered companies. Grubman and most analysts specialize in particular industries. In Grubman's case telecom was his area of focus.

Grubman came up three times in our story. The first time was when he introduced Nacchio to Anschutz. Second was when Grubman proclaimed XO as "the best opportunity ever in telecom history." Third was when Grubman sat alongside WorldCom's chairman Sidgmore and CFO Sullivan in front of Congress, testifying that he "always practiced . . . honestly held research and opinions." In effect, Grubman's excuse for making his disastrous recommendations was that he was simply one of the all-time worst analysts.

The truth was Grubman's opinions were not honest and didn't derive from research expected of an analyst. His opinions were caked in old-fashioned greed. And glowing praise of his Gold Rush darlings made him rich. Eventually, though, he would pay a price.

Grubman grew up an only child. For years Grubman told the story that he came from the working-class South Philly, the home base of singer Frankie Avalon and the neighborhood where the film character Rocky Balboa perfected his boxing. In fact, he grew up in the relatively prosperous middle-class Oxford Circle neighborhood of Philadelphia. In 2000, as the telecom industry began its meltdown, bringing heightened scrutiny by Congress and federal investigators, Grubman, the most prominent and highest-paid telecom analyst on Wall Street, admitted to a *BusinessWeek* reporter that he had been lying about his education for years. Grubman lied to his first Wall Street employer, PaineWebber, and in subsequent official bios and to the media, claiming that he had graduated from the prestigious Massachusetts Institute of Technology when, in fact, he received his undergraduate degree from Boston University.

He launched his career at AT&T before joining a Wall Street investment banking firm as a telecommunications analyst. The self-professed math genius failed the Series 7 exam, a required test to be an investment analyst, before passing it on his second try.

Grubman became an outspoken critic of AT&T and a loud cheerleader of AT&T's competitors. His AT&T experience gave him an edge over other analysts, and he carved out a niche by being bullish on the new entrants that attacked AT&T and its Baby Bell offspring.

In 1994 Grubman jumped ship to join Salomon, where he became the leading analyst on Wall Street. In 1997 he was number one in Institutional Investor's annual rankings. Wall Street insiders referred to him as the "ax," which informally knighted him as the sector's most influential analyst.

Grubman covered forty newly formed competitors to the Bell System, whose combined Market Value exceeded $1 trillion. His words fueled dramatic increases in stock prices, and he wasn't shy in how he characterized his influence. "I'm sculpting the industry," he proudly exclaimed.

"When Grubman said wonderful things about a company, it was like a narcotic—everybody wanted it" was how a former Salomon stockbroker characterized Grubman's influence. "He walked around like he was a god. And it was perceived in the industry that he was a god."

Grubman's influence led to lucrative engagements for his firm. Salomon collected over $1 billion from the clients covered by Grubman during the late 1990s and early 2000s. To reward him, Salomon paid Grubman $48 million between 1999 and 2001, by far the highest compensation of any Wall Street analyst.

The first time I met Grubman was in early 1999, when I was the group vice president overseeing the network organization for Level 3. I spent an hour with him in our Colorado offices detailing Level 3's execution plan. Grubman's focus was on pinpointing how to position Level 3 relative to other clients of his, such as AboveNet, Global Crossing, Qwest, and XO. As we conversed, it wasn't a substantive explanation that he was seeking. Grubman was searching for nuanced words he could use that would convey differentiation without dissing his other clients.

The next month Level 3 raised $1.4 billion by selling shares at $54 each, with Salomon engaged as Level 3's Wall Street advisor. On the day this offering was announced, Grubman raised his price target on Level 3 from $54 per share to $70. Four days later, Grubman published a forty-page report titled "Level 3 Communications—Optimizing a Layer of the Telecom Value Chain: The Intel Inside of Telecom." He touted that Level 3 was "a great play on Bandwidth" and "has a strong management team that is critical for success in this business." Grubman's influence was as good as advertised, as Level 3's shares jumped 34 percent to $72.

The expression "pay-to-play" refers to the linkage between an analyst's positive report on a client and the analyst's employer receiving lucrative engagements from that client. Grubman's positive report on Level 3, timed to coincide with Salomon being hired to support Level 3's stock offering, reeked of pay-to-play.

"It's buyer beware—many Wall Street firms have embedded conflicts of interest," said one analyst at the time.

"Does it look crummy? Yeah, it looks crummy. It's an appearance problem for the firm, but I don't know whether it's anything more than that," added a professor at New York University's law school. The laws at the time were unclear.

"Whenever [Grubman] picks up [coverage on] a company, I think the presumption is that there's a banking deal in the works," another expert chimed in.

Grubman was unapologetic, noting the sharp jump in Level 3's stock price: "I guess if anyone was particularly bothered by our report, it didn't really show in the market." When later questioned about the inherent conflicts in his pay-to-play antics, Grubman replied with his trademark swagger: "What used to be a conflict is now a synergy. Objective? The other word for it is uninformed."

WorldCom was the biggest example of Grubman's outsize influence. In October 1997 the *New York Times* published "For Salomon, as Adviser, Millions Plus Revenge," which focused on Grubman's critical role in WorldCom's successful acquisition of AT&T's biggest rival MCI.

> The only company with as much to gain from WorldCom's $30 billion bid for MCI as WorldCom itself may be [Salomon]. . . . WorldCom will pay Salomon $32.5 million in fees if the acquisition is completed.
>
> . . . The job of persuading Wall Street . . . falls to Jack B. Grubman, Salomon's senior telecommunications analyst. Mr. Grubman, who is renowned for his shoot-from-the-hip opinions and polemical research reports, has become a key deal maker at Salomon. He has known [Ebbers] since 1988, and has been an unflagging advocate of the company's stock.

McLeodUSA, a Midwest telecom company led by one of Iowa's wealthiest individuals, Clark McLeod, was another example. Salomon earned over $100 million in fees, including leading the McLeodUSA public offering in 1996 at $24 per share. Two weeks after the IPO, Grubman initiated a buy recommendation with a target of $40. "McLeod represents one of the truly great business models that will be executed in the new era of telecom," Grubman wrote.

Grubman was caught red-handed in a 2001 email message about Focal Communications. In February he published a research note on Focal reiterating a buy but noting some concerns. When bankers told Grubman that Focal's executives complained, he emailed the bankers: "If I so much as hear one more f-ing peep out of them, we will put the proper rating in this stock. . . . We lost credibility on [McLeodUSA] and XO because we support pigs like Focal." Grubman threatened to "put the proper rating" on Focal, which he said every smart institutional investor "feels is going to zero." On the same day as this email outburst, Grubman reiterated his buy rating, with the stock trading at $15.50 a share. He maintained the buy recommendation for five months, with the shares dropping to $1.24. Soon thereafter Focal went bankrupt.[49]

In April 2001 Grubman published "Don't Panic—Emerging Telecom Model Is Still Valid." In it, he recommended seven stocks, including Broadwing, Global Crossing, and AboveNet. By the end of 2001, the basket of stocks lost 58 percent of their value. Soon thereafter all but Level 3 went bankrupt.

The *New York Times* article titled "Telecom's Pied Piper: Whose Side Was He On?" focused on the conflict of interest among analysts who were rewarded with bonuses based on the fees the bank received from the companies they covered. Buy recommendations made for happy banking clients. Grubman, "one of Wall Street's highest-paid analysts," played the game.

The *New York Times* identified Grubman's culpability in the Boom and Bust: "The telecommunications mess stands out for another reason: One man is at its center—Jack Benjamin Grubman. . . . Mr. Grubman lured more investors into securities of nascent and risky telecom companies than perhaps any other individual."

Eventually, though, investors soured on the companies Grubman covered, and Grubman was exposed as a peddler of worthless stocks. "Even as [Grubman] rallied clients . . . to buy shares of untested telecommunications companies and to hold on to the shares as they lost almost all their

---

49 Bob Taylor, a former Ameritech and MFS colleague of mine, was the co-founder and CEO of Focal. In 2000 Taylor resigned and was replaced by Kathleen "Kathy" Perone, a veteran of MFS and Level 3. Focal filed for Chapter 11 bankruptcy in December 2002, completed reorganization in 2003, and was acquired by Broadwing in September 2004 for $189 million.

value, he was aggressively helping his firm win lucrative stock and bond deals from these same companies," the *New York Times* reported.

"Jack Grubman is the king of conflicted analysts," said a securities lawyer. "He used his picks to generate investment banking business for his firm and abused investor trust in his picks. He personifies the blurring of lines between investment banking and objective analysis."

Grubman wasn't alone. The analyst community in general shared in the blame. The *New York Times* article titled "Pied Piper" cited the questionable metrics that were used to value fiber providers.

> Because these companies had neither revenue nor earnings, telecom analysts devised new rationales, new metrics, to justify the purchase of already overpriced stocks. . . .
>
> The most popular method was to value the companies based on the money they had put into their networks—money from investors, not money that the company had earned. . . . [T]elecom shares were promoted by big brokerage firms as good values because they traded at prices that represented four or even five times what they had invested in plant and equipment.

In March 2002 the *New York Times* article titled "Telecom, Tangled in Its Own Web" tied Wall Street's pay-to-play behavior as contributing to the meltdown.

> At the center of this debacle stand the usual Wall Street enablers. Investment bankers raised money from investors for far more telecom networks than were economically feasible. Brokerage-firm analysts, eager to help their employers win ever more securities offerings, drummed up investor interest in untested companies.

The SEC didn't buy Grubman's "I-was-honest-but-incompetent" explanation. In 2003 the SEC filed a complaint against him for providing misleading research. Grubman paid $15 million in fines and was banned for life from the securities industry.

The destruction of the telecom industry led to one of the largest securities regulatory reforms in history. The new law, named the Sarbanes–Oxley Act of 2002, or SOX for short, was created by Sen. Paul Sarbanes and Rep. Michael Oxley. Among SOX's numerous provisions, it codified Grubman's behavior as unlawful.[50] Separation—typically referred to as a Chinese Wall—would be required to prevent bankers from colluding with analysts.

———————

As the mid-2000s neared, the worst of the Bust was behind us. Now it was time to transition to the next phase of the Boom/Bust cycle—the time when savvy investors rummage through the rubble and get their grimy hands on valuable assets for next to nada. The table was set for the phoenix to rise from the ashes.

———————

50 SOX also required that executive officers certify that they have reviewed the report, have read and understand it to ensure the report does not contain "untrue statement of a material fact or omit to state a material fact necessary in order to make the statements made, in light of the circumstances under which such statements were made, not misleading," and that the report fairly represents the company.

SOX makes it clear that signing officers are responsible for creating and maintaining effective internal controls to detect all material information that should be shared within the reporting period, including reporting deficiencies and material weaknesses.

SOX strengthened the independence auditors play in serving a company. An auditor would no longer be allowed to provide a client with non-audit work simultaneously. The lead audit for a company must switch every five years so that the possibility of fraud is decreased.

CHAPTER VI

THE PHOENIX RISES FROM THE ASHES

Be greedy when others are fearful.

—Warren Buffett, stressing the time to invest is when others are reluctant

No one could appreciate the size and scope of the Bandwidth mess. After all, it was the biggest Boom and Bust in human history.

Many of the iconic companies—AT&T, Genuity, PSINet, Touch America, and WorldCom—never recovered from the Bust. Others navigated through bankruptcy and regained momentum but were vulnerable prey hunted by predators seeking to rollup Bandwidth providers on the cheap. Qwest became part of a rural telephone company named CenturyLink.

Level 3 barely avoided bankruptcy and didn't sell out. "We talked about 'the pace of change' and about 'enormous revolution,'" Jim Crowe reflected in 2002. "I guess what I didn't appreciate is just how messy revolutions can be."

In this environment of fear and uncertainty, dozens of smaller fiber companies were sorting through their own bankruptcies or distress.

Forbes, in its 2002 article titled "Telecomeback," looked ahead.

By picking through this titanic wreckage, new fortunes will be made. . . . In the aftermath, . . . hard assets are selling for pennies on the dollar. That sets up the next telecom revolution, the next bout of wealth creation. It offers riches for the survivors and for investors bold enough, or crazy enough, to bet on the coming chaotic consolidation.

"The telecommunications industry is going through a period of unprecedented turmoil," said Level 3's Crowe to *Forbes* in its 2002 article titled "When Buffett Buys, a Bottom Must Be Near." Crowe added, "At the same time, the ongoing shakeout is creating extraordinary opportunities, as telecommunications companies, their network assets, and customer bases become available."

Forbes wrote, "The telecom sector remains shrouded in gloom, but clearly the potential consolidators are starting to raise their heads a bit and sniff around for opportunity. If Buffett is buying into a fiber-optics play like Level 3, that's a sign that telecom valuations finally have fallen low enough to tempt prudent value investors. Let the bottom-feeding begin."

Wired magazine was more brass in their 2002 article titled "Surviving the Fiber-Optic Fire Sale: "Incredible assets are up for grabs. . . . Vulture capitalists are circling, ready to pick the losers' bones. To win takes nerve, cash and luck. And the game could drag on for years."

While investors were frozen with fear, a new breed of entrepreneurs became greedy. I was one of them.

ICG—J. Shelby Bryan and Dan Caruso

I felt like a failure when I left Level 3 in 2003. Given the disastrous events of the early 2000s, I'm sure plenty of my industry colleagues did as well. Many never recovered, carrying the burden of their role in the meltdown through the remainder of their lives. Most of the leaders in the industry who remained never fully grasped the difficult lessons that needed to be learned.

I did. I deeply reflected on what had transpired in both the industry and throughout my career. I used my failures, and the failure of the industry, to carve a better path forward.

I applied the lessons I learned to achieve outcomes that earned me a Hall of Fame spot with nearly all my financial sponsors. For my first two investors, I achieved the largest internal rate of return (IRR) (>400 percent) and investment multiple (twenty-five times) in their combined fifty years of history.[1] These two sponsors backed me in my next venture, and they, alongside a few other investors, achieved the highest Absolute Return in their combined one-hundred-year history.[2] My investors, including the portion that was earned by management, earned $8.5 billion on their $1.1 billion investment.

My comeback began with a fiber provider in my home state of Colorado, whose backstory was colorful. In a 2000 article, the *New York Observer* described the leader of a fast-growing Bandwidth provider in Denver named ICG Communications:

> J. Shelby Bryan is tall, smooth and charming—seductive, even. He is an incomparable salesman. A blend of old Texas money and East Coast establishment gloss, the 54-year-old Mr. Bryan offers something for everyone: a few years working for Ralph Nader . . . , a stint as a Morgan Stanley mergers and acquisitions banker . . . and, when such a move was still considered daring, a run as a telecommunications entrepreneur.
>
> Vogue editor Anna Wintour left her husband for him. President Clinton schmoozed Manhattan Democrats in his Upper East Side salon. And on Feb. 29, cable titan and notoriously

---

1 Columbia Capital and M/C Partners were my Private Equity partners.

2 Columbia Capital and M/C Partners were joined by Battery Ventures, Charlesbank, Morgan Stanley, Oak Investment Group, and Centennial Ventures.

discerning communications investor John Malone took a $500 million stake in his high-flying . . . ICG.

ICG was born as Teleport Denver, a satellite operator in the mid-1980s that later rebranded to IntelCom Group. In the mid-1990s, with both its cash and prospects low, ICG recruited Bryan to become its CEO.

After receiving an MBA from Harvard University, Bryan launched his career as a Morgan Stanley investment banker. In 1979, he co-founded Millicom International Cellular SA, which had a Market Value of $1 billion when he resigned from his CEO position in 1994.

In his first year as ICG's CEO, Bryan raised $2 billion, much in the form of junk bonds. When Bryan observed Wall Street's reaction to AboveNet, MFS, Teleport, and others investing in metro fiber networks, he accelerated ICG's local fiber build-outs. The market reacted favorably. At the peak of ICG's share price, Bryan's stock was worth $88 million.

Bryan enjoyed the stature and perks of being a public company CEO, including raising capital and schmoozing with VIPs. However, he didn't much care for the aspects of the job that involved running the business. Bryan lived in New York City and spent minimal time in Colorado, the headquarters of ICG.

Bryan left the day-to-day operations to COO John Kane. In the late 1990s, Kane turned his attention to providing dial-up Internet access platforms for low-cost Internet Service Providers. In particular, ICG's customers were Microsoft, Earthlink, NetZero, which offered its customers free access to the Internet, and the discount retailer K-Mart, now bankrupt, for its BlueLight.com online service.

With ICG's customers offering free Internet access, traffic into the ICG network ramped up exponentially. The network wasn't ready for this traffic explosion, which led to an implosion of service quality. In 2000 the *Wall Street Journal* reported, "Key customers, including Microsoft Corp., NetZero and EarthLink, began screaming that users trying to connect to the Internet through the ICG network were getting blocked by busy signals or had connections cut off." The *Denver Post* piled on: "ICG's network was rife with troubles. It had experienced network outages, equipment failures, and technical difficulties. . . . Customers cut back on their business with ICG and threatened to cancel their contracts altogether."

During the early signs of duress, Bill Beans, a veteran of MFS and Teleport, replaced Kane as COO.

ICG's motive to aggressively offer this service was to exploit the access charges that the Baby Bells were required to pay for terminating this dial-up traffic. In 1999, $125 million of ICG's $479 million revenue was derived from this unsustainable regulatory framework. However, in 2000, the gravy train ended. Led by SBC, the Baby Bells delayed payments and forced a 90 percent reduction in the fees it was obligated to pay for terminating dial-up Internet traffic.

In the spring of 2000, before the customer challenges and rate reduction had fully played out, Bryan convinced John Malone's Liberty Media to lead a $750 million investment in ICG, which equated to an Enterprise Value at nearly $5 billion. Leveraged-buyout specialist Hicks, Muse, Tate & Furst invested $230 million alongside Malone's $500 million. The other $20 million came from investment advisor Gleacher & Company.

In August, in the midst of deteriorating service and shrinking revenue, ICG was on the cusp of combining with Alex Mandl's Teligent, which was also in Liberty Media's portfolio and was also faltering. The New York's Waldorf-Astoria Hotel was booked for an August 9 announcement. Mandl pulled out at the last minute as he better understood the magnitude of ICG's dire situation. Perhaps Mandl also knew that two wrongs don't make a right.

In September 2000 ICG slashed its estimates for 2001 revenue to half of its earlier forecast of $1.4 billion. The stock plunged to below $1 per share. With debt trading at a gigantic discount to face value, famed hedge fund manager Steve Feinberg bought a significant amount of ICG's high-yield bonds via his firm, Cerberus Capital Management. The $750 million invested in the Malone-led round just four months earlier was worthless. In Malone's exceptional career, ICG must have ranked as his worst investment.

Bryan was replaced by Liberty Media's Carl E. Vogel. A class action lawsuit claiming securities fraud against Bryan and others was settled for $18 million.[3]

With all hell breaking loose, a board meeting was held on September 15. Beans recalled fifteen board members and advisors, including Vogel and billionaire Thomas "Tom" O. Hicks of Hicks, Muse, Tate & Furst, were in the meeting when a blast struck the window of the sixth-floor boardroom at ICG headquarters.

"I was presenting, and I thought a bird flew into the window," Beans told me in 2024. The security people came in and said [it was] a bullet, not a bird." The police concluded the bullet was from a sniper, though the subsequent investigation didn't uncover the source. Vogel and Hicks resigned the next day, leaving Beans as COO, reporting to a thinned-out executive committee of the board.

Two months later, with the stock now worthless, ICG filed for bankruptcy. Randy Curran, who had restructuring but no telecom experience, was hired as the CEO. In October 2002 ICG emerged from bankruptcy as a public company with $205 million in debt and $94 million in cash. "This is a great day for ICG, our customers, and our employers," said Curran, CEO of ICG. Cerberus and W. R. Huff Asset Management emerged as the largest shareholders.

In 2003 Curran was named Denver's Telecom Executive of the Year by the Denver Telecom Professionals, who released the following statement: "Over the past year there have been very few telecommunications executives in the world who have been successful in turning their businesses around in this unstable environment, and Randy Curran is one of these executives."

Not all was as it seemed. By the end of the same year, Curran was no longer CEO. In the fall of 2004, just two years after emerging from

---

3 John Kane was also named in several lawsuits.

Chapter 11, ICG filed a statement with the SEC disclosing it would likely need to file for a second bankruptcy.

———————

While the ICG drama was playing out, I was wrapping up my time at Level 3. One of the hallmarks of Level 3's culture was convincing ourselves that we were the best operators in the telecommunications industry. Even in our survival years, we could point to Level 3 as being among the few that avoided bankruptcy or a desperate sale.

So when I left Level 3, I thought my phone would ring. It didn't. It turned out most investors didn't notice. Or perhaps they didn't care. Most of the telecom industry saw Level 3 as no different from our bankrupt rivals—like them, we also destroyed enormous amounts of stakeholder value.

Moreover, my rapid rise was fueled by a relentless desire to succeed no matter the cost. I had sharp elbows, low patience, and poor EQ. Colleagues had mixed feelings about me. They saw me as intelligent, driven, and successful—but also as ruthless, callous, and insensitive.

Lastly, the industry environment remained in peril. Too many executives were chasing way too few opportunities. Investors lost billions in telecom. Most were not ready to lose more.

I faced the unpleasant reality that my career had come to an end. I was only forty.

Just when I needed it, I received a big assist from a Level 3 colleague, Donald "Don" Gips. My relationship with Gips began in 1998, when he visited our Level 3 offices in Boulder. Gips arrived early for his scheduled meeting with Crowe, so I was asked to spend time with him. We introduced ourselves to each other, did the small talk thing, and slowly figured out that neither of us knew why we were talking to each other.

Gips was everything I wasn't—he had an undergraduate degree from Harvard University and an MBA from Yale. He was a management consultant at McKinsey & Company and then became chief domestic policy advisor to Vice President Al Gore. He served as chief of the International Bureau at the FCC, where he was responsible for wireless spectrum policy and auctions. He was sophisticated, well spoken, and worldly.

I was a schmuck from the blue-collar south suburbs of Chicago.

Crowe hired Gips to lead corporate strategy and development, which he did for ten years. In 2008 he resigned to serve in the White House, where he ran the Office of Presidential Personnel for President Obama. He was named as the US ambassador to South Africa.

In 2003 Gips sent this note on my behalf:

From: Gips, Don
Sent: Wednesday, February 26, 2003
To: James "Jim" Fleming
Cc: Dan Caruso
Subject: FW: Follow-up

Jim, I want to introduce you to Dan Caruso. Dan just left Level 3 and is one of the smartest people I have worked with here or anywhere. He is also a very good friend.

Don

Fleming was a senior partner at Columbia Capital (ColCap), a Washington, DC–based Private Equity company focused on telecom. John Siegel was Fleming's younger partner, and together, they were considering an investment in a distressed public company. Their investment thesis was to replace the CEO, but they didn't have a new CEO lined up.

M/C Partners, a Boston-based Private Equity firm that also focused on Technology, Media, and Telecommunications (TMT), was teaming with ColCap on the transaction. Peter Claudy was the lead partner on the deal. A fresh-faced MBA named Gillis Cashman was also involved.

A team of my former MFS colleagues were working with ColCap and M/C Partners on the deal. As the story goes, Claudy called up Siegel.

CLAUDY: Our MFS guys say they have the perfect person to be CEO of MPower.

SIEGEL: We have a name too. Don Gips from Level 3 is highly recommending him.

CLAUDY: That's a coincidence. The person recommended to me is also from Level 3. His name is Caruso.

SIEGEL: That is the same person Don introduced us to.

That deal didn't work out, so instead, Siegel and Claudy decided to back me as CEO in a bid to take ICG private. By this time ICG was burning $8 million a month and had $32 million of cash remaining. Do the math. They would be out of cash, according to their calculations, in four months.

They had $100 million in liabilities. The majority was a long-term lease on the corporate headquarters Bryan had built in Denver's Tech Center. The building was a lavish edifice of black marble, designed by an architect who was a college friend of Bryan's. When it was dedicated in 1998, Bryan boasted,

The building slants outward and upward, a little bit like our business. We are a relatively small company in a competitive industry, and in order to be successful, we have to be pretty darn aggressive. We wanted a dramatic-looking building that reflected what we were all about.

The building was described to reflect a bow of a ship, facing westward to signal that ICG was taming a new frontier. Instead, the new frontier tamed ICG and the bow-of-a-ship building became a lavish black marble anchor.

In the 2002 bankruptcy restructuring, ICG entered a long-term lease with the building owner. ICG's obligation was to pay $600,000 a month, which equated to about $60 million of the $100 million of 2004 liabilities.

While we were negotiating to buy ICG, I reached out to the building owner. A bankruptcy filing would have tied up their asset for several years. I offered to return it to them in eighteen months, in exchange for us paying them $240,000 a month. This eliminated over $55 million of liabilities with one stroke of the pen!

On July 18, 2004, with the revised building lease in hand, we entered the agreement to buy ICG for $0.75 per share, which equated to $5 million.[4] We agreed to invest an additional $3.7 million into ICG to provide them sufficient cash to get through the closing period, meaning our total investment was $8.7 million.

"When the ICG investment was announced, Jim and I received a call from a fellow investor who offered a sarcastic congratulations: 'I give you guys credit. At least you are going to die with your boots on,'" recalled Siegel. The phrase originates from the idea of soldiers or cowboys who would die in action, still wearing their boots, rather than surrendering or otherwise giving up. "This made me chuckle because we were confident in the team and their plan, but showed the abject fear in the market in the wake of the fiber meltdown," Siegel added.

A multi-month shareholder and regulatory approval process was required before we would be able to close the transaction. During this period we significantly cut ICG's cash burn rate, primarily by shuttering ICG markets that were not core to our Bandwidth strategy.[5]

In parallel, we negotiated the sale of ICG's California business. Three days after we closed our ICG purchase, we announced that MPower was

---

4 Gillis Cashman of M/C Partners recalled a conversation between the investors and me. I shared my analysis that I believed ICG's equity was worth at least $50 million. Peter Claudy of M/C responded that we could buy it for $5 million, to which I responded, "Holy crap. If so, I am going to make you a shit ton of money."

5 Several ICG markets, such as Boston and Philadelphia, had no fiber. We shut these down.

buying the California business for $38 million, consisting of $13 million in MPower stock and $25 million in assumed liabilities associated with a California fiber lease. The combination of the renegotiated headquarters lease and the California transaction reduced the $100 million of liabilities to $20 million—all within three days of owning ICG.

When MPower was sold in 2006, the value of our MPower stock grew to $20 million.

Like all Private Equity transactions, we needed a business plan to support the investment.[6] I winged a plan and put together numbers to support the thesis. I showed that the resulting businesses would be worth over $100 million, which would leave us with $40–$60 million of value—a return of two to four times the $15–$20 million we expected to invest. The clarity of the plan meant the entire team was in full execution mode from the moment we took control.

Each week we met with ColCap and M/C. We'd review progress against the plan, and in almost every case, we were way ahead of our targets. We sold nonstrategic markets.[7] We reduced the liabilities more and faster than we projected. We found hidden sources of value that we monetized into cash or equity.[8,9] Progress on expense reductions was well ahead of schedule.

The running joke was that whatever we achieved the prior week was old news. Quickly, the investor meetings were titled "What have you done

---

6 John Scarano, who was a colleague at MFS and Level 3, and Marty Snella from Level 3 were two of my initial partners. Rob Schmiedeler, a neighbor, was a third. Later, we were joined by Sandi Lewis (from MFS and Level 3), Dennis Kyle (from Level 3), and Matt Erickson (also from Level 3).

7 ICG's southern markets were sold to Xspedius for $18 million.

8 Part of the land that ICG owned at its headquarters was adjacent to Colorado's new light rail. We sold it for $2 million. If we held it another year or two, it would have been worth far more.

9 We sold ICG's voice business, called Voicepipe, for equity. Some of the equity was received from a Golden, Colorado–based company called New Global Telecom. When we sold ICG to Level 3, they didn't want what remained of our Voicepipe business and gave me the option of retaining Voicepipe, which we did. Later, we merged Voicepipe into Zayo for $3 million of Series A equity. Voicepipe's value grew by over ten times as Zayo's value rose, meaning Voicepipe alone became worth more than $30 million.

for me lately?" as we discovered additional value adders to celebrate. These were fun times for the management team and the investors.

Colorado was ICG's home turf. It owned one of the best competitive fiber networks in the state. We focused on the Bandwidth product. Revenue was growing while cash flow was being generated. We concluded this geography alone could be worth $130 million. Ohio was the second most important ICG geography. We turned this into a growth business as well and concluded it was worth $45 million.

Some, including me, wanted to keep Colorado and Ohio and use these operations as a consolidation platform. Others wanted to sell. One of our investors needed a big win to bolster their otherwise struggling funds. My management team had little to show for their Level 3 stock options and wanted to put money in their pocket. They were right. It was time to take chips off the table.

With Level 3, Qwest, and tw telecom all headquartered in Colorado, the Colorado network gave us leverage in the sale process. The result was Level 3 acquiring ICG for $173 million.[10]

We closed the transaction in mid-2006, two years after we took the company private. The outcome was phenomenal. Our total investment was the initial $8.7 million. Our cash balance never dropped below $10 million, which meant that we used their rapidly depleting cash balance of $30 million to buy and turnaround the company. The total proceeds were greater than $250 million, resulting in a Multiple on Invested Capital (MOIC) of 28.7x and an IRR greater than 400 percent.

———

ColCap and M/C rescued me from likely being an unnoticed asterisk in the story of the Bandwidth Bust. I helped the two post great returns in funds that were being devastated by the meltdown. ColCap and M/C became my

———

10 The $173 million purchase price consisted of $10 million in assumed profitability and the rest split between cash and Level 3 stock.

lead investors in my subsequent venture. I will forever be grateful to Siegel, Fleming, Claudy, and Cashman for their contributions and to Gips and my former MFS colleagues for their introductions to ColCap and M/C.[11]

Claudy was suffering from cancer during our two-year turnaround of ICG. He died soon after our exit. I was thankful that he was able to be there for the exit, as it ranked as one of his greatest accomplishments. Importantly, the exit locked in his portion of the value for him to leave to his family. Rest in peace, my friend.

Broadwing—David Huber

At the end of the Bust, Cincinnati Bell unloaded Broadwing and went back to being a regional telephone company. Broadwing seemed finished. It wasn't by a long shot.

Long before the Gold Rush, Dr. David Huber was a lab manager at GI in the early 1990s. Huber was working on commercializing a technology that was developed by British researcher David Payne and Bell Labs. Light would be routed through a prism to divide it into multiple wavelengths, also referred to as colors, so that each wave would carry a separate communications signal. The technology became known as dense wavelength-division multiplexing (DWDM), which became the enabler of fiber being able to carry vast amounts of data through communications networks.

However, GI was stubbornly uninterested in funding Huber's work. Instead, they agreed to license Huber the DWDM technology, conditioned on Huber securing funding of $3 million within eighteen months. Huber was able to close a $3.3 million investment just before the deadline. The new investors named a new CEO and rebranded the company as Ciena Corporation.[12]

---

11 Don Gips joined Zayo's board before our IPO and remained a director until we sold the company.

12 HydraLite was the initial name David Huber selected for his company.

Ciena went on a tear, which I suspect caused GI to shed some tears of regret. In February 1997 Ciena went public at a $3.4 billion valuation, the largest IPO valuation ever achieved by a start-up. However, by then, Huber had already moved on, departing the firm in 1995. Huber claimed his departure was because the board rejected his proposal for Ciena to develop an all-optical system.

Huber's Ciena stock became worth $300 million, which he used to launch another optical tech start-up, naming it Corvis Corporation.[13] This time Huber would remain CEO.

In July 2000, just before the transition from the Boom to the Bust, Corvis went public, raising $1.1 billion. The IPO price was nearly three times higher than the target. The stock price jumped 300 percent on the first day of trading and, after closing at $84, achieved a $28 billion valuation. Within days, the stock price increased another 30 percent, resulting in a valuation of $38 billion. Huber's 22 million shares were worth $1.9 billion.

This historically successful IPO occurred despite Corvis having only two customers, with Broadwing being the largest. Moreover, Corvis had yet to ship a product. Corvis's big balance sheet allowed it to provide vendor financing to its customers. Broadwing, for example, used Corvis's vendor financing to purchase Corvis's technology.

Soon after Corvis's IPO, the Bandwidth Bubble burst. Corvis's 2002 revenue was only $20 million; its losses were a staggering $507 million. And its largest customer, Broadwing, was under extreme duress. Its two other large customers were in bankruptcy. By early 2003 Corvis's stock price was $0.64.

Yet Huber didn't panic; instead, he played the only card he had up his sleeve: Corvis had nearly $600 million in cash remaining from its IPO. Perhaps he could leverage this cash while a host of others were reeling

---

13 Following the Bandwidth Bust, Ciena's annual revenue decreased from $1.6 billion to $300 million. In February 2001 the company raised $1.5 billion through a combination of stock and debt. Ciena used this cash to survive and acquire distressed rivals. In mid-2024, Ciena's Market Value was $6.6 billion.

from the meltdown. Huber teamed up with the founder of Cable TV company Charter Communications to buy Broadwing from Cincinnati Bell for a price of $129 million, plus operating liabilities of $375 million—a tiny fraction of the $3.2 billion Cincinnati Bell paid for Broadwing just three years prior.

"Times change," wrote an analyst, "[fiber] companies . . . are selling their assets for pennies on the dollar to get out from under their crushing debt loads."

Three years later Broadwing's board brought in a new CEO, Telecom industry veteran Stephen "Steve" Courter. Huber had decided to step aside, and just one month after Courter's arrival, Broadwing agreed to sell to Level 3 for $1.4 billion, including $744 million in cash and the remainder in Level 3 shares. So much for Broadwing being dead money. And Huber reaped a windfall for his shares. Later in 2006 he bought a $15 million home in the Port Royal section of Naples, Florida.

Some fairytales do have a happy ending.[14]

tw telecom—Larissa Herda

In the late 1980s, Larissa Herda's employer ran out of money. Not only was she left without a job, but her last paycheck went unpaid. The unemployed Herda, who lived in Washington, DC, stumbled onto ICC, the fiber upstart that also employed Ron Beaumont and was later acquired by MFS. In 1988 she convinced them to hire her as an account executive. "I knew nothing about telecom," Herda reflected in 2024. "When I sold a $15,000 per month deal to Marriott, it caught everyone's attention."

---

14 Steve Courter was a former US army captain and Desert Storm veteran who rose to lieutenant colonel. For only being in the CEO's chair for a month before the execution of the Level 3 transaction, Courter also had a financial windfall. He ran Broadwing for the next year during the closing process and reportedly distributed a big chunk of his proceeds from the Level 3 sale to some of Broadwing's senior managers. Courter remained in Austin, Texas, after passing the torch to Level 3 and became an award-winning professor for the University of Texas McCombs School of Business, education always having been his passion.

In 1999 Herda and her newlywed husband moved to Chicago. She used her ICC experience to land a similar position at MFS, becoming the first employee of the National Account Sales Group. She sold a $50,000 per month contract to a financial services company named Quotron, the largest deal for the company to date. This deal helped solidify Kiewit's conviction of their new investment in MFS.

Herda recalled another sale she closed, this one to a communications company. "I was in labor with my first child and closed Allnet from the hospital bed," she said, explaining that a female colleague of hers had returned from maternity leave and was told her job had been eliminated; Herda wanted to avoid a similar fate. Deservedly so, the sale earned her top salesperson of the quarter. During MFS's IPO road show, Royce Holland used the story of her closing the deal while in labor to illustrate the dedication of MFS's employees.

Herda and her husband didn't enjoy living in Chicago, so she asked to be transferred to Atlanta. Her boss, not wanting to lose his top performer, tried to block it. Herda went over his head and convinced Beaumont, now a president of MFS, for support. She was named general manager of Atlanta and turned it into one of MFS's fastest growing markets. Herda was promoted to regional vice president.

Immediately after WorldCom acquired MFS, Herda was recruited to join tw telecom, a unit of Time Warner, to be the senior VP of sales.[15,16] The tw telecom business unit was created in 1993 with the backing of Glenn Allen Britt and the support of his boss, Gerald "Jerry" M. Levin.

---

15 At the time Herda joined, the company was named Time Warner Telecom. The name was shortened to tw telecom in 2008 when Time Warner Cable was no longer a shareholder in the company. Just prior to Herda's arrival, the company adopted the Time Warner Telecom moniker. Previously it was branded Time Warner Communications. The entity that directly owned Time Warner Communications was Time Warner Cable, which was a division of Time Warner.

16 Herda was recruited to tw telecom by MFS veteran Steve McPhie. Herda met McPhie when she trained him for being her regional sales SVP counterpart in the Midwest region.

Levin was named CEO of Time Warner Inc. in 1993, and Britt became CEO of Time Warner Cable in 2001.[17]

By the time Herda arrived in 1997, tw telecom was operational in seventeen markets. Despite the market reach, 1996 revenue was only $24 million, and the company was bleeding $84 million a year in losses. Herda got to work. In her first six months, sixty MFS employees followed her to tw telecom. By the end of 1997, revenue had increased to $55 million, a 130 percent improvement over 1996.

Despite the momentum, Time Warner had lost interest in their tw telecom business unit. They tried hard to sell it but were unsuccessful. On her fortieth birthday, Herda received a call from a Time Warner executive offering her the tw telecom CEO position. She initially declined, recounting the conversation while on an entrepreneurship panel years later: "I told him he must be out of his mind because I had absolutely no idea what a CEO did, knew nothing about the capital markets, and I really wasn't the right person for the job."

His response, as recalled by Herda, was, "I've watched you for the past year. You're basically running the company already, and if you don't take the job, we're shutting the company down because Time Warner isn't going to fund the cash burn anymore." Herda accepted, feeling obligated

---

17 Perhaps the biggest blunder in business history was media giant Time Warner agreeing in 2000 to a "merger of equals" with dial up Internet provider America Online (AOL). The $350 billion combination was the largest in history. Soon after, it became abundantly clear that AOL's business model would be engulfed by the transition to dedicated, not dial-up, connections used by consumers to access the Internet. Eager consumers flocked to the more powerful, faster dedicated connections offered by the Baby Bells, Cable TV providers, and hundreds of upstarts. AOL rapidly lost market share, becoming increasingly irrelevant as it practically drowned in this transition's wake. The impact on Time Warner shareholders would be devastating.

Levin, who led the merger for Time Warner, was named number sixteen on CNBC's top twenty list of worst American CEOs of all time. Though not much of a consolation prize, AT&T's Bob Allen, WorldCom's Bernie Ebbers, and Enron's Kenneth "Ken" Lay topped him with rankings of twelve, five, and three, respectively.

AOL's shareholders, especially Internet pioneer Stephen M. "Steve" Case, made out like bandits. "I wanted to concentrate on my family," Case told the *Washington Post*. I'm sure his wife and five children wanted to spend more time with him and help him spend the many billions he pocketed from the merger.

to the loyal MFS colleagues that followed her to tw telecom. She immediately went on a road show, raising $400 million in three weeks to save tw telecom from being shuttered.

The reprieve didn't last long. On Super Bowl Sunday in 1999, Britt called Herda and revealed a tentative deal was in place to sell tw telecom as part of a bigger transaction with AT&T. Herda was irate. She convinced Britt and Levin to refrain from announcing tw telecom as part of the transaction. Soon thereafter, Herda forced their hand to allow her to take the business unit public under the threat that she and her executive team would otherwise resign.

tw telecom began trading on the NASDAQ on May 12, 1999, selling $250 million of Time Warner's shares to the public at $14 per share. By early 2000, tw telecom's stock soared to $89 a share, a sixfold increase from the IPO price.

In September 2001, in the market crash that followed the terrorist attack, tw telecom's stock dropped to below $10 per share. tw telecom's stock, again following the market, bounced back to reach $20 per share later 2001. The recovery was fleeting, with the stock dipping below $1 per share, and the debt traded at thirty cents on the dollar, in 2002.

When the bubble burst, tw telecom was feeling the chilling bite of telecom's nuclear winter. Eighty percent of Herda's revenue came from carriers, and most were filing for bankruptcy. WorldCom, in particular, was 14 percent of tw telecom's revenue.

Herda laid off a third of her workforce. She broke down in tears when she announced the layoffs to her employees. She recalled feeling not up to the CEO task because she let her emotions show. But showing that bit of humanity led to employees sending letters to her expressing their appreciation for her leadership and care.

Herda used her strong balance sheet to aggressively shift her enterprise customer revenue mix from 20 percent to 80 percent, significantly reducing her reliance on the rapidly shrinking carrier segment.

Prior to 9/11 and before the gravity of the Bandwidth Bust became clear, Herda acquired a struggling fiber property named GST Telecom.

John Warta, the founder of Electric Lightwave, launched GST Telecom in 1994. GST built fiber networks in multiple California markets: Tucson, Arizona; Albuquerque, New Mexico; and Houston, Texas. In 1997 he pursued a transaction to become the seventh telecom concession in Mexico for $30 million. His envisioned structure would allow him to build 2,200 kilometers of fiber along Mexico railroads, with an MCI deal in hand to fund much of the build in exchange for four fibers. In pursuing this transaction, Warta set up entities that were part owned by GST and part owned by entities associated with himself.

Warta pursued this venture without fully disclosing it to the GST board. When it surfaced in 1998, he departed GST and was named in a lawsuit filed by the GST board. Warta was accused of fraudulent transfer associated with the sale of a joint venture between GST and a Mexican company. GST also accused Warta of using GST assets to fund other telecommunications ventures. Warta's compelling perspective, which he shared with me in 2024, was he was being industrious in pursuing an opportunity that would create enormous value for GST shareholders. GST, which had a peak Enterprise Value approaching $2 billion, went bankrupt in 2000.

Herda seized on the opportunity, spending $690 million to acquire the assets of GST out of bankruptcy, which doubled the number of tw telecom markets to forty-four.

Five years passed before Herda did her next fiber acquisition. tw telecom purchased Enron's Portland, Oregon, fiber assets for $750,000 as part of a competitive auction in early 2004.

tw telecom's next acquisition was Xspedius Communications in 2006, which has a fascinating backstory of its own—one that also crossed paths with the heritage of MFS.

James "Jay" Monroe III was among the best examples of a clever entrepreneur buying distressed fiber assets in the wake of the Bandwidth Bust. Thermo Group, his family office, teamed with Brown Brothers Harriman and Colorado-based Meritage Funds to bankroll Xspedius. Jim Allen, the co-founder of Brooks Fiber and a former WorldCom board member, partnered with Mark R. Senda, a veteran of MFS, AT&T Canada (a.k.a. Allstream), and Brooks Fiber to form Xspedius.

Xspedius purchased the assets of e.spire Communications out of bankruptcy in June 2002 for $18 million in cash and $50 million in debt. At its Boom era peak, e.spire was worth over $1 billion and had 3,500 miles of fiber with networks in twenty states, including Georgia, Florida, Texas, South Carolina, Arkansas, and Alabama. In 1997 an analyst pumped up e.spire: "By almost any measure—revenues, network, etc.—[e.spire] is the fastest growing CLEC in our universe."

e.spire, which was originally named American Communications Services Inc. (ACSI), was started by MFS veterans Anthony "Tony" Pompliano and Richard Kozak. Pompliano was the original CEO of Chicago Fiber Optics, which became MFS. Kozak was the original leader of MFS Development who passed my resume along with the recommendation to ignore it to his successor, Kevin O'Hara. When e.spire declared bankruptcy in 2001, neither Pompliano nor Kozak were running the company.

"This is a very healthy business," said Senda. "Through the bankruptcy process, we were able to eliminate the debt [and receive] a business that is near net income positive."

In 2005 Senda and I struck a deal. Xspedius purchased five hundred miles of fiber assets from ICG in Birmingham, Alabama, Charlotte, North Carolina, Louisville, Kentucky, and Nashville, Tennessee. The purchase price was $18 million. "We think the fiber gods have smiled on us," Senda said. He was right. The value was far higher than the purchase price. Oh well—you win some, you lose some.

In 2006 Monroe sold Xspedius to tw telecom for $532 million, consisting of $213 million in cash and $319 million in stock. In total, the investors invested $200 million, half coming from Thermo. At the time of the transaction, the Xspedius investors more than doubled their value.[18]

───────────

Herda continued the steady high-single-digit revenue growth with strong profitability. By 2013 tw telecom posted $1.6 billion in revenue with profitability of $600 million. Herda, under the pressure of activists, was ready to listen when Level 3 approached tw telecom in 2014. Level 3 enticed tw telecom with a $7.3 billion purchase price, including the $1.6 billion of debt assumed by Level 3.

"Business was strong and growing, and it was actually fun," Herda explained to me in 2024. "In the end, as reluctant as we were to sell, we had a fiduciary responsibility to our shareholders." The $43 per share was three times the $14 IPO price in 1999. Given the fact that the division of Time Warner was basically worthless when she assumed the tw telecom CEO role, her value creation performance was phenomenal.

───────────

18 Monroe wasn't done. He rolled his $100 million of stock into tw telecom at $17 per share. When tw telecom sold to Level 3 in 2014, these shares were worth $41, meaning Monroe's $100 million Xspedius investment was now worth $350 million, including the cash Monroe received in 2006.

There's more. When Monroe sold Xspedius, he retained Xspedius Fiber Group, which owned fiber and conduit assets in California, Florida, Georgia, Maryland, Texas, Virginia, and the District of Columbia. He rebranded this entity FiberLight and focused the business on providing Dark Fiber solutions to carriers and large enterprises. Over the next seventeen years, they expanded to 18,000 miles of fiber. In 2023, Monroe sold it to a consortium of infrastructure investors for $1 billion. Assuming Thermo pocketed $300 million from this sale, the original $100 million Xspedius investment turned into a total of $650 million.

Herda was one of the most successful CEOs of the Bandwidth adventure. Without a doubt, she was the most accomplished woman. Kudos to Larissa and the entire tw telecom team.

XO—Carl Icahn; Verizon—Lowell McAdam

When we last left XO, famed buyout investor Teddy Forstmann, with the backing of XO CEO Akerson, was orchestrating a prepackaged bankruptcy. Corporate raider Carl Icahn had a different plan in mind.

Forstmann's proposed pre-pack would allow the company to complete a January 2002 deal that included Teléfonos de México (Telmex), Mexico's dominant telecommunications company, which was led by one of the world's richest, Carlos Slim. Forstmann and Slim would each invest $400 million in return for an 80 percent ownership stake.

The pre-pack was opposed by bondholders who believed that giving an 80 percent stake to Forstmann and Telmex left too little for the bondholders. During the bankruptcy, Icahn, the world's most infamous bottom feeder, accumulated a multimillion-dollar position in the unsecured bonds. In March 2002 Icahn made a competing offer to buy 50 percent of the equity for $500 million.

XO's management preferred the pre-pack over Icahn's offer, but when they attempted to move forward with Forstmann and Telmex, the two investors got cold feet, and in October 2002 each agreed to pay $12.5 million to get out of their offer. Forstmann, which owned about 24 percent of XO before XO's bankruptcy, also agreed to forfeit its ownership in XO.

The $1.5 billion Forstmann had invested in XO was worthless. Forstmann had many great successes during its thirty-five-year history, including Gulfstream Aerospace, Topps Playing Cards, Dr Pepper, and GI. XO was not one of them.[19]

---

19 Forstmann Little was also a major investor in McleodUSA. Learn how this turned out in appendix 3, "More Bandwidth Stories."

Sensing weakness, Icahn rescinded his restructuring proposal, blaming an inability to reach an agreement with XO's other creditors. In the meantime Icahn did what Icahn does best: he gained control of 85 percent of XO's senior secured debt. On January 17, 2003, XO emerged from bankruptcy with Icahn as the new owner.

Icahn told the *Washington Post*, "I'm hoping to see the company built up from here. I think it's got a great deal of potential." Icahn suggested this might be the beginning of an Icahn rollup of distressed Bandwidth companies.

XO reemerged as a public company with Icahn owning 83 percent of the equity. The $5.3 billion of prebankruptcy debt was reduced to $500 million. XO emerged with $554 million cash on its balance sheet. The equity holders lost everything.

XO's interim CEO, Nate Davis, who replaced Dan Akerson, commented to the *Washington Post*: "I feel like the rain and the dark cloud that have been following us around has been lifted. We've been telling people that while our balance sheet was sick, operationally we were strong, and this proves us right."[20]

Later in 2003 Icahn hired telecom veteran and former Global Crossing COO Carl Grivner as XO's CEO. "With its strong balance sheet and the combination of its wide-reaching broadband network assets and unique service offerings, XO is positioned to become a significant competitive force in the business telecommunications market," touted Grivner.

In 2004 XO was relisted on Nasdaq. XO also conducted a rights offering that was required by the bankruptcy reorganization plan. Other unsecured creditors exercised their rights under this offering, but Icahn took a pass—likely in error as I explain later—as this caused Icahn's ownership stake to drop to 61 percent. The rights offering raised $200 million and was used to pay down debt.

---

20 Dan Akerson succeeded Ed Whitacre, the former AT&T CEO, as CEO of General Motors, following Whitacre's successful turnaround of the auto giant.

Also in 2004 XO outbid Qwest and others in a $320 million acquisition of Allegiance Telecom.[21] *Forbes* quoted an unnamed source in a February 2004 article: "By owning the two largest [competitive phone providers] in the nation, it gives Icahn an unchallenged position to play a role in the inevitable restructuring of the whole industry. This deal guarantees him a place at the table." No, it didn't.

———————

Icahn's lieutenants on XO were New Yorkers Vince J. Intrieri and Keith Meister. Like Icahn, they knew little about operating a Bandwidth business. Their strategy was to just sustain the asset while Icahn sorted through how to get access to the net operating losses (NOLs).[22] XO meandered along, posting mediocre results.

———————

21 Allegiance Telecom was a competitive telecom carrier founded by former MFS COO Royce Holland at the same time Jim Crowe launched Level 3. The purchase price was $631 million, of which $311 million was paid in cash and the rest in the form of 45.4 XO shares. At XO's $6.80 share price, the company's shares were valued at $320 million.

Icahn was present throughout the overnight bankruptcy process auction. Each time Qwest outbid XO, Icahn would immediately top the Qwest bid. Qwest's team was not empowered to increase their bid without the approval of their CEO. So after each Icahn bid, Qwest would request a recess. . . This process continued throughout the night until Icahn wore Qwest into submission.

Genuity, which was acquired by Level 3 in 2003, was Allegiance's biggest customer. Level 3 sought to terminate Genuity's $25 million per quarter revenue commitment to Allegiance, as Level 3 could use its own network for this traffic. Level 3, citing service deficiencies, withheld payments. Level 3's John Ryan recounted what happened next. Holland escalated to Crowe, which led to Crowe asking Level 3's Corporate Development team what was likely to happen with Allegiance. They told him that bankruptcy was imminent. Crowe responded, "If the body is going to be found dead on the floor, do we really want to be standing beside it with a gun in our hands?" Level 3 released the payments.

22 Net operating losses (NOL) are the accumulation of net income losses that occurred in prior years. The NOLs are tracked on the balance sheet. If the company becomes profitable in the future, the NOLs may be used to shield the income from income taxes. Therefore, a dollar of NOL is worth 35 to 42 cents of avoided taxes, depending on the relevant state income tax rate. If a company with NOLs is acquired, the acquiring company can, depending on the circumstances, realize the tax savings.

In early 2006 I visited New York. I jumped at the opportunity when our banker offered to get me together with Intrieri and Meister. "Perhaps XO would like to buy ICG's Colorado and Ohio properties," my banker teased. I naively saw an opportunity for a reverse merger with me being CEO.

I met Intrieri and Meister at 767 Fifth Avenue, the lavish headquarters of Icahn Enterprises. The banker and I sat in a conference room for forty-five minutes after the scheduled start time when Intrieri and Meister finally showed up. The icy duo skipped past any pleasantries and acted like they weren't sure why they were meeting with me. Intrieri and Meister barely reacted as I explained the ICG turnaround, my past involvement with the XO-Level 3 fiber transaction, and the logic of combining ICG's remaining assets with XO. The meeting ended as abruptly as it started.

Later that afternoon my banker called me. Intrieri and Meister wanted to know if I was available to meet Icahn the next day.[23] "For certain," I beamed, if for no other reason than to meet the leverage buyout legend.

When I arrived, Icahn's assistant showed me into a large conference room. I scoped the large painting of a lion on a cliff with blood dripping from its mouth from the wildebeest carcass at its feet. I paused to look at the chess-like model of armies in full battle with Napoleon facing off against the Duke of Wellington. The large conference room had a large replica of a TWA airplane, a memento from Icahn's hostile takeover in 1985.

After twenty minutes of sitting alone, I was called into Icahn's private office. Icahn, Intrieri, Meister, and I sat at a small circular antique table.

"Why are you here?" Icahn blurted to initiate the conversation, as if I invited myself.

XO had $4.0 billion of NOLs, which had a value to XO of over $800 million. The US federal corporate tax rate was 35 percent in 2008, and the New York State corporate tax rate was 7.1 percent, bringing the combined tax rate to 42.1 percent. The $4.0 billion NOL would offset $4.0 billion of taxable income, which would reduce tax liability by as much as 42 percent. $800 million is 20 percent of $4 billion, reflecting a time value discount as the NOLs would take several years to consume.

23 Icahn was not available the next day. So I stayed over the weekend to meet early the following week.

"ICG turnaround blah-blah, XO/Level 3 Joint Build Deal blah-blah, XO blah-blah," I rambled as he couldn't have looked more disinterested.

"So let me get this straight. You think you could run XO better than my CEO, Carl Grivner," he barked.

I cited my investment thesis about focusing on bandwidth and not managed services.

Icahn interrupted, "See that phone over there?" as he pointed to a Polycom on the coffee table about three feet from his outstretched arm. "How about I grab that phone and I will listen while you explain to Grivner how you could run his business better than him?" he dared me.

I tried again to explain.

"I'm serious," he abruptly cut me off and began to scoot his antique chair toward the phone. "I am going to call him, and we will all be entertained by Grivner's reaction to your bullshit," he continued as he leaned in his chair and stretched for the phone.

"Sure, get him on the phone," I said, not sure I meant it. I watched as the tip of Icahn's fingers were inches from touching the speed dial button.

Then BAM! The chair leg broke, tossing Icahn flat on his back underneath the small table. Intrieri and Meister sat in their chairs, uncertain what to do, as it became clear Icahn had fallen and couldn't get up.

"Is this part of the intimidation play?" I thought.

We helped him up. Without missing a beat, Icahn grabbed the phone. The CEO was traveling, so Don MacNeil, the COO, took the call.

Icahn started to explain. "I got this guy here in my office. He's telling me all kinds of things about how he could run XO better than you guys."

MacNeil responded, "Who's the guy?"

Icahn pointed at me. "Hey, hey, what did you say your name was?"

We had a five-minute chat. MacNeil was kind, as he told Icahn that I was a credible guy in the industry. The call ended, and nothing came of it.

After the meeting I visited FAO Schwarz, the toy store made famous when Tom Hanks's character played the song "Chopsticks" by dancing on an oversize piano. I bought a big stuffed bear, which stands proudly in my Beaver Creek mountain house. I pet it every time I walk by.

The following week I sent a thank-you to Icahn with a bottle of glue. "I thought the Elmer's might be handy to fix the leg of your chair."

I received a written letter from him, which I still have.

Dear Dan:

Thanks for your kind note and the bottle of Elmers!

I enjoyed meeting with you and the opportunity to hear your thoughts on the telecom industry. And, I look forward to continuing our discussions in the future.

Please let me know when you will be in New York again as I would like to visit with you further.

Best Regards,
Carl

Icahn and I crossed paths a couple of years later—when I took a hostile run at XO.

XO NOLs would have a value of $800 million if Icahn could use them to shield taxable income. US tax law permits a corporate entity (parent) that owns greater than 80 percent of another entity (subsidiary) to use the NOLs of the subsidiary.

Post bankruptcy, Icahn, through a myriad of corporate entities that had taxable income, owned 83 percent of XO. Icahn entered a tax allocation agreement that resulted in him using $450 million of the tax shields in 2003 and 2004. Since he controlled both parties of the agreement, the favorable terms allowed Icahn to realize these tax shield benefits without properly compensating XO's other shareholders.

The 2004 rights offering dropped Icahn's ownership from 83 percent to 61 percent. The no-participation was likely due to a miscalculation by Icahn—perhaps he assumed other investors would not participate in the rights offering. Now below the 80 percent threshold, Icahn stood to lose the remaining $800 million of NOL value. He spent the next several years trying to rectify this.

Texas hedge fund R^2 Investments owned eleven million shares of XO's common stock, which it received alongside Icahn via the conversion of distressed debt R^2 purchased during XO's bankruptcy. When XO's stock price rose to $7 per share, R^2 was happy it rode Icahn's XO coattails.

R^2 was a lot less happy when XO's stock dropped to $0.3 per share in 2008, and 95 percent of their peak value was lost. R^2, which attributed the bottoming out to Icahn's mishandling of his fiduciary responsibilities as controlling shareholder, wasn't afraid to take on Icahn.

In 2005 Icahn signed an equity purchase agreement with XO that would have resulted in the sale of most of XO's assets and the NOLs for $700 million. R^2 filed a lawsuit challenging this self-dealing agreement, and after a preliminary hearing in March 2006, Icahn backed down.

In 2006 Icahn tried to sneak language into the proxy for the 2006 annual meeting that would provide him permission to consummate the 2005 transaction without shareholder approval. Icahn backed down again when R^2 commenced injunctive legal proceedings.

Icahn then decided to run the company into dire straits, which would provide him the opportunity to invest to rescue the company under terms that would result in his equity exceeding 80 percent. With debt maturities coming due in 2008, he ramped up cash burn to $10 million a month, which was draining XO of its remaining cash.

Several entities made unsolicited expressions of interest to purchase XO. In March 2008 a telecommunications company offered $1 billion to buy the whole company, a 50 percent premium over Icahn's offer.[24] Another inbound offer to buy XO arrived later that month.

Later in *Bandwidth*, Zayo's story will be told. Here, I will jump ahead to Zayo's second year when, in June 2008, I served a big fat softball to Icahn. My unsolicited offer was to buy XO's fiber business for $900 million to $1 billion while leaving behind the valuable NOLs in an entity that would continue to own the also valuable wireless spectrum.[25] Our offer included high-confidence letters from our financial advisors that Zayo could obtain financing. We indicated we could achieve a final purchase agreement within six weeks.

We signed a nondisclosure agreement and, on June 21, informed XO's banker that we were nearly complete with our due diligence. Two days later we firmed up our offer at $940 million and removed the financing contingency.

The next day XO's advisors summarized Zayo's offer to XO's board. The value of our offer was $1.52 per share, which was three times the $0.50 per share public stock price. Moreover, our offer allowed for XO to retain the value of the NOLs and wireless spectrum.

When XO didn't engage, I sent a letter on June 27 that documented the value of retaining the tax benefits as being an incremental $8.24 per share, bringing the total value to almost $10 per share, plus the value of the wireless spectrum. This represented an enormous premium over the $0.50 public stock price of XO.

---

24 The $1 billion translated to $2.25 per share, which was a 50 percent premium to the $0.79 to $1.27 price that XO was trading at in April 2008.

25 When Akerson became CEO of Nextlink, he envisioned a platform that combined wireless with fiber as solutions for its customers. To advance this vision, he paid $695 million to a company named WNP Communications, of which $153 million was used to fund the acquisition of wireless licenses from the FCC. He also paid $137.7 million to Nextel to acquire the 50 percent interest in a McCaw company Nextband. The other 50 percent was already owned by Nextlink. This gave Nextlink ownership of forty wireless licenses.

On June 30 I sent another letter questioning why the board was acting counter to its purpose of maximizing shareholder value by not responding to our offer.

On July 8 we expressed that we were prepared to increase our offer but only after having a face-to-face meeting. The response from XO's advisor was that we would need to raise our price to $1.3 billion and agree to a $25 million reverse breakup fee. On July 10, with XO's stock price now down to $0.35 per share, we offered to propose a purchase price materially higher than $940 million and stated that we did not object to a reverse breakup fee. However, to proceed, we reiterated our requirement for a face-to-face meeting with the actual decision-maker.

In Icahn's 2006 Elmer's Glue letter to me, he said he'd like to "visit with me further." Instead, XO cut off communications with us.

R^2 continued to put pressure on Icahn. In a June 12 letter, they wrote, "We believe you, as chairman of the board and the majority shareholder of XO, in clear violation of your fiduciary duties to minority shareholders, have been acting and will continue to act to advantage improperly your XO Debt to the detriment of XO's minority shareholders." Under threat of a lawsuit, R^2 requested (1) Icahn and those associated with Icahn resign from XO's board, (2) Icahn's stock be placed in a blind trust, and (3) Icahn refrain from exercising any managerial control or participation in shareholder, board, or management meetings.

Icahn ignored such warnings.

On July 24, 2008, XO executed a rights offering that resulted in Icahn achieving his goal of owning more than 80 percent of XO. The *New York Times* published an article titled "The Kettle? The Pot Says He's Black." The article focused on the hypocrisy of Icahn's reputation as a shareholder activist versus his behavior at XO, characterizing the transaction as follows: "Icahn paid $329 million and agreed to retire $450 million of the debt in return for a huge swatch of preferred stock. How huge a swatch? Enough to largely dilute all the other shareholders while pushing Icahn's stake above the magic 80 percent mark. Imagine that."

This transaction was not put up for a shareholder vote.

The *New York Times* went on to quote Powell of Telecom Ramblings: "Carl is very smart and acts very aggressively in his own self-interest. And if you get in the way of his self-interest, he will trample you."

A revised NOL agreement (tax allocation agreement) was also put in place under terms favorable to Icahn. The rights offering valued XO at $636 million, well less than half of the value of the Zayo alternative. The fairness opinion provided by XO's advisors included an extraordinarily unusual disclaimer.

> We express no view as to any other aspect or implication of the Transaction or any other agreement, arrangement or understanding entered into in connection with the Transaction or otherwise. Furthermore, we express no view on certain that Tax Allocation Agreement by and among [Icahn] and the Company nor are we expressing any opinion or making any judgment, evaluation or determination with respect to the value, characteristic, ability of any party to utilize or other attributes of the Company's tax position or of any net operating losses. In addition, you have not asked us to address, and our Opinion does not address, the fairness to the holders of any class of securities, creditors or other constituents of the Company.

In August R^2 sent a letter to Grivner, demanding the rights to books and records, citing their intent to "investigate possible mismanagement, improper influence or conduct, improper conflicts of interest and lack of due care on the part of XO's Board of Directors."

The next Icahn maneuver was to increase his ownership to 90 percent, which would thereby give him the power to execute a short-form merger to acquire the remainder of XO. In October Icahn disclosed that he increased his ownership in a private transaction by 3,995,000 common shares and 500,000 Series A preferred shares. This put him at 89.36 percent.

On July 9, 2009, Icahn offered to purchase the remaining 10 percent of shares held by others at $0.55 per share.

On August 4, 2009, R^2 sent a letter to XO's independent directors.

> Under Mr. Icahn's tutelage, we believe that this board is on the cusp of stripping almost all value from the minority shareholders. It has taken this board almost four years to find a way to give Mr. Icahn this company, but after a long, arduous process, the board has almost completed its apparent goal—finding a way to give Mr. Icahn all the assets and NOLs for as little consideration as possible.
>
> . . . We were truly aghast when we learned through discovery as part of our attached lawsuit that approximately one year ago one of the potential bidders valued its combined bid for the assets and the net operating losses at approximately $10 per share. The fact that the board rejected this offer in favor of the massively dilutive proposal from Mr. Icahn only causes us to question further the true independence of the "independent directors."
>
> As you well know, the attached complaint that we recently filed in the Supreme Court of the State of New York shows that this board dismissed proposals from FIVE different bidders that would have each likely garnered more value for the minority shareholders than anything Mr. Icahn is going to offer.
>
> Rest assured, we are going to do everything in our power to ensure that justice is served.

The R^2 lawsuit didn't frighten Icahn away from taking XO private. In July 2011 XO notified its shareholders that Icahn, as holder of more than 90 percent of XO's shares, had exercised his right to buy the remainder of the company and take it private at a price of $1.40 per share.[26] Two months later Icahn was the proud owner of 100 percent of XO.

---

26 The R^2 lawsuit was settled out of court at a cost that was not material to Icahn.

In February 2016 Icahn sold XO's fiber business to Verizon for $1.8 billion. The deal included an option for Verizon to purchase the wireless spectrum by the end of 2018. Between 2008 and 2018, Icahn was able to extract full value from the NOLs.

Given the $1.8 billion that Verizon paid, Zayo's offer of $940 million for the wireline business would have resulted in a good return for our investors. As I said previously, you win some, and you lose some.

Verizon, the proud new owner of XO's fiber business, was the Baby Bell powerhouse branded in 2000 after Bell Atlantic acquired GTE, and the Bell Atlantic-NYNEX-GTE triumvirate adopted Verizon as its new name.

During the Bandwidth Bust, Verizon was hit hard by the meltdown because of its financial ties to Genuity and AboveNet, as their bankruptcies resulted in Verizon losing billions of Market Value. Still, Verizon's strong wireless and traditional phone businesses helped insulate the telecom giant from the intense heat of the industry meltdown.

In 2005 Verizon bought the revived WorldCom for $6.6 billion, giving Verizon the fiber networks of MCI, WilTel, Brooks Fiber, MFS, and the UUNET Internet Backbone. Complementing all these assets were the significant ownership rights to fiber purchased as part of Verizon's Boom-era partnership with AboveNet. In 2016, Verizon added XO to its impressive inventory.

At the time of the WorldCom acquisition, Lowell McAdam was president of Verizon's wireless business. Six years later, in 2011, McAdam was named Verizon's CEO, and he spearheaded the acquisition of XO.

McAdam was excited with his XO purchase and concluded he wanted more. Zayo and I were in his crosshairs.

I haven't yet told the Zayo story, so I will be jumping ahead here to complete the Verizon story. By now, Zayo, in 2016, had been public for two years and had risen to be the biggest pure-play operator in the Bandwidth industry.

In late 2016 I received an unexpected reach out from someone I had never met—none other than Verizon CEO Lowell McAdam. "Hey, Dan, I will be flying over Colorado on my way to California next week. I'd like to meet you for lunch."

McAdam and I had lunch at the Ritz-Carlton in Bachelor Gulch. He told me how much he admired the company I founded. He then got right to the point. "I'd like to buy Zayo."

As McAdam looked forward, he saw the opportunity to leverage deep, dense fiber to simultaneously accomplish two important goals. First, the newest wireless architecture called 5G would need to leverage metro fiber to handle the massive increase in call and data load. Second, Verizon Business, which served enterprises, would be stronger if it leveraged its own fiber. McAdam knew he had great fiber assets while also being increasingly frustrated that Verizon struggled to leverage this fiber. Verizon's data records documenting the fiber assets were a hair-ball mess made worse by outdated systems and processes used to design and provision services. McAdam saw Zayo as a solution.

When I reported back to my board, the first question they asked was, "Could Verizon get the regulatory nod to do this acquisition?" I shrugged. "This didn't seem to be a hurdle to McAdam when I talked to him," I answered. This struck us all as peculiar. I followed up with McAdam, suggesting our regulatory attorneys chat on Verizon's perspective on regulatory approval risks.

The idea of Verizon acquiring Zayo was immediately shut down by Verizon's regulatory department. Convincing regulators to approve a Zayo acquisition immediately on the heels of XO was too much for them. Acquisition dead!

McAdam asked me to come back with other ideas. I presented a plan that would take Verizon's Brooks Fiber, MFS, and XO metro assets and combine them with Zayo's assets. We would carve out a layer of these assets for Verizon's exclusive use. We would leverage this layer both for Verizon's 5G deployments and to expand its fiber reach to enterprise buildings.

Verizon would become a 20 percent owner of Zayo—the maximum threshold for which it didn't require regulatory approval. Later, when the dust settled on the XO transaction, it could pursue an acquisition of Zayo.

McAdam liked the plan.

The next step was for me to present it to his team. We briefed his COO, John G. Stratton, and he assigned a team to lead the evaluation, using Georgia and Colorado as test cases.

We had a follow-up video call with McAdam, Stratton, and others. I could tell immediately from McAdam's body language that something was amiss. McAdam said great things about the work we had done and how excited he was to proceed forward. Then he turned to Stratton, who was connected via a video bridge, and said, "John, we've worked together for a long time. Whenever I see you scratch your chin like that, I know you have something on your mind."

As Stratton complimented McAdam for his highly perceptive observation, he began to express concerns that something didn't feel right to him about the analysis. I knew immediately, and said so in real time to my colleagues, that the idea was dead. The Verizon network organization would never let Zayo take over the fiber layer of its network.

Unfortunately, I unleashed a can of worms by going through this exercise by basically revealing the Zayo playbook to Verizon. This led Verizon to develop its own initiative branded as One Fiber. The Verizon executive team concluded they would take the Zayo proposal and do it themselves. Monopolistic thieves!

As I reflected later, I don't think I could have ignored the opportunity to form this strategic relationship with Verizon, our largest customer. However, it would have been better if I had, as I doubt Stratton would have been able to convince the Verizon board to accelerate funding of the One Fiber plan without us showing them the playbook. As the Verizon team executed One Fiber, Zayo's book of business with them steadily declined.

I waited a long time for my consolation prize. In the fourth quarter of 2023, Verizon booked a $5.8 billion impairment charge associated with its Verizon Business segment. This write-off suggested to me that Verizon's commercial performance in its fiber business has been a disaster, which is why I suggested Zayo could do it better. I got the last laugh.

CHAPTER VII

THE RECONSOLIDATION

No matter how great the talent or efforts, some things just take time. You can't produce a baby in one month by getting nine women pregnant.

—Warren Buffett

" What you have been seeing is the gradual reconsolidation of an industry that is highly capital intensive and that's perhaps best served by a relatively small number of large companies," an analyst surmised in January 2002. We are now fast-forwarding to the mid-2010s, with over a decade of his "gradual reconsolidation" behind us and the arc of consolidation gaining momentum.

Verizon had acquired WorldCom after the Bust and, in 2011, added XO after Icahn extracted his value from buying XO on the cheap during XO's bankruptcy. It would be Verizon's final major fiber acquisition.

SBC had acquired AT&T and then adopted the AT&T name. The acquisition of AT&T included Teleport, the Renegade AT&T acquired during the Boom, as well as the troubled intercity fiber providers AT&T gulped down during the meltdown. This was the last of AT&T's fiber meals.

All of which brings us to telling how the remaining national Bandwidth Infrastructure platforms were formed. We will complete the saga of Level

3, reveal the formation story of Lumen, and, finally, turn our attention to Zayo, the company I founded in 2007.

Grab a cappuccino and sit in your favorite recliner. I have several wild stories to share.

Level 3—Jim Crowe and Jeff Storey

Backspace to Level 3 in 2002, when I was still with the company. The stories of my ICG take-private, Huber's Broadwing acquisition, and Icahn's XO shenanigans had not yet happened.

Level 3's Jim Crowe had avoided bankruptcy by pulling two diamonds from the rough: one named Howard Diamond and the other Warren Buffett. This double bank shot enabled Crowe, like Icahn's XO, Huber's Broadwing, and my ICG, to become a phoenix rising from the ashes.

Level 3's rising had two differences from the rebirths of XO, Broadwing, and ICG. First, the threat of bankruptcy was still a cloud that hovered over Level 3 during its phoenix phase. Second, Level 3 would not be content with a good exit. Crowe wanted to finish what he started. He wanted to be so wildly successful that Level 3 would be viewed as a monopoly that needed to be broken apart. He knew, as the phoenix phase was beginning, that he had a long road in front of him.

"I've joked that before I retire I'd like an antitrust investigation," Crowe said. "But that's a problem for another day. Right now, everybody's got to keep from going broke. That's what we're maniacally focused on."

First, Crowe needed Level 3 to fix its day-to-day operations, as, despite the hype touting Level 3's team, the organization was painfully dysfunctional. Revenue was shrinking. Cash flow was negative. And the culture was a mess.

A decision was made that we needed to bring in an executive who knew how to run the business. Enter O. Lee Jobe III, the president of Concert, the BT, and AT&T joint venture. Hired to be Level 3's savior,

Jobe was named president of global operations, reporting to O'Hara. Jobe's responsibilities were, as I recall, the same as O'Hara's. He was, more or less, O'Hara's only direct report. Still following?

Jobe came in hot. Before "LOL" became an expression, he would laugh out loud while lecturing us on our pure incompetency. Though he had a point, his laughing also helped mask his substance abuse problem. After a late night NYC binge, Jobe showed up tardy and disheveled to Crowe's Wall Street analyst day. That was the last time I recall seeing Jobe. He went to rehab and never returned.

"Though receiving positive recommendations and completing a thorough interview process, the hiring of Jobe was a disaster," O'Hara recounted. "He was clearly struggling with what I assumed to be some type of substance abuse issues. I offered to send him to rehab while making it clear that rehab would be a condition of his continued employment. Jobe took me up on my offer. expecting to rejoin Level 3 after he completed his stint. I flew to New York once he was released and let him know his Level 3 position had been eliminated."

Jobe passed away in 2010, following a three-year battle with liver disease.

Before he left, Jobe brought on two of his Concert lieutenants—Neil Hobbs and Brady Rafuse. Putting aside Jobe's issues, these guys were cocky, direct, and hardworking. They knew their shit. Hobbs and Rafuse had little to lose as newcomers to the Level 3 mess. This was empowering. The Starsky and Hutch duo was particularly adept in guiding the sales and marketing side of the house. Hobbs and Rafuse were also a pile of dangerous fun to hang with.

In 2002 Hobbs, Rafuse, and I developed a special relationship. They concluded I knew the product, network, and financial side of the house—as well as where the dead bodies were buried from our first six years.

The three of us locked arms. Hobbs and Rafuse upscaled the go-to-market organization, something that was always lacking at Level 3. I used the product organizations to bring a commercial and financial focus to network, product, and service initiatives. No more drama. No more big strategic thoughts and colorful storytelling. We needed to sell the right services to the right customers at the right prices. For me, it was the first time where I could envision how to properly run a complex telecommunications organization, where emphasis was placed on value-creation outcomes.

We drove the entire company focused on the tactical realities of running a telecom business. The partnership between Hobbs, Rafuse, and me was going strong. Level 3 was turning the corner and steadily clawing its way back. Then my career at Level 3 came to an abrupt and confusing end.

I was invited by a Level 3 sales executive to participate in a large customer meeting at Level 3 headquarters. I spoke a lot, I guess, and the sales executive thought I was being disruptive to her agenda. During a break in the multi-hour meeting, I was asked to see my boss—Level 3's vice chair and former Bell South executive, Buddy Miller.

Before remote executives were a thing, our vice chair attended to his role from his quiet ranch near Atlanta. We'd see him every month or two. As best I could tell, he was my boss because Crowe wanted a buffer between O'Hara and me, as tension had developed between us in recent years.

When I sat down with Miller during the break, it was clear someone had instructed him to talk to me about my behavior in the meeting. My mindset wasn't ready to hear a lecture. I felt Level 3 had done me wrong during the 2000 reorganization and, post Jobe, failed to acknowledge I was a major reason the business was on an upswing.

My recollection was that I ended the conversation after about five minutes. "I'm done," I told Miller and left the office immediately and didn't return. It was a childish reaction, yes, but I was done with being underappreciated.

After a week I met Crowe for lunch. He asked me to return, ease back into Level 3, and all would be fine. I did, but I had no spirit left for the company. A couple of months later, I left for good.

Was the departure a result of poorly handled communications? Did I get fired? Were Miller and O'Hara reacting to me just being ready to leave and showing it? Or did I quit? Yes is the answer to all of the above. Also, I had been working with colleagues who weren't so keen on my increasing power in the Level 3 organization, so they perhaps saw an opportunity to clip my wings, which was not going to work for me this time.

Though devastating to me then, my cutting this cord was a true gift. If I had stayed, I wouldn't be the storyteller of this book.

In addition to fixing the day-to-day management of the company, Level 3 saw consolidation as the opportunity to cement its role as the industry leader. An industry executive took notice, saying in 2002, "I hope Level 3 is amazingly successful at buying other companies. It'll be a good thing to have fewer players."

Level 3 was amazingly successful as a driver of consolidation, and the company was fantastic for the industry.

In 2003 Level 3 acquired Genuity for $137 million. The $680 million purchase of WilTel 2 took place in 2005. In 2006 five acquisitions were completed: Broadwing for $1.4 billion and metro providers Progress

Telecom,[1] ICG Communications, TelCove,[2] and Looking Glass Networks[3] for a combined value of $1.5 billion. Level 3 added Global Crossing for $3.0 billion in 2011 and, five years later, tw telecom for $7.3 billion. The WilTel 2 acquisition was particularly noteworthy, as the company's CEO, Jeff Storey, would rise to significance in the aftermath of the merger.

The impact of Level 3's consolidation of fiber players was tremendous. A total of 30 percent or more of all fiber built out in the late 1990s and early 2000s was embedded in the Level 3 platform. This was the play Jim Crowe, and the industry, had so desperately needed. Yet Crowe, despite his enormous contributions, came close but didn't quite make it to the promised land. Crowe navigated the Bust and the rising phoenix eras to be a leader of the reconsolidation. He almost finished what he started but sadly not quite.

---

1 Progress Telecom was formed in 1998 by its parent, electric utility Florida Progress, to leverage its 1,100-mile fiber network. The telecom subsidiary of Carolina Power & Light was merged into Progress, which brought the fiber assets to 3,000 miles.

In 2003 EPIK Communications merged with Progress, which resulted in EPIK's investor Odyssey Telecorp owning 45 percent of the combined entity. EPIK, the telecom subsidiary of utility Florida East Coast Industries, invested $450 million during the Boom. During the Bust, they sold to Odyssey for $500,000. At the time of the sale to Level 3, Progress had nine thousand miles of fiber assets throughout the southeast. Level 3's $137 million purchase price was a great buy.

2 See appendix 3, "More Bandwidth Stories," to learn more about TelCove and its earlier incarnation as Adelphia Business Solutions.

3 Looking Glass Networks was founded in 2000 by my former colleagues Lynn Refer and Sunit S. Patel. Refer and I were together at Ameritech, MFS, and WorldCom, where he assumed my position when I departed to co-found Level 3. Patel was the treasurer of MFS when it was sold to WorldCom. He continued with WorldCom after its acquisition of MFS and served as its treasurer between 1997 and 2000.

Looking Glass raised $200 million from Madison Dearborn Partners and Battery Ventures. Refer and Patel added $275 million in debt, anchored by $100 million from Cisco Systems. Patel tapped into his Wall Street relationships, with support from each of JPMorgan Chase, Citigroup Salomon Smith Barney, Credit Suisse First Boston, Merrill Lynch, Deutsche Bank, and Barclays participating.

Looking Glass mimicked the early MFS business plan, with a narrow product portfolio of solely Dark Fiber and transport services. Armed with $475 million, they built metro networks in NFL markets. In 2006 Looking Glass sold to Level 3 for $165 million.

Level 3 struggled to deliver on its financial promises. This led to a strange disruption to Level 3's executive team, specifically involving COO and co-founder O'Hara and CFO Sunit Patel, that played out in the public spotlight. The outcome of this struggle paved the way for Crowe's successor to be a Level 3 outsider.

Patel was my MFS colleague. Like me, he stayed with WorldCom after it acquired MFS and served as its treasurer between 1997 and 2000. In 2003 Patel joined Level 3 and was named CFO. He played a central role in the fiber acquisitions, using his deep expertise and relationships to raise and restructure debt to enable these transactions.

In late 2007 Level 3 announced, with no forewarning, that it was launching a search for a new CFO to replace Patel. "As the company focuses on ensuring we take full advantage of the opportunities presented by our marketplace, we believe we need a CFO with skills and experience which emphasize both operational and financial management," Crowe explained. And I, from the outside looking in, scratched my head.

The news caught Wall Street off guard, as it viewed Patel as a strong CFO and a close confidant to Crowe. What was really behind this?

In October 2007 *Forbes* published "Level 3 Drops amidst CFO Mystery."

When is a demotion not a demotion?

Variants on that question are sure to be asked over the next few weeks to those who follow Level 3. The firm's shares stumbled Monday after it announced it will begin looking for a new CFO. The problem is that Level 3 believes its current CFO, Sunit Patel, is ill-suited to guide the projects he helped to start. But Patel won't step down until a replacement is found, and the firm insists it doesn't want him to leave. It just doesn't want him to be CFO.

An analyst provided some color: "Over the last two years he did a lot of financial engineering, and . . . made acquisitions. But he's weaker at projecting where the synergies are going to come from and what they're going to look like."

The analyst went on to explain why the stock was down 11 percent to $4.36 per share. "The person responsible for creating the guidance has announced that he's leaving." One week later Level 3 slashed its guidance for 2007 and 2008. The stock sank 25 percent to $3 per share.

The mystery became more mysterious when, five months later, Level 3 announced COO O'Hara would be departing and, in a 360-degree reversal, Patel would remain as CFO.

"Kevin O'Hara has worked closely with me for more than 20 years and this was obviously a difficult decision for both of us," Crowe explained in a statement. "At this time, however, Kevin and I have agreed that a different perspective will be of benefit to our company." Level 3 shares dropped 14 percent to $1.87 per share.

In 2024, O'Hara shared his reflections with me on the events surrounding his departure.

> In 2006, Crowe indicated to the board that he would like to transition out of the CEO role within a year or two. After an informal process, the board agreed that I would succeed Crowe. We began the transition planning, though the date remained uncertain. Around that time, we were going through a painful integration of seven significant acquisitions. I had grown increasingly frustrated with our finance organization, as the team were great at their external-facing responsibilities but they did not provide the operational financial support that I needed. Sunit Patel and I were frustrated with one another and, separately, tension had built between Jim and me.
>
> One day, Jim invited me to his office where he told me he had changed his mind—he wasn't going to retire for another few years. He asked if I'd be willing to stay on in my current capacity. I told him I would, though I stressed that we would

need to resolve the issues that were the source of the tensions. We agreed to organize our thoughts and get back together in two weeks. When we met, I told him I was all in so long as we could have an honest assessment, perhaps using a third-party executive coaching firm, of the senior team. Jim rejected this idea. Our meeting was at 8:00 a.m. By 8:15 a.m, I was unemployed. Like you, I'm still not sure if I quit or got fired. As a result of my departure, Jim no longer felt the need to replace Sunit and the search for a new CFO was canceled.

An analyst met with Level 3's executive team at the time of the announcement. His takeaway was that Patel and O'Hara had clashed over strategy. As support for Patel poured in from Wall Street, Crowe decided he should keep Patel as CFO and replace O'Hara.

"Given that most of the issues were operations-related, we never could understand why it was Sunit who was being replaced," the analyst wrote.

"There's a lot of disorganization," said another analyst, as he advised his followers to sell their Level 3 shares. "This whole incident just shows there's a lot of problems there."

Later that year Jeff Storey, who remained with Leucadia after it bought WilTel 2, joined Level 3 as COO. Storey's extensive operational background was seen as the medicine for Level 3's long COVID. For the next four years, Storey focused on integrating the assets from Level 3's acquisitions spree.

In late 2011 Level 3 resumed its fiber rollup by buying Global Crossing for $3.0 billion. Immediately following, a 1:15 reverse stock split was executed. As expected, the split caused the stock price, which closed at $1.55 on Nasdaq on Wednesday, October 19, 2011, to open at $22.23 on the NYSE the next morning.

In 2013 Storey succeeded Crowe as CEO.

Crowe acknowledged that the time was right to "get a good fresh set of eyes and ears in. . . . Any CEO has to ask, 'Am I still the best person, or is it time for a fresh point of view?' In my case . . . starting probably in '06, I thought as soon as Level 3 was financially strong and I had a good replacement, I should go do something else." Crowe also stepped down from the board.

"Jeff was the clear and unanimous choice of the board," said board chairman Walter Scott. "With 30 years of industry experience and his intimate knowledge of Level 3's customers, employees, and operating environment, Jeff is the right executive to lead Level 3 into the future." Scott retired from Level 3's board in early 2013.

"Of all the companies that I cover, the one company where I would have said if there were a change in CEO that would have the most positive impact on the stock, it would've been Level 3," commented an analyst in a poke at Crowe. "The reason for that was, at some point, Jim had just lost control of the company. He'd just become too far removed from the day-to-day operations to really understand what was needed to fix the company."

Level 3's stock traded at $42 per share at the time of Crowe's retirement announcement.

———————

Five years later, in February 2016, Storey took a two-month leave of absence to undergo heart surgery. In October 2016 Storey and I met for coffee in Boulder. This wasn't unusual, as we would get together about once a year, even if we didn't have anything in particular on our agenda. This meeting was just a catchup. We talked about our respective medical adversaries—mine cancer and his heart—with me blaming our health challenges on the stress we both carried as we navigated the Bandwidth Boom and Bust.

I told Storey that I recently purchased Level 3 shares, making me a shareholder for the first time in many years. I explained that a recent decline in Level 3's stock price seemed to me an overreaction to Storey's earnings call explanation of the challenges Level 3 faced. Storey stared deeply into my eyes before changing the topic. I was confused by his reaction. Though I didn't know it at the time, the stare was him wondering if I had insider knowledge about a pending major Level 3 announcement.[4]

On October 31, 2016, CenturyLink, the large regional communications carrier, announced it was acquiring Level 3 in a deal valued at $34.3 billion. The purchase consisted of $10.9 billion in assumed debt, $9.6 billion ($26.50/share) in cash, and 1.4286 shares of CenturyLink stock for each Level 3 share. When the deal was announced, CenturyLink's shares were trading at $28 per share. The combination of cash and CenturyLink shares meant Level 3 shareholders would receive $70 per share, a 49 percent premium to Level 3's closing price of $47 before the announcement. Level 3 shareholders would own 49 percent of the combined entity.

Level 3 was born on April Fools' Day of 1998. It died on Halloween of 2016. I was sad, perhaps because I thought I might be destined to be the CEO of Level 3 at some point.

An original 1998 Level 3 shareholder who held those shares for twenty years and then sold after the CenturyLink acquisition would have gained 87 percent, a paltry 3.1 percent IRR.[5] However, as we will soon

---

4 The SEC reviews stock purchases that occurred in the weeks before an acquisition announcement. My name caught their attention as a suspicious stock purchase. They contacted my financial advisor, asking for context. Fortunately, I sent an email to the advisor before making the purchase previewing my buy, citing I believed the stock market overreacted to a statement Storey made in Level 3's earnings call. They told me so long as I had no inside information, I was free to trade. My advisor shared this email chain with the SEC, and that was the last I heard from them.

5 When Level 3 was launched, its initial value was $2.5 billion, which equated to $5 per share. Adjusting for the 2:1 stock split in 1998 and the 1:15 reverse split that occurred in 2011, this equated to a value of $37.50 per share.

If recipients of these shares held them through to Level 3's sale to CenturyLink, the value received would have been $70 per original Level 3 share. They would have gained 87 percent during this twenty-year investment hold, a 3.1 percent IRR.

learn, long holders of Level 3, who retained instead of sold their CenturyLink shares, would see their returns turn negative.

In September 2021, Walter Scott Jr. died at the ripe old age of ninety. Though I met him only a few times, he was an inspiration.

In July 2023 James Q. Crowe passed away at the young age of seventy-four. His obituary read,

> In the 1990's, when web pages opened slowly and streaming video over the Internet was not viewed as realistic, Crowe hatched a vision of cheaper Bandwidth that would enable video streaming and the robust Internet experience we enjoy today. Crowe made his vision a reality.
>
> . . . Rather than have data move over traditional telephone networks, he envisioned voice, data, and video moving over a revolutionary new kind of network using fiber optic and Internet technologies. "A network for the eyes rather than for the ears" was how Crowe described it.
>
> Crowe formed Level 3 Communications in 1997 with exactly this vision. Together with his parent company, Omaha-based construction giant Peter Kiewit Sons', and Kiewit Chairman and Crowe mentor Walter Scott, Crowe raised $15 billion to construct a totally new network based upon the newest technologies and optimized for the Internet. They dug trenches and buried multiple conduits to carry optical fibers across the United States and parts of Canada, digging 16,000 miles along streets and railroads, installing electrical and optical equipment all along the path. They also built a similar network across Western Europe. Level 3 brought Bandwidth costs low enough to stimulate explosive growth in the Internet.

Early Level 3 customers such as Facebook and Netflix, among many others, took advantage of these lower costs to create the rich Internet experience the public knows today.

Crowe's obituary covered his transition from MFS to Level 3.

Crowe launched MFS Communications, the first company to bring robust local competition to the Baby Bells via fiber optic communication technology. He took MFS public in 1993 and built it rapidly into a Fortune 500 company, before it was acquired by WorldCom in 1996 for $14.3 billion.

A year later, Crowe founded Level 3. He was joined by a number of executives and engineers who had worked with him at MFS. His vision and generous nature attracted a loyal following, including but not limited to former MFS employees. American industry has many former Level 3 executives in leadership positions, virtually all of whom regard Jim Crowe as a major figure in their lives.

Count me as one of his loyal followers. He was a major figure in my life. The obituary ended with the following:

The Smithsonian Institution recognized Level 3 in a large public ceremony for its role in harnessing new Internet technologies to change telecommunications, stating: "Your achievement will inspire others to pioneer, and will provide a record for future generations that wish to understand how information technology changed our world."

Inspired by the memories triggered by Crowe's passing, I began to write this book in July 2023, just a few weeks after his passing. In many ways this book is a tribute to Jim Crowe.

Lumen—Glen Post and Jeff Storey

From the beginning of the book, we identified Lumen as one of the four remaining national Bandwidth Infrastructure companies. Lumen was the rebranded name of CenturyLink. This is the story of how Glen F. Post III turned CenturyLink into the largest operator of Big Bandwidth.

Post earned an accounting degree in 1974 and an MBA in 1976 from Louisiana Tech University. Immediately after graduating, he joined CenturyTel, a rural telephone company headquartered in Monroe, Louisiana, that served customers in thirty-seven states. Post spent his entire career with the business, culminating in a twenty-six-year reign as CenturyLink's CEO.

"As CEOs go, I would say he's more understated but very knowledgeable" was how Sprint's CEO described Post.

"There's no sharp elbow on the guy," a banker joked in 2010.

Post had a passion for duck hunting. "During the wintertime, [duck hunting] is his passion," the CEO of Progressive Bank said. "He's an avid hunter. He knows how to work the ducks and call them in. He's leading the hunt, and he's a great shot."

In perhaps the most unlikely twist in the Bandwidth adventure, this introverted, smooth-elbowed duck hunter turned his Monroe rural telephone company into the biggest consolidator of the fiber networks.

Before 2009 CenturyTel was a modest-size rural telephone company, with a Market Value of only $3.3 billion. The local telephone companies were struggling with declining revenues and margins. Post's plan was to get bigger through acquisitions. Sound familiar? This was Bernie Ebbers's plan when he took over the struggling WorldCom in the early 1980s.

In 2009 CenturyTel acquired its larger rival, Embarq Corporation, for $11.6 billion, half in stock and half in debt. Though CenturyTel was the acquirer, Embarq shareholders owned 66 percent of the combined company. CenturyTel traded at $26 a share when the Embarq transaction was announced. The merged company, which operated in thirty-three states and had revenue of $8.8 billion, was named CenturyLink.

CenturyLink was on the hunt. Qwest, through its merger with U S West, had a sizable traditional telephone business. This certainly played into CenturyLink's April 2010 justification for acquiringQwest, which cost CenturyLink $22.4 billion—$10.6 billion in stock and the rest in assumed debt.

CenturyLink's shares were trading at $36, up 40 percent from when Embarq was acquired, which implied a Market Value of $10.8 billion. The merger resulted in CenturyLink shareholders owning 50.5 percent of the combined entity and those of Qwest owning 49.5 percent. CenturyLink became the United States's third-largest telecommunications company.

"This combination will enhance our ability to deploy innovative IP products and high-Bandwidth services to business customers, expand broadband availability and speed to consumers, and offer superior, differentiated video products," proclaimed Post.

CenturyLink was both a customer and rival of Zayo. To develop a relationship with Post, I visited Monroe a few times in the 2010s. I'd sit with him in his boring office and have friendly one-on-one discussions on the industry. He was noticeably introverted, coming across as less sophisticated and far nicer than other telecom executives. I doubted he spent much time in front of high-tech customers. I surmised his comfort with technology was low. He was, nonetheless, a fine Southern gentleman.

I didn't view CenturyLink as an acquirer of Zayo, though in hindsight perhaps I should have. I suspect he was sizing up whether Zayo or our crosstown rival, Level 3, should be on his acquisition wish list.

Level 3 turned out to be his target. CenturyLink closed its acquisition of Level 3 in November 2017. The *Wall Street Journal*, in its article titled "Telecom CEO's Deal-Making Puts Louisiana Town at Center of Internet," was ready with dramatic words.

When Glen F. Post III took the top job at CenturyTel in 1992, the first website was little more than a year old, IBM computers dominated the workplace and mobile phones stood nearly a foot tall.

Some 24 years later, the internet and cellphones have supplanted phone lines as the telecommunications industry's main businesses, pushing many regional players into the arms of bigger rivals. But Mr. Post still leads the Monroe, La., company, now called CenturyLink, and is vying to turn it into one of the biggest internet providers in the world.

The 64-year-old executive is an unlikely skipper in the ego-charged media and telecom sector. He still lives in the small town of Farmerville, where he grew up and which is a half-hour drive from CenturyLink's headquarters complex on Monroe's outskirts. People who know him say he likes to hunt ducks, doesn't drink and rarely raises his voice.

"He's just a country boy at heart," said a former executive. ... "He knows where he came from but he also knows that he runs the largest company in Louisiana."

More than that, CenturyLink is poised to become one of the world's biggest internet carriers by adding Level 3 Communications, which has customers in more than 60 countries and thousands of miles of fiber-optic cables spanning four continents.

Not featured in the article was that CenturyLink's stock price eroded from $36 per share at the time of the Qwest transaction to $28 per share when the Level 3 acquisition was announced and fell again to $20 a share when CenturyLink closed its acquisition a year later.

CenturyLink's stock price decline lowered the acquisition price of Level 3 from $34.3 billion to $30.3 billion, implying Level 3 shareholders were receiving $57.50 per share instead of the $70 per share at the time the acquisition was announced.

Recall that Keith Meister was with me when his mentor, Carl Icahn, fell off his chair at Icahn Enterprises. He left Icahn to start Corvex Management, which specialized in activist investing. In May 2017, with the merger still months from closing, Meister disclosed he had accumulated 5.5 percent ownership of CenturyLink. In true activist fashion that would have made his former boss proud, he had a demand. He wanted Glen Post to accelerate his retirement and immediately hand the CEO reins to Jeff Storey. The transaction hadn't yet closed!

"Execution has been substantially weaker at CenturyLink than at Level 3 for many years," Meister said at the Ira Sohn conference in New York City. "Would you pay a premium to buy the New England Patriots and then not start Tom Brady?"

Hey, Storey. No activist investor ever said anything nice about me.

This pressure from activists, alongside a continued slide in Century-Link's stock price, hastened the transition of the CEO position. Storey became CEO on June 1, 2018, seven months earlier than planned.

In many ways the CenturyLink-Level 3 combination can be viewed as a reverse merger, with Level 3 being the surviving company. Storey was CEO, Patel was CFO, the executive team was mostly legacy Level 3 executives, and Colorado was the headquarters. It only lacked the name Level 3.

The collection of assets under CenturyLink was, and still is, astounding.

- The original CenturyLink
- Qwest by CenturyLink
- U S West Communications via the merger with Qwest
- Level 3
- Global Crossing via Level 3 acquisition
- Broadwing via Level 3 acquisition

- WilTel 2 via Level 3 acquisition
- Genuity via Level 3 acquisition
- BBN via Level 3's acquisition of Genuity
- tw telecom via Level 3 acquisition
- Xspedius via Level 3's acquisition of tw telecom
- GST via Level 3's acquisition of tw telecom
- TelCove via Level 3 acquisition
- ICG Communications via Level 3 acquisition
- Progress Telecom via Level 3 acquisition
- Looking Glass via Level 3 acquisition
- Embarq by CenturyLink

As a co-founder of Level 3, I felt pride. I also had jolts of jealousy that I wasn't part of the Level 3 victory lap.

As CEO of Zayo, I had mixed feelings about the merger. On the positive side, consolidation was a key part of Zayo's thesis. As the industry reduced to a handful of national platforms, the supply-demand balance would shift in favor of suppliers. Moreover, the merger left Zayo as the leading independent provider of Bandwidth Infrastructure.

On the negative side, I was intimidated by the sheer vastness of the assets under Lumen. If Lumen could get this behemoth under control, it could smash the rest of us. Then I smiled—Lumen is far more likely to be an utter mess than a fine-tuned machine.

Storey had a chance to put an exclamation point on the reverse takeover in September 2020 when CenturyLink changed its name. However, instead of Level 3, Lumen Technologies was introduced as the new brand. "Yuk" was my reaction as I yawned at the bland name.

Lumen CFO Patel resigned in September 2018. Storey retired in September 2022, with the stock price down to $10 a share.[6]

Industry outsider Kate Johnson became Storey's successor. She spent four years at GE before leaving to join Microsoft as president of North America. She left Microsoft in 2021.

I crossed paths with Kate Johnson in 2013 when she was an executive vice president at GE. As we prepared to take Zayo public, I sought independent directors for the Zayo board. After a few phone calls followed a banker introduction, Johnson and I were mutually interested in her taking the next step. Then Johnson abruptly pulled out. She told me her CEO, Jeff Immelt, blocked the idea and wanted her to focus on her responsibilities at GE.

Some ten years had passed since my interaction with Johnson. Now she was expressing unbridled optimism about joining Lumen and its mission. "I'm looking forward to leading this great company through its next chapter and helping customers leverage the Lumen platform to power amazing digital world experience."

Johnson's first job was to stave off a Lumen bankruptcy. The $10 per share stock price eroded quickly, dipping below $1 a share in September 2023, her one-year anniversary.[7]

This didn't stop Johnson from conveying confidence in Lumen with her pocketbook. In November 2023 she bought $1 million worth of Lumen shares, explaining, "I believe deeply in our strategy, our mission, the team that we've built and our progress so far, and so I went long in the stock. . . . I decided that this is the right thing to do, and I think it expresses my deep confidence in our business."

---

6 For those of you keeping score, the hypothetical person who held their original Level 3 share, which had a cost basis of $37.50 per share, would have a value of $53.50 per share, or 1.7 percent IRR, to show for their twenty-one-year investment. The $53.50 per share came from $43.50 per share in cash received from CenturyLink, plus the $10 per share of CenturyLink stock.

7 Okay, one more time we will visit our stubborn holder of an original $37.50 per Level 3 share. Their return is $44.50 per share, or 0.8 percent IRR. At least they didn't lose money.

The problems Johnson took on at Lumen are deep rooted and include operational deficiencies, legacy business overhang, and complex indebtedness. It might take five or more years—and perhaps a visit through Chapter 11—for Lumen to regain momentum. Still, the sheer assets buried in the Lumen platform should ensure its major player industry status for the duration.

Zayo—Dan Caruso

Zayo's origin began in the wake of the ICG 2006 exit. I had been recovering from double hip replacement surgery. To pass the long days, I googled fiber companies and reached out to their CEOs. Drawing on this dialing and discussion with CEOs, and my own very personal and qualified introspection on the state of the industry, the premise for Zayo emerged.

First, I noted the **Insatiable Demand for Bandwidth** would remain true for multiple generations to come. The daily widespread use of the Google search engine and Google Maps, the monumental adoption of streaming, and the continued expansion of cloud services and data centers were the most visible points of proof. Second, I explained that the problem was not a fiber glut; it was **Too Many Fiber Providers**. Too many fiber rivals fought one another for a finite set of revenue opportunities. Third, **Consolidation** would bring supply-demand balance, leaving the consolidated platforms that controlled the vast majority of fiber assets to flourish.

Fourth, I made the case that there was **Too Little Fiber**. As Bandwidth grew and use cases expanded, the sheer explosion in usage meant fiber would come to be woven into an interconnected fabric. Big customers and data centers would need a large quantity of fiber to handle burgeoning volumes of digital traffic that would pass among their facilities. And fiber demand would extend to small buildings, homes, and cell towers. The Internet of Things would lead to everyday devices such as farming equipment, vehicles, and watches being connected, driving more and more Bandwidth use cases.

Fifth, I suggested that **Now Was the Time to Invest.** The crash was so extreme and painful that most investors were ferociously determined to avoid fiber investments. To those who would listen, I likened it to a tequila hangover: "Remember your first overindulgence of tequila? You swore you would never touch tequila again. It took years before you took a sip again, realizing it wasn't the tequila. It was the pace, quality, and quantity of tequila you drank," I explained. I saw a window of opportunity to buy fiber properties at undervalued prices.

Sixth, I identified the unique post-Bust time in our industry. The upheaval of the 2000s brought an abundance of subscale forgotten fiber companies. I called them **Fiber Orphans.** The Fiber Orphans were owned by individuals, investors, or companies that weren't long-term investors. I termed these investors **Accidental Owners.** Some acquired the asset as a spin out of a distressed division of a bigger company. Others purchased the property through a bankruptcy. Still, others were financial owners turned Bandwidth operators as their CEO flamed out. My view: most would be eager for an exit.

Seventh and last, I observed that Accidental Owners ran tightly focused businesses. These businesses were laser focused on delivering Bandwidth solutions to customers using vast quantities of Internet to drive their own customer success and growth. When we launched Zayo, we coined the term **Bandwidth Infrastructure** to describe the business model.

This term was also personal to me. I would prove to myself, and to anyone in the industry who cared, that a Bandwidth Infrastructure business could be operated in a way that authentically created great outcomes for its investors.

Backed by my ICG sponsors, ColCap and M/C Partners, we[8] lined up a $225 million Series A and went hunting for Fiber Orphans.

---

8 John Scarano was my co-founder. Other key members of the founding team included Sandi Lewis, Matt Erickson, Marty Snella, Tim Gentry, Chris Murphy, and Scott Beer. Key executives who joined us a year later were CFO Ken desGarennes, Chris Morley, Fritz Hendricks, Glenn Russo, Hannah Wanderer, and Jason Tibbs.

Learning from past lessons of overzealous spending, we were frugal. Compensation was low—my salary, for example, was the minimum necessary to qualify for medical benefits. We were generous with stock options, firmly aligning management's interests with those of our investors.

We launched Zayo from a small office, which ironically was being paid mostly by Level 3, thanks to a sneaky move I pulled when we sold ICG to Level 3. We had been using a small office on Pearl Street in Boulder at the time for ICG and other purposes. This lease was in my name. As we were negotiating with Level 3, my team wrote a sublease that enabled ICG to lease half the space from me for 80 percent of the rent. With Level 3's campus ten miles away, the only Level 3 employees who would use the facility were legacy ICG executives, most of whom joined me for the Zayo adventure.

From our side of this small office, Zayo was born and would grow up to become Level 3's greatest rival. Was I driven by the opportunity to prove Level 3 wrong? To some degree, we all were, though most, if not all, of us were just having fun.

As we launched Zayo, blogging became a thing. Crowe's lesson about the Internet was burned into my consciousness. So I pledged to always stay current on the latest Internet-enabled trends.

I began blogging. And I gave more consideration to "the Bear," a nickname I was given while at Level 3. The nickname wasn't meant to endear me to colleagues. *Fur* sure. I was not a cuddly bear. I was more the grizzly startled awake from a deep hibernation. A whole lexicon rose around me. You're poking the Bear. Spray honey on your unsuspecting colleague to distract the Bear. Don't mess with the Bear's "Cubs," paying homage to my inner circle.

I embraced the Bear moniker, even naming my blog "Bear on Business." Like MFS's Royce Holland, howling in the trade rags in the early

1990s, I used Bear on Business and roared of Zayo's intentions to consolidate the industry.

We started great, acquiring multiple companies in our first year.[9] Then in 2008 subprime mortgages led the financial markets into a tailspin. Bear Stearns narrowly avoided collapse that March when J.P. Morgan Chase acquired it. In September Lehman Brothers, which had gotten the ball rolling downhill, escaped luck and went bankrupt. An international debt crisis was in full force. The US government concluded the banks were too big to fail, so it spearheaded a bailout. The Great Recession hit just as Zayo was entering kindergarten. Was our timing bad, or was this an opportunity?

CIT Group was one of our major lenders. We met in their NYC headquarters on our quest to raise debt to fund an acquisition. As we pitched our plan to a room full of bankers, few questions were asked, and no overt reaction was offered.[10] The mood was somber.

We learned the next day that the CIT team was anticipating mass layoffs. Instead, they received approval to provide the loan to Zayo. We were CIT's first post-crisis transaction that the bank approved, and this became a clear signal to the CIT team that CIT would get through its existential crises.

---

9 Our first three acquisitions were in Tennessee, Pennsylvania, and Indiana. We paid $9 million for Memphis Networx, an underutilized two-hundred-route-mile metro network that was a joint venture with the power utility and the Memphis business community.

PPL Telcom was a power utility spin out run by an industry veteran who treated it as a cushy semiretirement job. We paid $46 million and gained four thousand route miles of metro and regional fiber.

Indiana Fiber Works was a statewide fiber network that had metro networks in most of the major cities, including Indianapolis. We paid $22 million for it.

10 Thomas Westdyk and Joe Jenda were the two CIT bankers who supported Zayo. Among our memories is an after-midnight Ping-Pong game at the Standard hotel in NYC's Meatpacking District. Those were fun times.

During my ICG stint, I made a habit of extending my business travel for networking. When in Boston, I looked up Charlesbank and reached out to Michael Choe. We had a fancy breakfast together at Boston's Four Seasons hotel. We hit it off.

Choe was young, intelligent, curious, and ambitious. He was easy to spend time with. Though his firm had experience in telecom and tech, most of Charlesbank's investments were growth equity in traditional companies.

"Maybe we will find something to do together down the road," we both concluded.

When the mortgage crisis hit, Zayo had great momentum but was running out of dry powder for more deals. I reached out to Choe, who saw this as an opportunity. "Zayo is poised to accelerate its rollup in an environment where owners will be anxious to sell and other investors will be fearful to buy." Charlesbank led our Series B, providing $128 million of additional dry powder to continue our rollup. Morgan Stanley also participated in the Series B.

In our first five years, we completed over twenty acquisitions and gathered fiber assets scattered throughout the United States. Among my favorites were Onvoy,[11] AGL Networks,[12] and American Fiber Systems (AFS).

---

11 Minnesota-based Onvoy was one of my favorite acquisitions, solely because of the friendship I developed with their COO, Fritz Hendricks. From the moment I met Hendricks, I knew he was special. He was exceptionally intelligent, could go toe-to-toe with any telecom CTO, and his business acumen was extraordinary. Hendricks ran the business even though someone else held the CEO title.

Zayo purchased Onvoy for $72 million, and our focus was primarily the Bandwidth business. We noted, however, that Onvoy had an underappreciated wholesale voice business with durable revenue and cash flow. Hendricks detangled the Bandwidth business, which was combined with Zayo, from the wholesale voice business. Hendricks ran the wholesale voice business using the Onvoy brand.

Onvoy became the gift that kept on giving. Hendricks steadily grew revenue and consistently delivered cash flow. Eventually, one of our investors bought Onvoy Voice Services. Zayo's overall return on Onvoy easily exceeded four times. Thank you, Fritz.

12 AGL Networks was a favorite for two reasons. One, it was primarily a Dark Fiber business, which gave us a full appreciation of the strength of the Dark Fiber business model. Second, it was how we won the deal. On the last day of bidding, our CFO, Ken

AFS's CEO, Dave Rusin, was previously the president of Rochester Tel, later rebranded as Frontier. His LinkedIn profile lists his accomplishments as having "transformed the 125-year-old telephone monopoly into the 1st open market local carrier in the world" and "Catapulted Frontier Communications to be the 4th largest telecommunications company in the US." It's his LinkedIn, so he is entitled to embellish.

Rusin's clone was Vizzini, the bumbling, bully, mastermind thug played by Wallace Shawn in *The Princess Bride*. Rusin was not a fan of mine, showing frustration as he watched me raise money and acquire fiber companies. "He stole my idea," he must have told his investors.

Eventually, I convinced his largest investor, Jeffrey "Jeff" Drazen, who represented Sierra Ventures, to sell AFS to Zayo. When I held an AFS all-employee session, Rusin had his HR lead ask three questions, submitted anonymously in true Vizzini fashion.

The first was, "We understand Level 3 had a lot of problems when you were there. Did Level 3 get on track after you were fired?" I told them Level 3 was now doing fine.

The next question was, "AFS has a great brand in the industry with a name that captures its Bandwidth mission. Zayo is a complete unknown with a name that stands for absolutely nothing. What will the name of the combined company be?"

"Zayo," I answered.

Then the third diabolical question came. "AFS is led by a CEO that has an incredible track record as a longtime industry CEO. You have far

---

desGarennes, told us we were outbid. When I pressed him on how much, he went back to the bankers and twisted their arms. We raised our bid $1.2 million and agreed to sign within twenty-four hours.

My phone rang. A CEO friend with limited acquisition experience was seeking my advice. "But before I do, I need to ask you a question. Did you execute a deal to purchase AGL Networks?" I answered carefully, "No, I did not execute an agreement to purchase AGL Networks." I had not yet signed it, so my answer was technically true. She went on to tell me that they had the deal locked up, and then, while routing for final approval, they were outbid by $0.5 million. "It's almost like they knew what we bid. Is there anything we can do?" she asked.

We quickly signed. Nice job, Ken!

less experience and a questionable track record. Who will be CEO of the combined company?" I answered that I would be the CEO without feeling a need to justify.

AFS, which had accumulated metro fiber assets in the rising tech markets of Boise, Marietta, Las Vegas, Minneapolis/St. Paul, Nashville, and Salt Lake City, was a great acquisition for Zayo.

———

AboveNet had grown steadily under Bill LaPerch's leadership post meltdown, exploiting the incredible assets that were built during the Boom times. That was when LaPerch decided to explore a take-private transaction in which management would co-own the company alongside private investors.

In late May 2011, the financial press caught wind that AboveNet hired investment banker JPMorgan. AboveNet was in play!

In June 2011 the *New York Post* published "NY Fiber Firm Buy in Works," reporting AboveNet "has been shopping itself for several months, received a binding offer in the last few days from a private-equity firm . . . [and] has been asking for a minimum bid of $80 a share." At the time AboveNet was trading at $73 per share, a 71 percent gain in the past year. The *New York Post* added, "[S]ources believed it would be a challenge to get anyone to meet that price."

Hmmm. Opportunity was knocking hard at my door.

My instinct was to insert Zayo into the process, knowing that as a public company, AboveNet could not ignore our expression of interest. I contacted LaPerch and asked for Zayo to be included. LaPerch went to his board and informed me that he could not respond to market rumors but that they would consider an offer assuming we had appropriate financial backing. I didn't have the dinero at the time; there wasn't much I could do.

We learned later how the bidding war played out in 2011—and how close AboveNet was to being acquired in early 2012.

Jan: JPMorgan presented to AboveNet's board that the prospects were high that AboveNet could be sold at a premium value to its stock price.

Feb: JPMorgan was engaged to explore with an eye toward an exit.

Mar: Five Private Equity firms submitted indications of interest at prices of $71–$80 per share. The three highest bidders were invited to engage in diligence.

Apr: A sixth Private Equity firm requested permission from AboveNet to speak to Zayo about a joint bid. AboveNet denied its request.

On the same day as my call to LaPerch, a CEO of a public company expressed interest and received permission to participate in the AboveNet process, but this bidder withdrew from the process in June.

Later in April a third CEO reached out to JPMorgan, only to withdraw in May.

Interest in AboveNet grinded on. In May AboveNet provided draft merger agreements to the three Private Equity firms and one of the strategic bidders. This latter bidder offered $85 per share, of which 30 percent would be in the form of shares of its stock. Yet another strategic bidder reached out to JPMorgan to express interest in AboveNet and submitted a bid for $80 per share.

On June 11 a financial bidder submitted a bid at $74 per share. The board rejected the offer as inadequate. Both of the strategic bidders lost interest. By November the board concluded it was time to move on and ended the process.

LaPerch watched as Zayo gobbled up more than a dozen fiber properties. Despite AboveNet having a fifteen-year head start, Zayo had become nearly as big. With the take-private process now terminated, LaPerch, who shied away from M&A, was ready to join in the consolidation game.

In October 2011 Reuters published an article titled "AboveNet CEO Keen to Make His First Acquisition," quoting LaPerch: "It wouldn't be out of the realm of reality for us to make an acquisition if there is a good strategic reason to do it, and it makes good financial sense."

LaPerch set his sights on 360networks.

Unfortunately for LaPerch, I had been laying the groundwork with 360networks for years. Between 2008 and 2011, I visited Greg Maffei once a year at Liberty's Denver headquarters to see if he was ready to sell 360networks. I also checked in with his executives Rick Coma and Chris Mueller twice a year. 360networks's assets were concentrated on the West Coast and would make an ideal complement to Zayo's growing Pacific Northwest regional network.

In 2011 my persistence paid off. Maffei's team approached us and said they were ready to sell. We worked on an agreement. Then as we were approaching signature, AboveNet emerged, determined to outbid us. Despite AboveNet's Hail Mary bid, 360networks picked Zayo as its dance partner. Our M&A track record gave Maffei confidence that we would close without any hitches.

In 2024, LaPerch shared with me his inside story. "Zayo and AboveNet coincidentally ended up with the same final bid. When I found out, I called Maffei and offered him $10 million more. I still remember his response. 'Bill, you run a great company and get great returns on your organic investments. Dan Caruso has done sixteen acquisitions, and I feel more comfortable that a sale to Zayo will close quickly. I am going to take the Zayo offer, even though it's less.' That one really hurt me."

The $318 million price exhausted our cash resources, but my coveting for AboveNet only intensified.

Zayo was rolling. In 2011 my investors were ready to take Zayo public. A 2012 exit would have been lucrative. Investors were five times in the money and quite eager to cement an outcome. But I wasn't on board. And my vote trumped all of theirs; Zayo needed me as its CEO to ring the NYSE opening bell.

While all this was going on, it became clear to me that to continue with our consolidation, we needed more cash. My investors were tapped out, as most of them had already invested their maximum. Despite being unable to invest more, they were reluctant to welcome in a new investor, as this would require a meaningful additional voice in the board room, thereby complicating governance. A new investor would also reset the investment clock, as time would be needed for their investment to grow in value before an exit.

We compromised. My syndicate gave me the opportunity to find a Series C lead, but we'd need to strike a balance on governance and exit time frame.

On the hunt for a Series C lead, I accepted an invitation from Stephens, a boutique investment bank in Little Rock, Arkansas, to attend its annual TMT conference. My assigned seat at dinner was next to Phil Canfield, a senior partner at GTCR in Chicago. We struck up a conversation, during which Canfield explained GTCR's approach: "Each January we select four macro themes that will guide our investment strategy for the year. We then identify the best platform in which to execute each theme."

Canfield, I learned, was also a University of Chicago MBA. His analytical approach resonated. He continued, "Fiber infrastructure was the leading theme for GTCR this year. After a few months of research, we identified Zayo as the best platform to execute on this theme."

I'm not the fastest swimmer in the pool, but I caught on that it wasn't a coincidence we were sitting next to each other. Canfield told me of

GTCR's success in sector rollups and explained its leader's strategy: "All PE firms give lip service to the importance of the CEO. At GTCR, we learned the CEO is by far the most important factor in value creation. We only back the most capable CEOs."

I enjoyed my ego being stroked.

Over the next few weeks, we solved the governance and exit dynamic and entered a handshake agreement with GTCR for a multihundred-million-dollar Series C. Zayo had our third hunting license.

Canfield asked me which fiber company we wanted to target. I answered AboveNet. He responded, "I was hoping you'd say that."

Armed with more ammo and AboveNet in my crosshairs, I headed to Miami's South Beach to attend Metro Connect, one of the last remaining annual industry conferences from the Boom era. Since the launch of Zayo, I leveraged this annual conference, which was held at Eden Roc in March, as a personal hunting ground to spread the word on Zayo's ambitions.

The first year I went was 2008. It was fun, with CEOs from many of the Fiber Orphans in attendance. A few investors and bankers also showed up, but overall, attendance was sparse.

The subsequent year I hosted my first annual invitation-only after hours at the Forge, a trendy wine bar in South Beach. Handpicked CEOs, investors, and bankers were invited in advance, which drove up attendance at the conference. Soon, the after-hours invite was seen as a golden ticket.

I began giving the keynote address as the kickoff to the conference. Instead of giving flowery updates on Zayo, I gave thought-provoking speeches to create a buzz. Examples were "Reflections of a Serial Entrepreneur," "Telecom's Boom, Meltdown, and Resurgence," "Batman versus Superman," and "Reflections of an Anxious Chairman."

The most infamous experience was when I plopped my wallet on the table during a panel discussion. My spontaneous reaction, with Bill LaPerch sitting next to me, was the first in a series of steps that rapidly culminated in Zayo's acquisition of AboveNet. Here is how it played out.

I wasn't the only one who enjoyed the energy of Metro Connect. LaPerch enjoyed its resurgence as much as I did. AboveNet was the conference's most prominent company attendee; its metro networks were the envy of us all. Court would be held as Bill sipped whiskey at the bar. Vendors sought to sell him equipment, carriers wanted access to his fiber, and bankers pitched to be his advisors. Plus, LaPerch was friendly, witty, and easy to be around.

The featured panel to kick off Metro Connect was the fiber big dogs offering their predictions for 2012. John Purcell, the CEO of Fibertech Networks, was closest to the moderator.[13] I was in the middle, and LaPerch was to my left. Purcell and LaPerch were the two most respected statesmen of the industry; me—not so much so. I was the corporate-raiding interloper. The room was packed with over one thousand analysts, bankers, and fiber executives.

The first question went to Purcell, the softball the Metro Connect crowd was hoping for. "John, a lot of consolidation activity has accelerated over the past two years, much of it driven by the guy to your left. What do you think will happen this year?"

Purcell was both articulate and entertaining. He began dryly, "I don't know why you posed that question to me" and explained how Fibertech has grown 100 percent organically. "I don't even understand how acquisitions work."

Turning his wit toward me and touching my shoulder, he claimed to have no idea how I've convinced my investors to go on such a wild

13 John Purcell spent thirty years at Rochester Telephone, which rebranded to Frontier in the 1990s. In 2000 Purcell partnered with colleague Frank Chiaino to launch Fibertech. Their simple strategy was to be the underlying fiber infrastructure for carriers and Internet providers. Undeterred by the Bust, Purcell steadily built Fibertech and created a fantastic outcome for his investors, selling to Lightower in 2015 for $1.9 billion. Purcell, among the most likable and competent executives in our industry, passed in 2017 at the young age of seventy-three.

spending spree. "It appears everyone in the room will eventually be bought by Zayo, and we will all work for Dan someday. I keep waiting for him to drive his dump truck to Rochester and share that green with me."

His eloquent ramble then segued to Zayo, which, according to his math, was unfortunately out of dry powder. He must have had a banker do some work because he cited the size of the prior round, the outlays on acquisitions, and our debt ratio: "Unfortunately, Caruso's wallet is now running on empty."

Instinctively, I reached into my back pocket. He couldn't see me, as he was turned to the audience. I pulled out my wallet and dropped it on the coffee table in front of us. When it was my turn, I declared, "Don't you worry, John. I have plenty of dry powder."

Several of my investors were in the audience, and I spotted them wincing at my wallet drop. The rest of the audience howled, as did Purcell when he pieced together what just happened.

LaPerch chimed in, "Dan that is only a wallet. You are going to need a wheelbarrow of wallets if you are interested in AboveNet."

As I dropped my wallet, my attorneys had been putting the final touches on an offer letter for Zayo to acquire AboveNet. I asked LaPerch to meet with me on Metro Connect's final day.

The next day, the final day of Metro Connect, we sat near the pool. LaPerch seemed unusually nervous. I got right to the point. "Bill, Zayo is making an offer to buy AboveNet for a healthy premium over its stock price."

To my surprise, LaPerch was ready with an answer. "Dan, I appreciate your interest and respect you for your industry leadership. However, my board would be uncertain on how to value your equity, and given the quality of AboveNet's assets relative to Zayo's, we would, of course, place a much higher multiple on our profitability than yours." He was right on the comparative quality of our respective assets.

I caught on to what was happening. LaPerch was ready for the conversation but was coached that we would propose a merger where some or all of Zayo's purchase price would be in the form of stock in the combined public company.

I answered, "We are offering 100 percent cash. We have the equity and debt fully committed. We can get through diligence and be ready to sign the deal by late March."

I dropped a manilla envelope on the table and slid it toward him. "The offer is in this envelope. Feel free to open it here, or you can take it with you to open it later." He sat quietly for a minute. He didn't have any questions. He politely said he would get back to me, leaving the envelope untouched.

Within an hour LaPerch texted me and asked for the envelope to be delivered to his hotel room.[14] When LaPerch opened it, he saw an all-cash offer backed by debt and equity commitments for $80 per share.

The next day we heard back from AboveNet's general counsel, requiring us to sign a confidentiality agreement that included a standstill provision as a condition to having further discussion. A standstill outlines specific restrictions on the potential acquirer's ability to take certain actions such as launching a hostile takeover bid or engaging in other aggressive actions. Requiring a standstill before any conversation was certainly an aggressive response.

We concluded we had little choice but to comply, so we agreed.

On March 4 AboveNet's banker, JPMorgan, provided us with a draft merger agreement and access to a data room.[15] They set a deadline for us to complete diligence, return a markup of the merger agreement, and submit our bid along with signed debt and equity commitments by March

---

14 Glenn Russo, who was part of my Zayo executive team, had the task of delivering the envelope to Bill LaPerch's hotel room. After sliding it under the door late in the evening, he panicked, wondering if he had the room number right. Russo texted me that evening, but I didn't see it until the next morning. He recalled the sleepless night he had.

15 Marco Caggiano and Fred Turpin from JPMorgan were AboveNet's bankers.

11, 2012. We would have one week to complete a transaction to buy the multibillion-dollar public company. This was unprecedented.

When we pushed back on the crazy tight time frame, JPMorgan's answer was, "If you don't want to proceed forward as a result, that choice is available to you." No doubt the response was scripted. But why?

We learned later that AboveNet was hours away from completing an LBO with Providence Equity as the sponsor. Providence reapproached LaPerch in December 2011, expressing a renewed interest to acquire AboveNet. Immediately after the 2011 holidays ended, a fast-track process to take AboveNet private picked up steam.

Jan: Providence offered a price range of $75 to $80 per share. Later in the month, they firmed up its price to $77 per share.

Feb: JPMorgan was reengaged to represent AboveNet. A counteroffer of $79 per share was provided to Providence. Providence increased its bid to $78 per share. AboveNet provided Providence with a draft merger agreement. Providence responded with a markup of the merger agreement. AboveNet responded that the markup was unacceptable and would allow the exclusivity agreement to expire on February 27.

On February 23 Providence received approval from its investment committee and expressed a willingness to show more flexibility on the terms.On February 29 Providence submitted a revised markup of the merger agreement, addressing many of AboveNet's concerns.

My meeting with LaPerch and delivery of our all-cash offer for $80 per share was March 1, the day immediately after they reached agreement with Providence! Though accidental, our timing could not have been better.

This wasn't good news for the AboveNet team. Providence would have retained the management team, giving them the opportunity to continue to run the platform. The recapitalized AboveNet would have dry powder to do acquisitions. Our offer, even though higher by $2 per share, was not a welcomed consolation prize. They were at the one-yard line of a management buyout, and I was interloping on LaPerch's parade.

As a public company, though, AboveNet couldn't just shoo us away.

AboveNet's management, including LaPerch, made themselves available to meet with us but only for a limited amount of time. We flew to New York and met the AboveNet team in a private hotel room. LaPerch had his GC, COO, and CFO with him. I had Ken desGarennes, along with a couple of our investors and investment bankers.

We also had a conference bridge open, where we had a nearby war room of twenty-five Zayo executives complimented by the deal teams of our investors. They poured through AboveNet's online data room and listened in as we asked our questions, some of which they prompted via texts to desGarennes.

Joe Valenti from Barclays, one of our investment bankers, recalled the diligence session. "The meeting tone began to shift to probing for reasons not to do the deal. You called a side bar and told everyone from our team to refocus, stressing the need to see the forest through the trees. 'Acquiring AboveNet is a slam dunk from an industrial logic and a value creation perspective. Let's shift our focus on how to get it to the finish line.' It was a great leadership moment"

LaPerch and his team politely answered our questions. As we approached lunch, LaPerch excused himself, saying, "I think my team can handle the rest."

We prepared a one-hundred-page diligence report for our investors. However, the premise was simple. We were paying $2.2 billion, which represented a tenfold presynergy Profitability multiple. Its Profitability margin was 44 percent, which we were confident we could get to at least 55 percent. AboveNet had 20,600 miles of the absolute best metro fiber assets serving four thousand customer buildings including most datacenters in the most important geographies. When combined with ours, its assets would be even more valuable. The icing on the cake was AboveNet's stellar customer base.

The lengthy diligence readout was not needed to be enamored with this deal. However, the comprehensive deck paved the way for rapid approval by our investors.

We didn't quite realize we were in a bidding war, though we knew something was amiss, which made us thoughtful in each move. We hadn't sniffed out that AboveNet was so close to a management buyout backed by Providence.

On March 12 Zayo submitted a markup to AboveNet's proposed merger agreement with a confirmation of its $80 per share price. The next day AboveNet's leadership reported to their board that Providence was making progress on reducing the number of open issues in the purchase agreement, and except for price, Providence's terms were more favorable than those of Zayo. To force Zayo's hand, AboveNet provided us a one-day deadline to respond to whether Zayo would accept AboveNet's revised agreement and, if so, at what price. Zayo's advisors pushed back, saying they would not respond unless AboveNet provided clarification on price and other value-related terms.

On the following day, March 14, AboveNet's attorneys resolved most of the remaining outstanding issues with Providence and reported to their board that the agreement was nearly ready to be signed. The board instructed JPMorgan to communicate that Zayo must provide its best price and terms and that if Zayo failed to respond by 7:00 p.m., AboveNet would cease negotiations.

We still were without a clue that we were in a real-time bidding war.

At 7:30 p.m., Zayo provided a revised markup and increased the price to $81 per share. The message sent by increasing the price by $1 was that we had more upward flexibility; the uncertainty about how much we'd be willing to pay prevented AboveNet from being able to conclude that another deal was superior.

AboveNet's lawyers continued to negotiate with Providence in an attempt to resolve the remaining issues. A final draft of the purchase agreement would be ready later that evening. The board instructed JPMorgan to inform Providence of the existence of an unsolicited offer at a price of $81 per share. Note that AboveNet still was silent with us about another bid; by sharing our bid price, they were clearly steering the deal to Providence.

AboveNet also reported to the board that Zayo's markup resolved most of the issues but still had less favorable go-shop and breakup fee provisions. The board instructed JPMorgan to express disappointment in Zayo's proposed terms.

We sensed AboveNet was looking for an excuse to kill our deal and knew they were running out of reasons to do so. We concluded we needed to remove this final objection. On March 15 we conceded the go-shop and breakup fee provisions. This prompted JPMorgan to plead with Providence that it needed to increase its bid, or it would lose the deal to another bidder.

Providence increased its proposed price to $78.50 per share and communicated that this was its best and final offer to AboveNet. We were told later that Providence simply didn't believe that another bidder was at the finish line with a higher bid. They feared they were being tricked into bidding against themselves.

AboveNet communicated to Zayo that it would enter a transaction with Zayo at a sale price of $84 per share and that it would require a response by the next day, March 16, at 5:00 p.m.

On March 16 Providence signed the merger agreement and set a deadline of March 18 for AboveNet to countersign. Unbeknownst to us, any misstep by Zayo would have allowed AboveNet to complete the management buyout with Providence.

Just before the March 16 at 5:00 p.m. deadline, Zayo informed AboveNet that we were prepared to accept the $84 price and were ready to execute the merger agreement. We were the highest bidder, and AboveNet, whether its management team liked it, was obligated to accept our deal. On March 18 Zayo and AboveNet executed the merger agreement.

On the morning of March 19, Zayo's acquisition of AboveNet was announced, just seventeen days after I first delivered the unwelcomed envelope to LaPerch at the Eden Roc pool in South Beach, Miami. I suspect this is a record pace for a multibillion-dollar public company hostile acquisition.

In 2024, I learned from LaPerch that Craig McCaw, whose original investment in AboveNet traced back to 1997, remained a shareholder through its sale to Zayo. According to LaPerch, McCaw made a thirteen-times return on this fifteen-year investment.

In the two years after AboveNet, we completed several more acquisitions,[16] including multiple in Europe that complemented the AboveNet assets.[17] Each of the acquisitions bolstered the depth and reach of Zayo's increasingly impressive collection of fiber assets.

With the success of our acquisitional hunt, we were ready to enter prime time. Those who chose sports as a career aspire to make the NFL or NBA. Career politicians aim for a governorship, Congress, or the White House. Entertainers strive for an Oscar or a Grammy. For CEOs, the equivalent is taking a company public on the NYSE. It was time to reward our early investors with liquidity. Management, too, was ready for a payday. It was IPO time for Zayo.

---

16 A favorite during this period was Fiberlink LLC, which had a fiber and conduit route that connected Denver to Chicago. This link was critical as we slowly assembled a nationwide fiber network. Chris Jensen, who ran Fiberlink, became a good friend.

17 We purchased GEO, a fiber network that covered the UK and Ireland; NEO Telecoms in France; and Viatel Technology Group that had a fiber footprint tying together the major cities in western Europe.

We had our IPO kickoff meeting in early 2014. Several dozen bankers, attorneys, accountants, and Zayo investors filled the room.[18] First on the agenda was for me to give the Zayo story, noting that most in the room were familiar with it.

To break the ice, I told the following story.

> My mom belongs to a church group. The first Tuesday of every month, they meet to share their life updates with one another.
>
> It was my mom's turn. "What is new in your life, Penny?" they asked.
>
> "Well, my son is doing an ippo (pronounced like 'hippo' without the *H*)," she proudly answered.
>
> "What's an ippo?" her elderly lady friend asked.
>
> "I don't know, but it sure sounds important," was her answer. The church ladies nodded their approval.

This set the tone for the meeting and reflected our confidence that the Zayo IPO would be smooth sailing.

The IPO road show is an eleven-day process that starts on a Monday and culminates, if successful, with the IPO being finalized on Thursday evening of the following week. Before the road show is launched, a meeting occurs after the markets close on the preceding Friday to assess the conditions. Is investor interest high? Are the markets receptive to an IPO? Depending on the assessment, a go-no-go decision is made.

During our go-no-go call on Friday, October 3, 2014, our bankers shared concern. "Markets are showing signs of instability. Some pundits are fearful the IPO markets might be closing. Other IPOs are being delayed until the markets settle down." Despite these warnings, their overall recommendation was "Demand for Zayo is strong. Though the choppy markets are concerning, our recommendation is to go."

---

18 Our IPO investment banking team was led by Jason Rowe of Goldman Sachs, Jon Yourkoski of Morgan Stanley, and Joseph "Joe" Valenti of Barclays.

Zayo's board agreed. We were a go. We flew to NYC over the weekend to launch the IPO road show on Monday.

Over the next seven days, we covered NYC twice, Boston, London, Chicago, and San Francisco. As I had my morning cappuccino in the city by the Bay, I was alerted that MoffettNathanson initiated coverage of Zayo. Its provocative headline was "Dark Fiber versus Towers: Batman versus Superman, or You're No Jack Kennedy," written by its talented young analyst, Nick Del Deo.

The business model of Towers—led by Crown Castle, American Tower, and SBA—was loved by investors for a good reason. Del Deo wrote, "Towers are almost universally viewed as the best business model in telecom, if not one of the greatest business models of all time." The Tower model was a steady revenue growth, extraordinarily sticky customers, and strong cash flow. Crown Castle, American Tower, and SBA enjoyed valuations that were twenty-five times profitability, which was double the eight to twelve times typically ascribed to Bandwidth companies.

Was Dark Fiber similar in strength to Towers, like Batman to Superman? Or was the comparison unwarranted, as Sen. Lloyd Bentsen famously told Sen. Dan Quayle that he was no Jack Kennedy in the 1988 vice president debate?

Del Deo's conclusion was "close but no cigar." Dark fiber, in isolation, had the same attributes and was perhaps deserving of the Superman comparison. However, the business model of Dark Fiber at scale was never done. Zayo, though the closest to a Dark Fiber pure play, had significant other Bandwidth revenue streams. Bandwidth, in general, certainly was no Jack Kennedy.

In any case this was positive for our IPO. Even being close but no cigar to business models that traded at twenty-five times profitability was a strong endorsement. With newfound optimism, we completed the day in the Bay and headed to Los Angeles.

———————

When we made the go-no-go decision, the S&P 500 index was nearly 2,000. Each day of the road show, the market environment was worsening. When we started our day in Los Angeles on October 15, one day before the IPO was to price, the S&P closed at 1,800, down 10 percent. The market was in no mood for an IPO.

Other IPOs were being pulled. The feedback our bankers received from Zayo presentations was positive, but investors were hesitant to commit to buying shares in the shaky market environment. The order book, which is the commitment from investors to buy shares offered in the IPO, was alarmingly low.

That evening, in Los Angeles, we began talking about the real possibility that we would need to pull our IPO. The bankers were noncommittal. The order book was weak, but verbal interest was encouraging. The general theme was, "Investors are focused on the losses in their current portfolios, not IPOs—but they like Zayo nonetheless."

The plan for Thursday's IPO pricing day was to drop down in Kansas City to visit American Century Investments and Waddell and Reed, and then continue to NYC. We'd arrive shortly after the markets would close and spend the next two hours with our bankers and board to execute, hopefully, the IPO.

I pushed back on the Kansas City stop. "Shouldn't we head straight to NYC? We need to drum up interest in the IPO."

The bankers explained, "That's the job of the sales teams at Goldman, Morgan Stanley, and the rest of the syndicate. Your focus needs to be on Kansas City. Strong interest from one of the big Kansas City Funds could make the book."

As we headed to the Los Angeles airport, we read the front page of *USA Today*: "Oil, Europe, Ebola Spook the Markets." The flanking articles were "Signs of Trouble in Ukraine" and "Obama Seeks Fast Acting 'Swat Team' from CDC."

Were we just going through the motions at this point? Was the IPO dead on arrival?

The bankers played the role of cheerleaders. "It's premature to give up. Let's nail it in Kansas City and hope the markets stabilize today."

On the plane from Kansas City to New York, we crossed our fingers and knocked on wood, which seemed to work as the markets responded with a modest bounce back.

We arrived in NYC and beelined to Morgan Stanley's headquarters. I insisted we relocate from the quiet private client floor to the trading floor. "If we have to pull this IPO, I at least want to be surrounded by high-energy traders." We relocated. Much better.

At about 4:00 p.m., our bankers began giving us updates on the book. Orders were light, but interest remained. Investors were using the market softness to push back on price. The $21–$24 range we signaled during the road show was not going to happen. However, if we were willing to show flexibility on price, we had a chance.

"$20?" we asked. "Maybe" was the banker's reply.

Every twenty minutes, they provided updates from the sales teams at each of the banks.

"Interest is sufficient at $15–$16 per share. We don't yet have enough interest at prices above $16."

Our board agreed with management. "$16 is too low. Tell your sales teams that the price needs to be $20, or we will pull the IPO." If we pulled the IPO, that would be costly to the bankers. Whether we meant it, the threat motivated them.

Canceling the IPO would also be devastating to management. We would have to wait several months and then start the process all over again. Moreover, the markets would ask, "Did they pull it because investors didn't like Zayo? Or was it really just the market environment?"

More than an hour passed. The book strengthened. "Good thing we didn't listen to you," one of the bankers joked with me. "We just got a big order from Kansas City at $20 per share."

The bankers were now signaling we could get the IPO done but only if we accepted $19 per share. "There just isn't enough demand at $20."

In hindsight $19 versus $22–$25 wasn't a big driver of value. We were selling about $600 million of stock, which would be about 10 percent of our total Market Value of about $6 billion. The primary value resided in the 90 percent we didn't sell.

Nonetheless, the $19 felt like we were selling ourselves short. Emotions were high.

Some of our investors favored pulling the IPO. One selfishly told us, "We are not selling at that price. The rest of you can do what you want." Their stubborn and self-centered position poked us in the eye. We pushed back, yet they wouldn't budge.

At about 5:30 p.m., Zayo's CFO, Ken desGarennes, had his best Zayo moment. His financial team was exhausted from the IPO preparation. The road show was grueling for the two of us. The entirety of Zayo had been working twelve hours a day since we founded the company seven years ago. He spoke candidly to the investors: "The board can pull the IPO if it wants. If you do, I hope you have better luck next time with your next CFO."

Now that was laying down the gauntlet. I nodded to him in appreciation and motioned for the bankers to come into the room. "How much time do we have before we make the final decision?"

Surprisingly, the bankers from Morgan Stanley, Goldman, and Barclays weren't sure. They huddled outside the door for a few minutes. They walked back in and pointed to the clock. "It's 5:45. If you don't pull it by 6:00 EST, you will be a public company on the NYSE whether you sell shares or not."

We were all shocked. How could it be that we didn't talk through this scenario in advance?

"What's the minimum amount we can sell?" we asked. At a low price, we could mitigate dilution by reducing the number of shares. The trade-off is that Zayo needed enough shares in the public market, which is referred to as float, to allow sufficient liquidity for healthy trading activity.

"You can reduce the volume by about 20 percent. If you need to reduce it further, we would recommend you pull the IPO instead."

"How's the book looking?" we asked, knowing that updates were continuing to come in from the sales teams.

"We can get it done at $19 per share, especially if you reduce the shares by 20 percent."

DesGarennes's speech made us all realize that pulling the IPO because of $19 was an overreaction. We reduced the volume of shares being sold from the planned twenty-nine million to twenty-one. To accommodate investors who didn't want to sell at the reduced price, we shifted to a majority of the shares being primary versus secondary.[19]

At 5:58 p.m., we gave the bankers the thumbs-up to fill the orders. We raised $400 million, with $100 million distributed to investors and $300 million staying with Zayo.

We rang the bell to open the NYSE the next morning. We watched as the first trade was made at just over $21 a share on the way to a closing price of $22. We met a much bigger team of Zayo colleagues, investors, and family that evening at Buddakan in NYC's Meatpacking District. The backdrop of the markets in turmoil made Zayo's IPO that much more special.

───────────

In the two years following the IPO, most of my original executive team moved on. They all made millions on their Zayo equity, and most were entering the primes of their careers. The team was understandably tired, as the first decade of Zayo was a work-around-the-clock experience. It didn't help that most lived in Colorado, where cycling, hiking, and skiing were their alternatives to another grinding year at Zayo.

In the aftermath of the IPO, we continued the consolidation. We bought AT&T Canada, which had been rebranded as Allstream and

───────────

19 Primary are newly created shares that are sold by the company, where the company retains the cash. Secondary are shares sold by existing shareholders of the company, where the cash is to the benefit of the selling shareholder.

provided us fiber assets that stretched from Vancouver through Calgary, Edmonton, Toronto, Quebec City, Montreal to Halifax.[20] We acquired Spread Networks, made famous from Michael Lewis's book *Flash Boys: A Wall Street Revolt*.[21] We purchased Electric Lightwave (ELI) in the Pacific Northwest, which provided Zayo with ten thousand miles of fiber in thirty-five markets that both densified existing footprints and added several new markets.[22] We even acquired a Vancouver Canada Dark Fiber provider named Optic Zoo.

The combination of 360networks, Allstream, and ELI—on top of the deep big city networks of AboveNet and Zayo's other acquisitions—made Zayo the only legitimate nationwide rival to Level 3. However, as each year passed, execution of our plan became more difficult.

Zayo's acquisition of AboveNet and subsequent IPO drew the attention of large institutional investors. They realized Zayo's $1 billion of invested equity was now valued at over $6 billion. They wanted a bite of our Bandwidth pie.

"We are hunting for a fiber platform, similar to Zayo, that we can use to participate in the consolidation of the space" was the late-to-the-party banter of dozens of institutional investors. They locked arms with bankers, lenders, and consultants, convinced they could do what Zayo did, but far better. Acquisition multiples rose, while commercial deals became more contested.

I had opened Pandora's box!

---

20 To learn more about Allstream, see appendix 3, "More Bandwidth Stories."

21 To learn more about Spread Networks, see appendix 3, "More Bandwidth Stories."

22 Electric Lightwave was the combination of multiple companies in the Pacific Northwest and the North: Integra Telecom founded by Dudley Slater, the original Electric Lightwave launched by John Warta, Minneapolis-based Eschelon, and opticAccess. After Slater's thirteen-year reign ended abruptly in 2011, a revolving CEO door included Global Crossing's Tom Casey, Level 3's Kevin O'Hara, TelCove's Robert "Bob" Guth, and tw telecom's Marc Willency. Zayo paid $1.4 billion for Electric Lightwave.

A subcategory of these growth investors was infrastructure funds, whose source of capital was sovereign wealth and pension funds. These infrastructure investors swarmed aggressively on the opportunity and viewed that their advantage was a low cost of capital, as the ridiculously low-interest environment of the 2010s suggested modest single-digit equity returns would be sufficient for their investors. This directly led to infrastructure funds being able to justify higher values for Bandwidth Infrastructure companies.

Over the next several years, a dozen growth equity and infrastructure funds acquired ownership stakes in fiber companies throughout the Americas, Europe, and Southeast Asia.[23] Profitability multiples ranged from the midteens to midtwenties.

Zayo's IPO road show and the provocative "Superman versus Batman" analyst report drew the attention of the tower industry. If Dark Fiber had business model attributes akin to tower, perhaps the tower companies should enter the market. Pandora again!

Tower companies traded at twenty to thirty times Profitability and, given the strength of tower business models, received favorable terms on

---

23 In 2014 Ridgemont Equity Partners funded Cross River Fiber. In 2015 Antin Infrastructure Partners bought EuroFiber. In 2016 Partners Group acquired Axia NetMedia, which operated networks in Canada, Massachusetts, and France. In 2017 KKR bought Australian-based Vocus and bought 40 percent of Telefonica's fiber spinout, Telxius. Tiger Infrastructure Partners funded Crosslake Fibre.

In 2018 TPG Capital acquired RCN cable and rebranded it as Astound. Antin purchased FirstLight from Oak Hill. Goldman Sachs's West Street Infrastructure Partners partnered with Antin to take the UK's CityFibre private. AMP Capital from Australia bought Everstream, a fiber operator in the Midwest. EQT funded a Europe rollup of IP-Only, GlobalConnect, and Broadnet. Antin bought Ufinet Spain. Infracapital purchased 50 percent of SSE Enterprise Telecoms in the UK.

In 2019 Macquarie Infrastructure Partners acquired Bluebird Network. EQT purchased inexio in Germany. Grain Management rolled up Hunter Communications, Ritter Communications, and Summit Broadband.

There were numerous transactions in 2020. Stonepeak also purchased Level 3's South American business, branding it as Cirion. Brookfield Infrastructure prevailed in a bidding war with Macquarie to acquire Cincinnati Bell. Digital Colony acquired a majority stake in Beanfield. SDC Capital Partners acquired FatBeam. Swedish-based Polhem Infra purchased Telia Carrier.

debt. Would these advantages result in tower companies being able and willing to pay a multiple of twenty times or more for fiber properties? If so, Zayo would be priced out of future acquisitions. The higher multiples also might lead to tower companies offering commercial terms that would be difficult for Zayo to match.

Tower companies were set up as real estate investment trusts (REIT), which provided them tax advantages over typical corporate structures. REITs are not subject to corporate income tax on the distributed earnings. If fiber and wireless infrastructure qualified as real estate under the REIT regulatory framework, tower companies could use REIT status to underscore their advantage over Zayo and other Bandwidth companies.

The tower industry in the United States is dominated by three companies: American Tower Corporation, Crown Castle, and SBA. Two of them—American Tower and SBA—decided to stick to their knitting. Why roll the dice on an unproven business model when your core business, to quote "Superman versus Batman," is "the best business model in telecom, if not one of the greatest business models of all time"?

The third, Crown, reached the opposite conclusion. In 2014 Crown received a favorable private letter ruling from the IRS confirming that fiber networks would be permitted revenue in its REIT structure. Emboldened, Crown outlaid over $11 billion between 2014 and 2020 toward acquiring fiber companies, including several that I was targeting.[24] They paid substantially more than what Zayo could rationalize.

In Zayo's second-quarter 2015 earnings call, an analyst asked me, "With Crown Castle entering the game and Profitability multiples rising ever higher, has Zayo's rollup finally run into a steep enough hill to slow it down?"

My rambling answer conveyed uncertainty.

---

24 The biggest company acquired by Crown Castle was Lightower, whose CEO was the very capable Rob Shanahan. See appendix 3, "More Bandwidth Stories," to learn more about Lightower. Crown also acquired FPL Fibernet run by CEO Carmen Perez, as well as Dark Fiber provider Sunyses and Wilcon, a data center and fiber provider in Los Angeles. I tried multiple times to acquire each of these companies.

It's good to see . . . these [fiber] properties are worth a whole bunch of money.

We're always asking ourselves where that next set of opportunities are, not looking backwards. . . . The one thing that won't change is that we're focused on an infrastructure strategy. Infrastructure is what we're all about. Infrastructure is how we think about the world. We love investing in assets that have decades worth of life to them. That's not going to change.

I expect us to continue to be acquisitive. That's our nature, we're good at it, and I think you'll see us continue to be effective.

Was Crown's rollup good or bad news for Zayo? Arguments could be made on both sides. Bad news was that Zayo would no longer be able to roll up fiber properties at attractive prices. Any fiber property that matched up well with Crown's plan would be priced beyond Zayo's reach.

The good news was that Crown consolidated many of Zayo's pesky rivals and, better, was repurposing these companies to be primarily focused on mobile infrastructure. Crown would be increasingly disinterested and disadvantaged in the many other ways to leverage fiber. For example, financial services, healthcare, and tech firms would be reluctant to use a tower company for their mission-critical service. Verizon and AT&T would be reluctant customers, as Crown was now their Bandwidth competitor.

Most important, Crown was a tower company. Could they really effectively operate a Bandwidth Infrastructure business? Probably not was my assessment.

Crown's fiber foray had one other implication: was Zayo severely undervalued? Crown was willing to spend $11 billion for a collection of assets that were 35 percent of what Zayo assembled. Does this mean Zayo was worth over $20 billion, instead of the $12 billion implied by our stock price? Maybe it was time to find out.

In 2008 I heeded Buffett's advice by being "greedy when others [were] fearful." In 2017, with the tables turned, I needed to comply with the other half of Buffett's wisdom and "be fearful when others are greedy."

Zayo's window was closing. Our opportunity to generate outsize returns was shuttered by investors who were comfortable with mediocre returns.

In the mid-2010s, I met investor Marc C. Ganzi for the first time at a small investor conference hosted by a boutique investment bank at the Deer Valley Resort in Utah. When I introduced myself to him, he asked with a beaming smile, "You know what investors call me?" I shrugged. "They call me the Dan Caruso of Towers." He paused, smirked, and then added, "And you are known as the Marc Ganzi of Fiber."

One of my investors used a baseball reference when I told him of this conversation: "Yeah, except you started out in the grounds crew. Ganzi was born on third base."

Ganzi is the grandson of Walter J. Ganzi Sr., an Italian immigrant who co-founded the Palm Restaurant Group. When Ganzi Sr. passed, his family had a public legal fight over ownership control of Palm. I suspect this took a toll on Ganzi.

Ganzi married Melissa Potamkin, the granddaughter of Victor Potamkin, who built a $1 billion-a-year automotive empire. In 2012 the *Palm Beach Post* dubbed them the "royal couple of polo" for their ownership of two separate polo teams, as well as Wellington's Grand Champions Polo Club. They also own Aspen Valley Polo Club.

Ganzi's father, Ted Ganzi, was a successful real estate developer, which influenced Ganzi down the real estate path. After receiving a business degree from Wharton, Ganzi bought a distressed real estate company, which led, eventually, to his focus on the real estate of towers.

Ganzi co-founded Global Tower Partners in 2003 and sold it for $4.8 billion to American Tower in 2013. The timing of the exit coincided with my Deer Valley encounter with Ganzi. Ganzi co-founded Digital Bridge Holdings, LLC (DigitalBridge) in 2013 to focus on the broader Communications Infrastructure sector, including Bandwidth Infrastructure. DigitalBridge merged into the public company Colony Capital in July 2019.[25] Ganzi became the CEO of Colony Capital in 2020 when its founder, Thomas J. "Tom" Barrack, got into legal trouble.[26]

Returning to my budding relationship with Ganzi, he invited me to his home in Aspen, Colorado, in July 2018. We sat on his patio, which overlooked the Roaring Fork River, and shared a bottle of wine. Ganzi expressed interest in partnering with me to take Zayo private. As a CEO of a public company, I responded appropriately. Zayo would do what is best for its shareholders, and my involvement should not be viewed as a condition of a transaction.

I was open to a sale and would have considered staying on as CEO. However, I knew Ganzi and I were cut from the same control-freak cloth. At this point in our respective careers, we both needed to be the top dog. I couldn't imagine a scenario where I'd work for him. And his personality was such that he would not be a hands-off owner who stayed in his investor lane, especially considering Zayo would be his biggest investment. He was Jerry Jones, not Robert Kraft, and I was Mike Ditka, not Dick Jauron.

The financial services firm Cowen and Company hosted an annual Communications Infrastructure Summit in Boulder.[27] The combination of Zayo's headquarters being Boulder and my brand as the industry's outspoken leader was what led them to select Boulder in the mid-2000s. This conference remains a major tour stop for the players in our industry.

---

25 When DigitalBridge merged into Colony Capital in 2019, the stock traded at $22 per share. After dipping below $10 in early 2023, it bounced to $12 a share in August 2024.

26 The billionaire Tom Barrack, an early and avid supporter of Donald Trump, was indicted in 2021 on nine charges stemming from his alleged lobbying for the United Arab Emirates as an unregistered foreign agent. In November 2022 he was found not guilty on all charges.

27 The Boulder gathering is now branded the TD Cowen Communications Infrastructure Summit.

Ganzi continued his overture in a one-on-one discussion at my house during the 2018 Summit. Though no price was offered, he reiterated his desire to partner with me to take Zayo private. I assured him we would respond appropriately to any offer while again stressing the importance of my representing the interests of Zayo's public shareholders, which wouldn't be possible if I were to partner with him.

In October Ganzi called me again. He said he was working with other investors, and they were soon to make an offer.

Private Equity firm Stonepeak, with $50 billion to invest, also was showing interest.[28] Brian McMullen, Stonepeak's head of digital infrastructure, was the lead partner in its investment in a specialized fiber company.[29] Ganzi was also an investor, with Ganzi and McMullen anchoring the board. The strained relationship between the two of them was said to be entertaining.

In October 2018 McMullen visited me in Boulder, which led to a meeting in NYC a week later. McMullen expressed interest in acquiring Zayo and hoped to have an offer submitted within two weeks.

Ganzi and McMullen weren't the only ones interested in acquiring Zayo. EQT Partners, a Swedish-headquartered infrastructure investor with over $200 billion under management, was also salivating. The primary investors that backed EQT were European pension and sovereign wealth funds, which were at the time earning near-zero interest rate returns. Infrastructure was seen as moderate risk. Therefore, they surmised a mid-to-high, single-digit return would be a prudent risk-return

---

28 Stonepeak led an investment in euNetworks, backing euNetworks's CEO and my former colleague, Brady Rafuse. Columbia Capital bought a controlling interest in euNetworks soon after we formed Zayo with a strategy that mimicked Zayo. I was an investor and assisted in bringing Rafuse on board as CEO. Under Rafuse's leadership, the investment performed exceptionally well.

Stonepeak also partnered with Columbia Capital to fund Cologix, backing another Level 3 colleague of mine, Grant Van Rooyen. It, too, had a great outcome for investors.

29 Stonepeak partnered with DigitalBridge to purchase Extenet Systems, a Chicago company focused on fiber infrastructure for wireless carriers. The original lead investor in Extenet was Columbia Capital.

trade-off. This cost-of-capital advantage led to purchasing assets at prices that were materially higher than the norm.

In the early 2010s, EQT named Jan Vesely to lead telecommunications infrastructure investments in the Americas. Vesely led EQT to substantially outbid others in acquiring two southeastern fiber providers for a combined $1.45 billion in early 2018.[30] Vesely, who was in his thirties, had a decade of post-business school experience that was entirely in investment banking. Despite having limited experience in operating businesses of scale, his approach as the investment owner was to be very hands on. If Ganzi was the Dallas Cowboys' Jerry Jones, Vesely was the Las Vegas Raiders' Al Davis. Later in 2018 Vesely turned his attention to Zayo.

———

With interest in Zayo heating up, I got prepared. The responsibility of a public company CEO is to maximize value for the shareholders. The CEO is personally liable for behaving in a manner consistent with this responsibility.[31] If fraud is involved, the CEO is criminally liable.

Every material public company acquisition was followed by ambulance-chasing attorneys organizing lawsuits aimed at the CEO and board. The lawyers will look for anything that can be spun to look suspicious and claim that the CEO behaved in a manner that did not maximize the outcome for the shareholders. An acquisition price for Zayo would be unusually large—the second-biggest LBO since the 2008 financial crisis.

---

30 Lumos and Spirit were merged together and rebranded as Segra.

31 Insurance can be used to mitigate the financial exposure. However, extreme situations could result in liability that is not covered by insurance.

It would draw a lot of attention from unscrupulous lawyers who sought to profit from any transgression.[32]

To protect from such claims, best practices are as follows. The CEO surrounds himself with extraordinary legal and professional advisors. Every step and communication are documented and shared with the attorneys. The CEO avoids conversations that do not include another representative of his company to ensure there is a witness to attest to what was discussed. Before taking any action, advice from lawyers and bankers is sought and documented. The board is genuinely engaged and formally directs each major decision.[33] For a deal doer like me, who thrives on acting autonomously and unpredictably during the negotiation dance, this required an adjustment.[34]

On November 6, 2018, Zayo had its quarterly board meeting. During the meeting, Ganzi called to tell me he was working on providing a written, fully financed proposal at a price between $41 and

---

32 A public company is required to file a "Proxy Statement Pursuant to Section 14(a) of the Securities Exchange Act of 1934," also known as a Schedule 14A. In this proxy the details of all material events that led to the transaction are shared by the company with its shareholders. This Schedule 14A would be scrutinized by corporate litigation attorneys as they search for claims that the CEO screwed something up.

33 I am grateful for my capable Zayo deal teams, which included CFO Matt Steinfort, General Counsel Mike Mooney, Corporate Development Lead Rachel Stack, Investor Relations VP Brad Korch, Jason Rowe from Goldman Sachs, and the dynamic duo Marco Caggiano and Fred Turpin from JPMorgan. Zayo's external attorney, Kenton King, partner at Skadden, Arps, Slate, Meagher & Flom LLP, represented us and was a godsend in keeping us focused and on a safe track. Zayo's board, especially Don Gips, Linda Rottenberg, Scott Drake, and Yancey Spruill, was excellent throughout the process.

34 As I tell how the acquisition played out, I will exclude the specific involvement of the others on my team. This is not to diminish their important roles and contributions or overinflate mine. It's only meant to improve the storytelling and firsthand authenticity of the events.

As is clear in the Schedule 14A, all my actions were the result of direction from Zayo's board, and at least one member of my executive team joined me in each meeting and call. We abided by these best practices, and I was thankful we did. The transaction held up to scrutiny, and though a big class action lawsuit was filed as expected, all but one count was dropped at the onset. The single remaining count was trivial and led to a settlement that was immaterial for the new owners.

$42.50 per share, which I scribbled on the whiteboard while my directors watched. At the end of the board meeting, Phil Canfield, who led Zayo's Series C round for GTCR, resigned as lead director.[35] After his resignation was official and after he heard of Ganzi's price range, Canfield pulled me aside to tell me he was considering participating in a potential acquisition of Zayo. That seemed unusual, but I told him I'd be supportive.

Michael Choe of Charlesbank, who led Zayo's Series B round, called me later on November 7 with the message that he was hearing chatter in the markets about Zayo and assured me that he had a favorable view of the company. I don't know if his message was a comforting reassurance as a friend or if he was fishing for a reaction.

Zayo announced its third-quarter financial results, which were slightly lower than expected. We also announced that we were evaluating a plan to spin Zayo into two companies—one focused on infrastructure and one focused on services. The pressure was getting to me, and my rushed announcement of a possible spin was clearly a mistake. The share price dropped from $30 to $23 on November 8. Not fun!

On Friday Stonepeak's McMullen contacted me to reaffirm his firm's interest in acquiring Zayo. During the following week, we learned Canfield and Choe were teaming with McMullen. On Friday, November 16, we received a written offer for $33 per share from a consortium consisting of heavyweights Blackstone, I Squared, Kohlberg Kravis Roberts (KKR), and Stonepeak partnered with Zayo's biggest private investors, Charlesbank and GTCR, with the latter two partnered with Stonepeak as the consortium leaders.[36]

On Sunday I communicated to McMullen that the offer was insufficient. The next day Bloomberg News published an article that Zayo was

---

35 Don Gips replaced Phil Canfield as lead director, and Yancey Spruill was appointed to replace Canfield on the board.

36 The lead for Blackstone was Greg Blank; Gautam Bhandari represented I Squared; and Brandon A. Freiman was on point for KKR.

in play. Clearly, the consortium was leaking to the press to put pressure on us. This seemed cheesy.

On Tuesday they revised their offer to $34.50 per share. When McMullen asked for feedback a few days later, I told him this offer wasn't meaningfully different from the prior offer and therefore still too low to engage. They went quiet for a week. I supposed they were going to see if anyone on our side would reach out to them. We didn't.

In the first week of December, we heard from two activist stockholders. They encouraged us to consider selling Zayo in the low to mid-$30 range. The consortium's pressure tactics now included leaking to the activist community, as was clear when the consortium timed its delivery of a revised $35 per share offer later that same week. Two days later I responded that although we would not sell Zayo at that price, we would support them in the due diligence process, including hosting them in Boulder.

On December 16 McMullen requested and was granted permission to include five former Zayo executives in the diligence process, including ex-CFO Ken desGarennes and my former co-COOs Chris Morley and Matt Erickson. Alongside Choe and Canfield, a majority of my most important Zayo partners were colluding to buy Zayo.

Zayo was likely the most lucrative investment ever made by Choe from Charlesbank and Canfield from GTCR. Both personally made millions of dollars from their involvement with Zayo. The ex-Zayo executives pocketed millions of dollars, in some cases in excess of ten million. Very few, if any, had net worths of more than a million before Zayo, and none of them had yet established themselves as industry executives. In no small part, they owed some credit for their success to me.

The consortium, as will become clear as I share the play-by-play, appeared to use hostile tactics to force us to sell Zayo to them at bait-and-switch prices. This put my former investors and executives, most of whom were close friends, in an adversarial position with me. Did my friends start out with the right intentions—to help me exit successfully from the public markets? Or was greed their motivation, seeing another potential run at wealth creation through Zayo, regardless of whether this would be abrasive to me? Indeed, as the transaction hostility intensified, it was unclear whether my former investors and executives were friends or foes.

I felt embarrassed. How would those around me—my board, Zayo employees, and the entire industry—view this? It certainly seemed to me an act of disloyalty. I took the situation personally. I was angry.

I also was disadvantaged, with the fight lining up to be a 10:1 battle in their favor. And they had the secret copy of my negotiating tricks play-book. Yet they underestimated me. I knew that Zayo was successful because of me, not despite me. Over my dead body were my former executives and investors going to force me to sell Zayo on the cheap. A 10:1 battle was, in this case, an even match up. As Michael Buffer bellows before a championship boxing match, "Let's get ready to rumble!"

Aesop told us about the Sun and the Wind. One day the Sun and the Wind had an argument about who was stronger. To settle the dispute, they decided to have a contest to see who could make a passing traveler take off his coat.

The Wind went first, blowing with all its might to send strong gusts of cold air at the traveler. However, the harder the Wind blew, the tighter the traveler clutched his coat to his body. The Wind's forceful efforts were in vain.

Next, it was the Sun's turn, who smiled warmly and began to shine brightly. As the Sun's rays gently touched the traveler, now feeling warm and comfortable, he willingly removed his coat and continued his journey.

The Stonepeak consortium played the role of the Wind, and I was the traveler. I needed to buy time and find an investor to be the Sun.

Wind requested to have its attorneys work with ours to work on the merger agreement. We denied this request, emphasizing the need to first reach agreement on price. By not working with them on the merger agreement, we would make it difficult for them to put legal and activist pressure on us to accept an offer that we viewed as too low. Without a negotiated merger agreement, Stonepeak consortium's pressure tactics would be thwarted.

Wind requested permission to talk to sixty-three potential co-investors. We denied this over-the-top request, knowing they would lock up each investor with standstill agreements to further reduce the likelihood of a competitive bid. Nasty.

On the same day, DigitalBridge contacted us that they were considering a joint bid with EQT at a price of $36 to $38 per share. Perhaps the two of them together would emerge as the Sun. Encouraged by the prospect of a bidding war, we hosted Sun in Boulder on December 20.

Wind must have caught wind that DigitalBridge and EQT were teaming up. On December 28, McMullen, Canfield, and Choe called me to again propose that the lawyers begin work on a merger agreement. I reiterated the need to first reach agreement on key terms, including price. They were frustrated.

Though I didn't disclose it to Wind, I directed our attorneys to complete a markup of the merger agreement we received from Wind. If we agreed on price and other key terms, I wanted to make sure we were in a position to move quickly. I also sought to have an agreement ready for Sun to speed them up.

On January 9 we hosted Wind for an in-person diligence meeting in Boulder. I felt very awkward with my former colleagues peppering me with questions. The tables felt like they were turned on me. Undoubtedly, they had a large group of former Zayo employees on the other end of a conference bridge, repeating the same tactic that I used in the AboveNet transaction.

On January 12 McMullen requested permission to add DigitalBridge to Wind to ensure they had sufficient levels of funding. Luring DigitalBridge into the Wind consortium would cut me off at the knees. When McMullen told me that Blackstone already initiated a conversation with

DigitalBridge, I cited this as a nondisclosure violation. We denied Wind's request in hopes of preserving the competitive tension. However, damage was already done, as DigitalBridge was now doubting the pursuit of Zayo was worth its time.

I had a call the next day to deliver the no message to McMullen. I harped on their violation of the nondisclosure by reaching out to DigitalBridge without our permission. I relayed the message that we already provided Wind with sufficient access to our data. Now it was time for them to bring the dance to an end. We needed to see a firm offer. In anticipation of a bait and switch, I warned McMullen against coming back with a lower price offer.

Instead, Wind continued to pepper us with questions. I told McMullen that we would no longer respond to requests until the price was firmed up, noting that we were interpreting Wind's messaging that they would be lowering their offer price. I expressed concern about leaks to the press, conversations with shareholder activists, and nondisclosure violations, which we interpreted as hostile tactics designed to pressure us to accept a lower price.

In the meantime we observed that Sun was not active in the data room. We terminated its access to the data room, which they seemed to not notice. Ganzi and Vesely appeared to have lost interest. We were cornered.

The South Beach Miami Metro Connect conference was the following week. Nearly all the players would be present. With all the press leaks and activist activity, all eyes would be on me. For the first time, I headed to Metro Connect not as the Roaring Bear but as the vulnerable prey.

I liked it better when I was the predator. So I decided to go hunting, specifically aimed at getting Ganzi and Vesely back into the bidding. I set up separate meetings with each of them in locations that would be highly visible to the curious onlookers.

On the first day of Metro Connect, I met with Vesely over lunch in the poolside venue. Nearly every table was occupied with telecom bankers, investors, and executives. Vesely noted that though he was very interested in acquiring Zayo, he was concerned that Wind had already convinced funding sources that the deal was locked up. He expressed that he did not want to waste EQT's resources on diligence if there was no real prospect to complete a deal. In response, I emphasized we would open a window for them to get through diligence. "Submit a bona fide offer and we will grant you exclusivity to catch up," I assured him.

Later that day I met with Ganzi in the courtyard that also was very visible to the early cocktail hour crowd. I emphasized the window of opportunity to submit a bid and complete diligence. Ganzi also expressed concern that funding sources were convinced that Wind had the transaction locked up. Ganzi also shared with me that Blackstone, citing his source as a former Zayo executive, told him that I would never agree to sell Zayo to him. Ganzi also shared that Blackstone told him that Wind would allow DigitalBridge to invest $1 billion if Wind was successful. This further violation of the NDA was a bribe to reward DigitalBridge to stay on the sidelines. Ganzi emphasized that he did not want to be used by me as a stalking horse.[37]

The next day I met with McMullen, at his request, in his private hotel suite. To my surprise, his bid partner, Gautam Bhandari from I Squared, was in his room. They orchestrated a clumsy good cop / bad cop routine, messaging that they were close to submitting a revised proposal but likely at a lower price.

The following day Zayo received a letter from an activist shareholder who encouraged us to accept an offer for an acquisition in the low to mid-$30s per share. A day later another activist sent a letter, encouraging us to accept an offer for $27 to $32 per share. Clearly, Wind was tipping off these activists to deliver body blows that would compel them to accept a reduced offer.

---

37 The expression "stalking horse" in this context means that a company gains negotiating leverage by enticing a secondary bidder to join the process, thereby prompting the primary bidder to strengthen its bid.

Two days later Wind played the anticipated card. They lowered their offer to $31.50 per share. Given the content and timing of the activist letters, we concluded that a Wind insider was violating the nondisclosure. We decided it was time for some head banging.

Through our bankers, we requested that all members of Wind, along with its legal and investment advisors, join a call to hear our feedback on its proposal. Wind must have been surprised when the only attendees representing Zayo were our legal and investment advisors. At our direction our attorney read a stern warning conveying that leaks to the press and to activist stockholders were unacceptable. Also, the revised offer price of $31.50 was insulting. Afterward, as we received reach outs from Wind's leaders, we had a stress-relieving belly laugh at their "we would never do anything like that" denials.[38]

The next night McMullen and Choe flew to Boulder for dinner with me. Their first message was that no one from their consortium leaked information or violated the confidentiality obligations. I conveyed my reaction with a smirk. Second, they informed me that one of their members dropped out of the consortium. They asked for but were denied permission to contact additional investors. I emphasized that Wind's track record left us wary of its tactics and bellowed that they were not the only bidder.

In the meantime momentum was building with Sun. On February 12 Ganzi called me to report they were making substantial progress with EQT. Three days later we had a call with Vesely and Ganzi, during which they conveyed they would send an offer within a week.

A week later, on the same day, both Wind and Sun made their moves. Wind submitted a "best and final offer" with a purchase price of $32 per share, while Ganzi verbalized that Sun would be submitting a written offer at $35 per share. Whew. Ganzi and Vesely sent this message just in time.

---

38 We don't know for certain if any members (or, if so, which specific individuals) of the Wind consortium were leaking confidential information to the press or activist investors. We did know the leaks were harming Zayo and gave the warning in hopes of quieting the source.

We received Sun's written offer four tension-filled days later. With the offer in hand, we delivered the message to McMullen, Canfield, and Choe that we would not transact for $32 per share. When McMullen asked me if I could provide feedback on the $32 per share, I had fun with my answer: "Our board took at face value your written statement that $32 was your best and final offer and, as such, did not provide any additional feedback." I knew this would piss Wind off. The tables were turning.

On March 5, 2019, we entered a thirty-day exclusivity agreement with EQT and DigitalBridge, which would prohibit us from having any conversations with Wind. Sun would have time to catch up; Wind would be left stewing at being on the outside looking in.

A strategy in Texas Hold'em is to get the opponent to be pot-committed. During the early betting rounds, a player is prodded to push many chips into the pot, making it difficult to pull out even as the player begins to question the strength of his hand.

Wind already spent millions on consultants and lawyers. If they lost this deal, they would have to write it off as a loss. They were pot-committed.

Sun signaled that its diligence would be extensive, involving lots of consultants, accountants, and lawyers. The deeper they got into diligence, the more leverage we'd have. When Sun threw waves of MBAs, CPAs, and JDs at us, its chips accumulated in the middle of the betting pile. They quickly became potcommitted.

If we played our cards right, we could end up with two bidders, neither of which would want to go home empty handed. For the first time in a long time, I was happy.

We extended Sun's exclusivity for two additional ten-day periods. This gave them time to lock up the $14.3 billion of equity and debt to fund the deal. It also provided the partners of Wind more time to seethe.

A couple of days into the exclusivity period, the activist shareholder, Starboard Value, released a public letter urging the company to sell but did not specify a price at which it would support a sale. We smiled, knowing that its informants were now at an information disadvantage.

A few days later, I received a call from Neil Hobbs, my close colleague from Level 3, who was an advisor to Stonepeak. Fishing for information, Hobbs asked, "Does it make sense for McMullen to continue working on the deal?" I stuck to our public disclosures, which was a clear hint that we were under an exclusivity with another party but still considering options. I did not comment when Hobbs asked if there was any concrete feedback concerning steps that McMullen could take. I imagined Hobbs reported back in his infamous colorful tongue, "I told you dumb fucks that you want Caruso on your side of the table, not lined up against you."

In parallel, more as a ruse than a contingency plan, we launched a process to sell our data center business. This noise would keep both Wind and Sun on their toes. Two weeks into the exclusivity period, McMullen emailed me, asking for a call. When I told him I could only speak to him if the topic was the sale of our data center business, he confirmed data centers was the topic of the call. During the call, as expected, McMullen focused on his bid for Zayo: "Once the exclusivity period ends, Stonepeak would like to reengage, suggesting he would advocate for Wind to increase its offer to $35 per share." He blamed others in his consortium for the bait and switch. My response was simply that this is not a topic that I was permitted to discuss. I wish Zoom was the norm back then so McMullen could see my childish grin.

The McMullen reach out was also an assist with Sun, as the NDA obligated us to disclose the McMullen reach out, prodding Ganzi and Vesely to avoid a bidding war by getting to the finish line quickly and with no retrades.

At Ganzi and Vesely's invitation, we scheduled a dinner. I picked Corrida, a Spanish Steakhouse that overlooks the flatirons of Boulder. I wanted the atmosphere to convey partnership, which made Corrida a natural choice. Their goal was to begin discussions around my commitment to remain as CEO. My message was that if Sun was successful in its bid, they could count on my support in any way they needed—whether or not they wanted me to remain as CEO. I was not looking for any commitment in exchange. I was open to rolling over a portion of my equity, which gave them the assurance that I'd have skin in the outcome.

As we transitioned to mid-April, I received almost daily reach outs from McMullen. "Are you in a position to discuss our bid?" he'd ask. I answered him by ignoring the emails.

Finally, on April 24 the exclusivity period expired. I spent the prior two days with Ganzi and Vesely hammering out the remaining terms of the merger agreement. We were close to done but not quite there. Ganzi and Vesely pleaded for another extension. My message was clear: "I'm getting daily emails from the other suitors. You need to get the deal done or risk losing it."

Like clockwork, McMullen reached out the next morning. He was clever in his search for a scoop. "We would be willing to join another consortium at $35 per share to fill an equity gap," he stated. My reaction, he hoped, would reveal that the DigitalBridge and EQT team were stuck. Instead, I had fun by playing dumb. "I'm not aware of another consortium with an equity gap" was my retort. "Are you?"

McMullen noted that he would follow up further with the other members of Wind and get back to us. Two days later Wind submitted an offer, which did not specify a purchase price. Were they still fishing? Or was there friction within Wind on what to bid?

We declined to engage, letting them know we were mindful of our prior experiences with them. Our message was clear: if we give you another shot at this, stop jerking us around.

Later that day I received a call from Canfield. "We shouldn't have sent you an offer without a price. We remain interested and will come back with a price." Six days later they came back with $33.50 and also noted Blackstone dropped out of the Wind consortium.

Believing we were days away from signing with Sun, we had our bankers pepper Wind's investors one last time: "Why do you remain below the $35 price?" Their response was that $33.50 was the highest consensus they could get to with their current group. To stall for time, we provided them a draft merger agreement but without a reaction to the $33.50 price.

In parallel, I focused on the final issues in the Sun agreement. With all substantive matters behind us, our board authorized me to have discussions with Ganzi and Vesely about my commitment post the acquisition. In a video call on May 6, they expressed their interest in having me continue as CEO. I emphasized that I didn't want to delay the transaction by negotiating the details of an employment agreement. With Wind in the backdrop, Ganzi and Vesely were also wanting to avoid a delay. To get them comfortable that I'd be aligned with them, I agreed to roll over $105 million of ownership.

With this behind us, we were ready to get to the finish line. I had my team come to my house in Boulder, as I wanted us to be together when we signed the agreement. At 2:00 a.m. on May 8, we exchanged signatures.

The fight was over. Sun, whom I wanted to win, prevailed. My clean exit was near.

The saga was not quite over. Telecommunications transactions require an extensive closing process. Approvals are required from each country, state, and city. The process often takes a year or more.

My responsibility as a public company CEO was to ensure Zayo got to closing as promptly as possible while avoiding any risks that the transaction would be derailed. DigitalBridge and EQT were responsible for raising the capital to close the transaction, which included nearly $8 billion in debt. Other than ensuring Sun had the support they needed, I didn't pay much attention.

In February the debt raise took a twist. My CFO stopped by my office. "Vesely asked if you are joining the road show next week." A road show, similar to the IPO shindig, was required to sell the debt offering to investors.

I was confused. "I wasn't planning on it." I wasn't asked for my opinion on the financial plan. Given the behavior during the first six months of the closing process, I assumed they'd be bringing in a new CEO. I had no reason to think they wanted me to be part of the debt raise.

"I think Vesely is going to call you and encourage you to join them." When Vesely called, he told me it is typical that the CEO participated in the debt raise. I agreed to join the road show.

I was ten minutes late to the New York City kickoff of the debt raise. The room was packed with fifty professionals from DigitalBridge, EQT, their bankers, and their law firms. Most were sitting around a huge table, and all eyes were on me. I looked for a seat against the wall, thinking I was an observer who might answer a few contextual questions. Instead, I was directed to the center of the table.

Vesely said, "Dan, the team would like you to walk them through the presentation you will use on the road show."

Huh? I am the one raising the debt? When was this decided and by whom? And why didn't anyone give me a heads-up?

Whatever. I did as I was asked and took the deck, which I was seeing for the first time, and danced through my pitch. Most of the material was from my Zayo story decks, so it wasn't hard. I turned it over to Vesely and my CFO to explain its financial plan.

We used Ganzi's jet to bounce between New York, London, Boston, and San Francisco. As expected, the international representation in the meetings was extensive, as the story had appeal to Asian and Middle Eastern investors. In each meeting I did 75 percent of the talking.

The road show was fun. I enjoyed the time with Ganzi and Vesely. Investors liked the Zayo take-private story. Debt markets were frothy. Not only did we generate more than enough demand, but DigitalBridge and EQT were also successful in negotiating fantastically favorable terms.

Something else was in the air in February 2020—COVID-19. I spent the second half of February in bed begging my wife not to ship me to the hospital, as I feared what they did to people with this mysterious disease. I had visions of a young John Travolta as Bubble Boy. After a week of lying in bed shivering with a fever and sucking oxygen from a tank, I got better.

With the debt raise done and all regulatory approvals in hand, we were ready to close the transaction. On March 9, 2020, the sale to DigitalBridge and EQT was official.

In the press release, I said, "We are excited to launch this new chapter of Zayo as a private company under the ownership of a consortium led by two highly experienced infrastructure investors who have a deep understanding of our business and bring significant value to Zayo. . . . As a private company, we will have greater flexibility to pursue our long-term strategy and leverage our fiber to fuel global innovation for our customers." These were more than just the right words. It conveyed what I believed to be true. I still do.

I had minimal discussions with Ganzi and Vesely as we approached closing. When my executive team asked me if there would be changes post close, "I have no idea" was my answer. If I were being forthright, I would have told them, of course, changes were soon to come.

Soon after closing, Ganzi and Vesely informed me they were bringing on a new chairman, Verizon's former COO, John Stratton. As mentioned in the previous chapter, my last encounter with Stratton was a few years prior, when he threw cold water on the discussions I was having with his boss, Verizon's CEO Lowell McAdam. After McAdam left Verizon, Stratton was passed over for an outsider CEO. Overall, Stratton was a likeable guy. However, the thought of Zayo (and me!) being guided by a career Verizon executive was amusing.

Stratton told me that if I played my cards right, I could continue as CEO. "Geez, thanks," I sarcastically thought. I was perplexed why he thought I'd be interested in continuing as CEO in this environment. I was way past the point in my journey where I'd work for a chairman picked by owners without my involvement. I had one year to wait before I was fully vested, so I played along.

The board was staffed with nine people, and no spot was left available for me. "Eleven more months until vesting," I thought.

Consultants of all types arrived. The cost consultants declared Zayo was inefficient. The customer experts advised our customer focus was too narrow. The service experts cited concerns about the health of the network and the quality of customer service. The system experts declared our systems and data needed an overhaul. The talent consultant concluded the leadership team needs to be up-leveled.

The consultants applied the consulting playbook. Their phase one findings led to an expensive phase two engagement. When phase two didn't show the results, the consultants teed up an even more expensive phase three.

DigitalBridge and EQT purchased Zayo. It was their company now. They bought the right to do whatever they thought best for the company. My job was to get out of their way and wish them the best.

Six months after we closed the transaction, I left Zayo. Steve Smith, the long-tenured CEO of Equinix and a friend of mine, took over the reins of CEO.

———

Zayo's sale to DigitalBridge and EQT became the second-largest LBO since the 2008 financial crisis. My thirty-five-year journey as the Bear of Bandwidth had an appropriate and positive ending. For those of you keeping score, $1.1 billion was invested into Zayo's equity. The value of this investment grew to $8.5 billion. The MOIC of eight times and the IRR of 35 percent were extraordinary for an investment of this magnitude.

Zayo was, and still is, the only national pure-play provider of Bandwidth Infrastructure.

More than a dozen members of Zayo's management team became C-level leaders at other Bandwidth companies and many were backed by

Zayo investors. Most of them assembled teams that were anchored with Zayo veterans. I am most proud of this portion of my legacy.

I proved to myself, and less importantly to anyone else who cared, that a Bandwidth Infrastructure business could be run in a way that produced a fantastic long-term outcome for its stakeholders. Warren Buffett, I think, would be impressed. Jim Crowe, for certain, would be proud.

CHAPTER VIII

A BRIGHT GLIMMER OF LIGHT

In the short term, reality underperforms the hype.
In the long term, our imaginations fail us.

—Jim Crowe

Artificial Intelligence (AI) is already causing an explosive need for data center capacity and, with it, demand for Bandwidth Infrastructure is increasing.

Quantum technology will soon plunge into our world with a forcefulness that will spark another societal revolution. It will collaborate, not compete, with AI, causing an exponential explosion in Bandwidth requirements.

Quantum computing will be dramatic in its own right, but quantum technology goes well beyond computing. Quantum sensing will spur a quantum Internet of Things that will span across quantum positioning systems, autonomous vehicles, and the discovery of rare minerals on Earth and beyond.

Quantum radio frequency will be the enabling technology for 7G mobile networks. Its ability to detect minute changes in radio frequency fields with greater precision will open up new bands of spectrum and unimaginable applications.

When the Internet was first commercialized, it ran on old-fashioned phone networks using copper and circuit switches designed to carry voice. Within its first decade, newly designed networks using fiber and Ethernet violently replaced the old technology. The same will happen with quantum.

Quantum communications and networking will enable quantum computers and sensors to connect natively to each other, meaning that information will remain in a quantum state as it traverses networks. Fiber will remain essential, though new fiber types that are more effective at carrying quantum signals might be introduced. Certainly, new photonics and chip technology will be required to support quantum networking.

AI and quantum together will enable robotics and augmented reality to find their way into our everyday world.

Starlink and soon Kuiper are revealing the increasing role of satellites in communications. These will complement, not replace, terrestrial fiber. SpaceTech, in general, will create large quantities of data and communications requirements that will put more demand on fiber networks to carry the load.

The future of Bandwidth Infrastructure is a bright glimmer of light.

The industry that was once just AT&T, and then exploded to hundreds of rivals, is now largely reconsolidated down to four major national providers: AT&T, Verizon, Lumen, and Zayo, with Zayo being the only pure play. All four are facing challenges with their Bandwidth platforms.

AT&T has the most modest presence in Bandwidth Infrastructure. The depth and breadth of its commercial fiber assets are materially less than those of the others.[1] Most of AT&T's fiber is used to support its residential, wireless, and enterprise business. AT&T is not a major supplier of Bandwidth Infrastructure to external customers. AT&T's stock

---

1 The statement excludes fiber assets that are primarily used to serve AT&T's residential customers.

price trades at near its thirty-year low, which is half of the high achieved during the Boom years.

Verizon has a vast reservoir of fiber assets. The combination of World-Com assets, including WilTel, MFS, Brooks Fiber, MCI, and XO, along with the fiber it built on its own, puts its inventory at comparable levels to Lumen and Zayo. Yet Verizon has failed to develop a healthy Bandwidth Infrastructure business model, as evidenced by its $5.8 billion write-off in late 2023. Verizon's stock price is near its fifteen-year low.

Lumen finds itself in a vulnerable position, with the risk of a bankruptcy filing remaining acute. In mid-2024, Lumen's stock hovered just over $1 per share, which equated to a Market Value of only $1.3 billion. Lumen's $20 billion in debt trades at a steep discount to its face value. Though Lumen has a few years before it is required to pay back its debt, it is unclear if a solution to its precarious predicament will be found. Lumen's CEO, Kate Johnson, remains confident, as demonstrated by her purchase of a second tranche of $1 million of Lumen stock in 2024.[2]

Zayo, though in a stronger position than the others, faces its own challenges. At the time Zayo went private, Zayo had $2.6 billion of revenue with profitability of $1.3 billion. Four years later the revenue

---

2 In early August 2024, just as we were finishing the final edits to *Bandwidth,* Lumen's stock soared to nearly $8 per share. The trigger was Lumen's announcement of new contract revenue bookings of $5 billion, alongside active discussions of an additional $7 billion. Lumen's disclosures inferred that Microsoft was a main customer and AI was a catayst.

CEO Kate Johnson explained: "To summarize what's happening, the dramatic rise in AI innovation springs explosive growth in data center buildouts, and data centers simply have to be connected. . . . [W]e're repositioning the company for the future in the growing market of AI."

On the same August day, Lumen announced an agreement with glass manufacturer Corning Incorporated to procure a "substantial supply of next-generation optical cable." Lumen positioned this "fiber-dense cable" as a technology that would support "major cloud data centers racing to stay ahead of AI workloads." Zayo also reported record sales bookings—citing AI as a major driver.

Perhaps the critical role of fiber as digital infrastructure, the consolidation of the fiber industry, and the excess conduit that was deployed to house advanced generations of fiber technology are on the cusp of driving a valuation boom among the Bandwidth leaders.

remained at $2.6 billion, while profitability declined to about $1.1 billion. The 50 percent profitability margin declined to 42 percent of revenue. Zayo's indebtedness increased from $7 billion when it went private to nearly $9 billion in 2024.[3]

An important difference is worth noting between Zayo relative to the others. AT&T, Lumen, and Verizon each have structural problems that are intertwined with their core businesses. These structural challenges compromise their abilities to execute their Bandwidth Infrastructure strategies. Separating their Bandwidth businesses from their core businesses is impossible for AT&T and Verizon and extraordinarily difficult for Lumen. In essence, these businesses are stuck in deep mud.

This leaves Zayo in a far better position, as it is the only pure-play, North America–centric national Bandwidth Infrastructure platform. Zayo doesn't have the deeply rooted structural problems that are encumbering its rivals. Execution is the primary challenge, which is on the path to being addressed.

With AT&T, Lumen, Verizon, and Zayo distracted, might another national rollup pick up momentum? The next largest collection of fiber assets sits inside Crown Castle. However, the status of Crown's Bandwidth business is not good. In late 2023 activist Elliott Investment Management concluded Crown's entry into fiber was a colossal mistake, citing Crown's lagging performance relative to its rivals, American Tower and SBA. CEO Jay Brown, who led Crown's fiber rollup, was forced to resign, and Crown announced a comprehensive review of its fiber business. Two independent directors were named to its board by Elliott, one of whom was former Level 3 CFO Sunit Patel. Crown has hung a For Sale sign on its fiber business. Akin to Lumen, separating Crown's fiber business from its core tower business will be very difficult.

---

3 After Zayo was taken private, the new investors sold its data center business. $600 million of the proceeds were distributed to investors and lowered net indebtedness. Acquisitions and negative cash flow from operations increased net indebtedness.

In mid-2024, Zayo separated Europe from the North America business, presumably preparing to sell it and use the proceeds to reduce its debt.

Other material pockets of Bandwidth Infrastructure exist. Dave Shaeffer's Cogent, which acquired PSINet for a pittance and purchased the fiber assets of Sprint from T-Mobile in 2023, could be a starting point for a fifth national platform. Lightpath, led by Zayo veteran Chris Morley, also has substantial fiber, which is mostly concentrated in Northeast assets. Significant pockets of fiber assets are buried in the distressed public companies Windstream and Uniti Group; they are on a track to be unified into one company. Other regional clusters of fiber assets remain independent. However, all of them combined would be about the size of Crown's fiber business. And most have deep structural problems of their own. Another major rollup platform is possible but not likely.

One other possibility is a play by Big Tech. Significant fiber assets have been accumulated by Alphabet, Meta, and Amazon—and to a lesser degree by Microsoft and Apple. They have participated and, at times, spearheaded joint builds. The fiber has typically been used to connect important data center locations. Not unlike Amazon's AWS play, one of the Big Five might reach the conclusion of owning a vast fiber platform as a strategic addition to its portfolio.[4] The Bandwidth platform could be viewed as a means of extending the Google Cloud, Microsoft Azure, or AWS platforms to the customers' edge. AI, quantum, augmented reality, and robotics—all areas of keen interest by Big Tech—will drive an ongoing need for fiber. A $20 billion price for Zayo would be a drop in the bucket for Big Tech. Though unlikely, a bold move by one of the Big Five is plausible.

A leader of the Cable TV industry such as Comcast, Charter Communications, or Cox Enterprises might consider a move.

Lastly, T-Mobile is a plausible strategic suitor. Verizon and AT&T each own and operate dense fiber networks, with one of their major purposes being to provide Bandwidth connectivity to support their mobile services. T-Mobile, whose services now include providing wireless Internet

---

4 Big Five refers to Alphabet (Google), Amazon, Apple, Meta (Facebook), and Microsoft.

access to homeowners, might determine that owning Zayo or Lumen might open up more opportunities to expand or differentiate their product offerings.

The most likely outcome, though, is Zayo regains its footing and continues on with its consolidation—with the acquisition of Crown's fiber business being a leadership-cementing achievement—and relists as an NYSE or Nasdaq public company. In parallel, Lumen continues its slow crawl-back, perhaps enabled by a Chapter 11 restructuring, from its financial abyss. And perhaps before the end of this decade, Lumen and Zayo will merge into a mighty Bandwidth provider.

CHAPTER IX

FINAL REFLECTIONS

The value of a business is "the present value of the stream of cash that's going to be generated . . . between now and doomsday."

—Warren Buffett

ow did I achieve a financial outcome far better than the norm?

When I launched Zayo, I decided I must document my core management philosophy and beliefs, which would be my guardrails for not repeating mistakes of the past. The result was eventually published as the Zayo Owner's Manual. A modified 2014 version is included in the appendix.

I observed the industry didn't have a reliable methodology for measuring value creation. Standard accounting approaches were more than unhelpful; they were destructive. So I developed my own approach to measuring and holding ourselves accountable for value creation. I consider this my most important innovation, and I hope to see it adopted by others. The appendix also includes my "Equity Value Creation" methodology.

Lastly, I was innovative in my use of systems, Salesforce.com, in particular, and data, which is all important in operating a Bandwidth business. Together, systems and data allowed us to achieve industry-shattering

Profitability margins of 50 percent while quickly and fulsomely integrating dozens of acquisitions.

I would be remiss if I didn't include the other factors that led to success. I was an insecure badass, a rough-around-the-edges, southside Chicago guy who wouldn't tolerate a life of being average. I was the Bear of Bandwidth.

The big takeaway from the Bandwidth fiasco was that the leaders were not focused on durable cash flows. Accounting gimmicks, misleading disclosures, and other bad behaviors were used to fool investors. Eventually, places to hide were too hard to find, and a steep price was paid.

Buffett likens a bubble to a fairytale: "Akin to Cinderella's Ball . . .," a Boom is an elaborate party. Bernie Ebbers found himself on the cover of *BusinessWeek* with the headline "Telecom Cowboy." Jim Crowe was showcased in *USA Today* alongside Warren Buffett, Bill Gates, and Walter Scott. The *New York Observer* glorified ICG's Shelby Bryan as "tall, smooth, and charming—seductive." In 1999 I danced to a live performance by KC and the Sunshine Band at a Level 3 Christmas party while driving around in a yellow Ferrari with Level 3–inspired license plates.

Buffett continues his analogy by saying the party guests delay their departure because "they hate to miss a single minute of what is one helluva party," while investors "continue to speculate in companies that have gigantic valuations relative to the cash they are likely to generate in the future."

Just as Cinderella knew her carriage would turn into a pumpkin and her horses into mice, the Bandwidth Boom executives and investors knew their party would come to a devastating end. However, unlike the Bandwidth CEOs, Cinderella knew an important detail—she was able to watch the hands of the clock as they ticked closer to striking midnight. "The giddy participants all plan to leave just seconds before midnight.

There's a problem, though: They are dancing in a room in which the clocks have no hands," Buffett explained. The Bandwidth Boom partiers didn't know when the Bandwidth Bust would hit.

In the wake of a Bust, a new opportunity, like the proverbial phoenix, rises from the ashes. Savvy entrepreneurs jump in, buy assets on the cheap, and through fundamentally sound business practices, create new fortunes.

Bandwidth wasn't the first and won't be the final Boom, Bust, and Resurgence cycle. The smashing together of AI, quantum, blockchain, augmented reality, genomics, and robotics might already be fueling a new Gold Rush. The pattern of Boom, Bust, and Resurgence will repeat itself.

Heed the words of the Oracle from Omaha. Be greedy when others are fearful. Be fearful when others are greedy. Know there are no hands on the clock. Leave the party before midnight.

And always remember, Cash Flow is King.

ACKNOWLEDGMENTS

After nine months of intense work, *Bandwidth* was a meandering mess. Then I was introduced to Steve Aranguren, who knew little about technology but was exceptionally knowledgeable on the topic of storytelling. Steve became my creative editor and shaped *Bandwidth* into a story that is an enjoyable and riveting read.

Within days of meeting Steve, I had an introductory lunch in Boulder with Marcus Brauchli, who, I learned as we enjoyed Maine-inspired fish and chips, was a twenty-four-year veteran of the *Wall Street Journal,* serving as the managing editor. Between 2008 and 2012, Brauchli was the executive editor for the *Washington Post.*

Brauchli leaned into my *Bandwidth* project, offering to introduce me to John J. Keller, a journalist on the *Wall Street Journal* team who covered the Bandwidth industry in the heyday of the Boom and Bust. Keller jumped right in as my editor and readied *Bandwidth* for prime time.

Brad Feld is a Venture Capital icon in Colorado and well known in the global venture community. He is also a friend. I asked Feld, who wrote and published several books of his own, to take a gander at my *Bandwidth* manuscript. Feld instead dove in deep. He edited the draft, provided guidance on how to tighten the story, and offered colorful anecdotes from his own Dot-Com journey.

Courtney Caruso and Emily Notaro of the Caruso Ventures team were instrumental in the research, project management, and execution of this ambitious undertaking.

Steve, Marcus, John, Brad, Courtney, and Emily—thank you. I am immensely grateful for the passion, enthusiasm, and expertise each of you applied to help me complete my memoir.

OWNER'S MANUAL

Owner's Manual for Zayo Investors

**The Business Principles That Guide the Decision-Making
of the Zayo Management Team**

**Published in January 2014 by Dan Caruso, chairman
and CEO of Zayo Group**

Introduction

In 1983 Warren Buffett drafted a set of business principles; in 1996 he published them in an Owner's Manual as part of Berkshire Hathaway's annual report. Every publication of the annual report since has included the Manual.

The content of Buffett's principles has shaped my business thinking and has been ingrained in the culture and practices of Zayo. Buffett's decision to publish his Owner's Manual struck me as an appropriate method to set expectations with existing and prospective investors.

With respect and deference to Buffett, I decided to publish an Owner's Manual for Zayo Investors. The Owner's Manual applies to both debt and equity investors, whether private or public.

Principle 1: Treat investors as long-term business partners, and view management as managing partners.

An enterprise must be managed with the perspective that investors, not management, own the assets contained within the company. Management, on behalf of its partners, is the steward of the company's assets. Management's responsibility is to act, to the best of its capabilities, to maximize the value of the assets.

Principle 2: Be transparent (i.e., clear, open, and honest) in communications with investors.

Whether the news is good, neutral, or bad, management must provide sufficient information to enable investors to understand the ongoing performance of the business. Information should be provided in a useful, consistent, and unbiased fashion, enabling Investors to gain insight into the results and trends that impact the value of the enterprise.

Management, when sharing information with investors, should focus on the operational metrics management uses when operating the business. By doing so, management is providing insight into its decision-making process as well as the data itself.

Heed the following wisdom of Warren Buffett: "Elsewhere triumphs are trumpeted, but dumb decisions either get no follow-up or are rationalized. This behavior finds its way into the culture and operations of a company. High-risk business decisions are encouraged because if they pay off, the rewards are high and if they don't, the ramifications are less than they should be. Though human nature might resist, management must be candid in its self-assessment, even when it comes up short."

Principle 3: Understand the meaning of Intrinsic Value and make maximizing Intrinsic Value the basis for business decisions.

Warren Buffett refers to Intrinsic Value as an all-important concept. He defines it as the discounted value of the cash that can be taken out of a business during its remaining life. Maximizing Intrinsic Value must be the guide for all business decision-making. Buffett says: "If Intrinsic Value increases, the stock price will eventually follow."

I look at Intrinsic Value as the true value of an enterprise, which in turn is captured by the following statement. An enterprise will be worth what its free cash flows, discounted at a risk-free discount rate, are really going to be. Risk-free rate is used in this context because the rate is applied to actual cash flows. Of course, forecasts of cash flows reflect uncertainty— the higher the degree of uncertainty, the greater the risk for investors and accompanying discount rate. My point is Intrinsic Value will follow from what free cash flows actually turn out to be.

On one hand, this is the tenant of Net Present Value—and hence the third principle might be viewed as extraordinarily basic. On the other hand, I've observed too many instances where management teams and investors have sidelined the importance of this basic tenant. Instead, decisions have been made based on achieving revenue, EBITDA, or capital targets or on maximizing metrics such as eyeballs or growth rates.

Of course, the free cash flows resulting from decisions are rarely known with certainty, and often the level of uncertainty is high. Good decision-making is a special art, not a simple math problem. Principle 3 provides clarity to the objective that should guide decision-making. First, understand Intrinsic Value will result from free cash flows. Second, use the maximization of Intrinsic Value as the basis for making decisions.

Note that maximizing Intrinsic Value does *not* equate to maximizing the price that a buyer might pay for an enterprise at a particular point in time. University of Chicago, of which I am an alumnus, originated the "efficient market theory" (EMT), which states that the most accurate estimate of the Enterprise Value of a public company is reflected in its stock price. Buffett likes to point out that if EMT is correct, he would more likely be a panhandler than one of the world's richest people.

Said differently, the stock market's view of Enterprise Value is often out of sync with Intrinsic Value. Like Buffett, I believe management's focus must be squarely on Intrinsic Value, not a point-in-time market-based estimate of Enterprise Value. Proper long-term decision-making should be driven by the quest to maximize what free cash flows are really going to be while applying appropriate discounts for the time value of money.

Principle 4: Investors should be informed of the company's true Intrinsic Value, not more and not less.

Buffett points out that Intrinsic Value is easier to define than to accurately calculate. Management's goal (to the best of its ability) should be to help investors gain an accurate understanding of the Intrinsic Value.

This principle could be a source of conflict with the common-held belief that management's goal is to maximize stock price (in the case of a public company) or the sale price (in the case of a private company). If a stock price (or sale price) reflects an Enterprise Value that differs from Intrinsic Value, the result is that a selling Investor will benefit at the expense of a buying investor, or vice versa. For example, if the valuation implied by the stock price exceeds the Intrinsic Value, new investors will be overpaying exiting Investors.

The goal is to be fair to all investors. To accomplish this, management's goal must be that the perceived value of the firm (whether the entity is a public or private company) aligns with Intrinsic Value.

Investors should never lose sight that the measurement of Intrinsic Value is elusive for both investors and management. Importantly, management does not necessarily have greater insight into Intrinsic Value than a particular informed Investor. Principle 4 emphasizes that the foundation of management's communications with Investors is to help them gain insight into the firm's Intrinsic Value.

Principle 5: The principles in the Owner's Manual also apply to debt holders.

As members of the investor group, debt holders should be treated as partners and in a manner consistent with the entirety of the management principles. Through the application of the articulated principles, management's intention is for debt holders to receive a fair return for their level of risk as lenders.

Management's strict responsibility to the debt holders is to comply with the obligations set forth in the debt facilities agreements. In satisfying this responsibility and while applying these principles, management will seek to optimize the company's overall cost of capital.

Principle 6: Management's objective relative to equity holders is to maximize Equity Value Created

Principle 4 centered on management's responsibility to accurately portray Intrinsic Value. Principle 5 addressed management's obligation to debt holders. Equity Value is the portion of Intrinsic Value that remains after all debt obligations are satisfied. The maximization of Equity Value is achieved by maximizing Intrinsic Value while using debt to optimize the overall capital structure. As such, principle 6 is the direct outcome of principles 4 and 5.

Equity investors seek to achieve an appreciation in the value of their equity investment. The minimum rate of appreciation that equity investors would view as adequate is referred to as the Hurdle Rate. The word "hurdle" is quite descriptive. If the rate of equity return is at or above the Hurdle Rate, management is satisfying or exceeding the expectations of its equity holders. If it is below the Hurdle Rate, management is falling short of expectations.

Note that principle 6 isn't simply to exceed the Hurdle Rate. That is, management shouldn't make decisions with the objective of simply clearing the Hurdle Rate. Nor is it the goal of principle 6 to maximize the rate of appreciation of equity, which I refer to as Equity IRR. Instead, principle 6 emphasizes the goal as maximizing Equity Value Created.

Management, and its equity holder partners, must appreciate the interrelationship between Hurdle Rate, Equity IRR, and Equity Value Created, while acknowledging the overriding goal is to maximize the amount of Equity Value Created.

I will use an example to illustrate these interrelationships. In the example, I will make the unrealistic and flawed assumption that Intrinsic Value is known and accurately measurable by all. Another unrealistic assumption I will make is that the Hurdle Rate is a fixed and precisely knowable number. (In reality, investors will have varying views of risk and return thresholds, and capital markets are volatile. As such, Hurdle Rate is never known with certainty.)

Let's assume an equity holder owns 10 percent of the equity of a company that has an Equity Value of $1 billion. The equity holder's value is $100 million. Perhaps this investment was made that very day, by buying 10 percent of the outstanding equity of the company. Or perhaps this investment was made long ago and, by equating to 10 percent of the $1 billion of value, is now worth $100 million. Either way, this holder is looking to make an incremental return on his $100 million of value.

Let's assume the Hurdle Rate is 12 percent. If the equity holder remains an investor for a year, the minimum appreciation of the investment would need to be 12 percent, or $12 million, for the equity holder to be minimally satisfied. If the holder's investment was worth less than $112 million, disappointment would be anticipated; at greater than $112 million, satisfaction would be expected. The level of excitement or disappointment would correlate with the degree of under- or over-performance.

Equity Value Created is the amount of Equity Value gain in excess of the amount required to achieve the Hurdle Rate.

Continuing with the example, assume the actual value of the holder's investment grew to $118 million during that year. The amount in excess of the Hurdle Rate would be $6 million, hence the Equity Value Created relative to that particular equity holder. The holder would have earned an Equity IRR of 18 percent—6 percent above the Hurdle Rate. The equity holder should have been pleased that Equity IRR was meaningfully in excess of the Hurdle Rate; substantial Equity Value was created.

Principle 6 is maximizing Equity Value Created, not maximizing Equity IRR. Why the distinction? From the perspective of this individual equity holder, a higher Equity IRR is a direct result of higher Equity Value Created. However, from a company perspective, decisions might be available that would increase Equity Value Created while decreasing Equity IRR. Principle 6 emphasizes management should choose the one that maximizes Equity Value Created, even if Equity IRR is thereby not maximized.

Principle 7: Since maximizing Equity Value Created is a principle, Equity Value Created should be measured and used as a tool in operating the business.

Most executive teams state that maximizing value created is the key financial objective. However, few attempt to measure it and even fewer use it as an operating metric. Public companies might use stock price as a proxy. When the stock price is up, they tend to cite it as a validation of management performance. However, when the stock price is down, they often attribute the disappointing stock trend to external factors—either macroeconomic impacts or the fickle misgivings of stock market traders.

Buffett points out that Intrinsic Value is easier to define than to accurately calculate. Since the measurement of Equity Value Created requires Intrinsic Value to be calculated, a dilemma is encountered. Notwithstanding this reality, principle 7 emphasizes that measurements should be made and then used as tools in operating the business.

Intrinsic Value cannot be precisely measured, but it can be meaningfully estimated. Most equity investors—whether private or public—estimate the Intrinsic Value of a business when making decisions to buy or sell an ownership stake at a given price. When they perceive a gap between Intrinsic Value and price, they see an opportunity. Likewise, most Private Equity investors use a mark-to-market methodology to track and publish, on a quarterly basis, the value of their equity investments.

Using similar techniques, a methodology around Equity Value Created can, and should, be employed in the management of a business. The methodology should be used in conversations between the investors and management to assess the performance of the business. Likewise, the Equity Value Created performance of each segment of the business can be monitored via the methodology.

The important takeaway from principle 7 is that Equity Value Created can and should be estimated on a regular basis, both at the enterprise level and by each business segment. Though the measurement is imprecise, it is effective in driving appropriate decision-making and in understanding the overall performance of the enterprise.

Principle 8: The communication and management of risk profile requires particular attention of management.

Risk profile permeates its way through many of the principles.

In the discussion of Intrinsic Value, I emphasized that forecasted cash flows are estimates that reflect risk and/or uncertainty. As such, estimates of Intrinsic Value will reflect a higher than risk-free interest rate. Higher levels of risk result in higher Hurdle Rates, which will dampen the estimate of Intrinsic Value.

Debt holders, when pricing the terms of a debt facility, are assessing the likelihood that debt will be repaid in full, and conversely the risk that it won't.

The amount of debt relative to equity alters the risk profile of an enterprise. The optimization of debt costs, and the Hurdle Rate for equity holders, are directly impacted by the debt-to-equity ratio.

Management must recognize the importance of risk profile in its overall relationship with its investor business partners. Embedded in principles 1 and 2, management must be transparent regarding its understanding of the risk profile of the business. If external factors cause changes in risk profile, management should provide insight to its investors.

If internal management decisions are being contemplated that substantially alter the risk profile, these decisions should be considered in the context of prior understanding management has with its investors. Sudden changes should be avoided. If management believes changes are appropriate, the contemplation of decisions should be shared with investors well in advance (if practical) of the decisions being made.

Principle 9: Forecasting competency should be viewed as a strategic capability.

Accurate and thorough forecasting, which I'll refer to as forecasting competency, can reduce risk profile without adding material cost to a business. Forecasting competency bolsters Intrinsic Value in multiple ways.

First and most importantly, strong forecasting competency often leads to better operational decisions. Decision-makers are aided by having reliable estimates of future financial performance. They learn about the likely impact of prior decisions sooner and more thoroughly. This improved and accelerated feedback loop leads to stronger ongoing decisions.

Second, a strong forecasting competency will result in fewer unexpected events.

Third, the cost of debt is likely lower for companies with a strong forecasting competency (all else being equal), as debt holders will price less risk into a debt facility. The reduced cost of debt directly leads to increased Intrinsic Value.

Fourth, the Hurdle Rate is also likely lower for companies with a strong forecasting competency because the reduced risk lowers the threshold return required by an equity investor.

For all these reasons, I view a strong forecasting competency as among the more important operational competencies of an enterprise. Quite often, improving forecasting competency is a low-cost way to improve decision-making while removing avoidable risk within a business. As such, operational metrics inclusive of reliable forecasts centered on Equity Value Created measurements are paramount to the operations of the business.

Principle 10: A long-term and healthy interrelationship between customers, employees, and investors is essential to maximizing Intrinsic Value.

The Owner's Manual, by its nature, focuses on the principles management will apply in satisfying its obligations to its investors. The principles would be incomplete without emphasizing the interrelationship of customers, employees, and investors, who together are the stakeholders of a business.

Principle 3 emphasizes that management's decision-making should be based on maximizing Intrinsic Value. I fundamentally believe that Intrinsic Value can be maximized only if an entity truly enamors its customers. Generally speaking, the more value a firm provides its customers, the more Intrinsic Value it will create for its investors. Management should never view that it is dividing a fixed amount of value between its customers and investors; instead, enamoring customers increases the total amount of value—that is, the size of the pie becomes larger—and both customers and investors are better served.

Equally as fundamental is my belief that the needs of customers and investors can be fully realized only if employees are motivated and inspired to deliver outstanding execution. The day-in and day-out exceptional efforts of committed and capable employees are necessary to enamor customers and deliver exceptional investor returns. To attract and retain the best employees, these team members must believe they are being treated in an appropriate manner. For optimal investor results to be achieved, employees must be made to feel they are part of the partnership with investors. The culture should be such that employees feel they are being treated fairly (and then some!) on a holistic basis that includes compensation and work environment, relative to their contributions, capabilities, and alternative employment options.

When employees feel excited with their occupation, proud of their company and its mission, and appreciated by what they receive in exchange for their contributions, the firm will perform better. The capability to enamor customers increases, and customers will in turn reward the firm with their loyalty and an expanded business relationship. The Intrinsic Value of the enterprise gets larger. The key is to find a harmony between investors, customers and employees/management—such that all are better off because of the effective partnership with one another.

EQUITY VALUE CREATED

"My Most Important Innovation"

Dan Caruso

Introduction

M anagement's primary responsibility is to create Equity Value at a pace that exceeds their investor's return threshold for an investment of a similar risk characteristic. Despite universal agreement, private companies have no method of measuring performance relative to the company's Equity Value creation responsibility. And public companies use the stock price as this barometer, which clouds performance as stock prices are heavily influenced by factors outside the management's control.[1]

Yet Private Equity investors track the value of each of the investments in their portfolio. Moreover, they report the value to their clients, usually

---

1 When stock price increases, CEOs don't hesitate to highlight this as an achievement of the management team. However, when stock prices are in decline, CEOs remind their investors of the unreliability of the stock price due to stock market factors that are outside the control of the CEO.

on a quarterly basis. How are Private Equity investors able to track value creation while CEOs do not?

This bothered me—a lot. For many years. If you don't keep score, how do you know whether you are exceeding value creation expectations? If creating Equity Value is the goal, the ability to measure Equity Value creation is paramount.

I was deeply unsatisfied with the norms to get around this. Enormous efforts would be put into establishing a budget, often involving negotiations to allow management to lower, and then beat, expectations. Rewards and repudiations depended on who made and missed their budgets. This struck me as an enormous waste of time, as the outcomes were focused on actuals versus budget instead of how much value was created or destroyed.

Accounting methodology does not focus on measuring value creation. Net income, cash flow, and net assets are tabulated, but nowhere does the methodology attempt to address whether value creation goals are being met.

Economic Value Added (EVA) was created as a method for addressing this issue. Though EVA is a meaningful step in the right direction, it has not been adopted by the Private Equity industry. In my opinion, EVA is not an appropriate approach for good reasons.[2]

My solution flowed naturally from how the Private Equity industry values its portfolio companies. I adapted this approach to how

---

2 Economic Value Added (EVA) has three shortcomings. First, it doesn't directly incorporate cash flow into the calculation. Instead, it uses depreciation as a proxy and change in net asset value. The all important role of cash flow is obfuscated.

Second, EBITDA—which is a primary valuation metric used by investors in many industries, including telecom—is not used in EVA. Though EBITDA is not a GAAP-defined term, it is used because it is the best approximation of cash flows generated from operations.

Third, EVA relies on the accounting definition of value—the balance sheet. Material gaps often exist between the stockholder's equity as tabulated on the balance sheet and the perceived value of the business. I put weight on the perceived value, as this value is very real to the investors who are buying and selling ownership positions in companies. Investors report perceived value to their investors in a process sanctioned by the SEC. They "mark to market" when communicating value to their investors instead of citing stockholders' equity.

management, as responsible stewards of its investors' capital, track and report its value creation performance.

My Private Equity sponsors loved it. They saw the calculation of value creation each quarter, which aligned with how they measured the performance of their investment. The discussions validated that management was focused on the right levers of value creation. If we excelled, the quantitative discipline of the methodology would be the evidence. If we fell short, it would quickly reveal itself in the numbers. Our interests were aligned. We talked in each other's language.

We didn't waste time negotiating budgets. We didn't rely on management storytelling as the basis for evaluating management's value creation performance. We relied on the financial calculations of value creation.

The methodology proved to be versatile. It applied to periods that included acquisitions, new equity investments and distributions, increases or decreases in debt, and major capital outlays.

Some of my Private Equity investors tried to get other portfolio companies to embrace the methodology. Some of my management team took the methodology with them to new ventures. To my knowledge and disappointment, none have resulted in sustainable traction.

I view Equity Value Creation methodology as my greatest accomplishment. It might die as a result of me hanging up my CEO spikes. I hope someone who reads this book takes an interest.

Value Creation Calculation Dilemma

Management's responsibility is to create Equity Value at a pace (Equity IRR) that exceeds the investor's return threshold for an investment of a similar risk characteristic (equity hurdle rate). Assuming the equity hurdle rate was 15 percent, the investor's level of satisfaction at various outcomes is illustrated in the table below.

Equity IRR Performance	Investor Reaction
> 30%	Extremely Satisfied
18%–30%	Satisfied
12%–18%	Neutral
0%–12%	Dissatisfied
< 0%	Extremely Dissatisfied

The methodology is rooted in this simple concept. How does the Equity Value at the end of the period compare with the Equity Value at the beginning of the period, and how much time has elapsed? Based on the calculation, how satisfied would the investor be in the Equity Value creation outcome?

For example, if an investor invests $1 in a company, and later the investment is worth $1.30, how satisfied would the investor be? The answer depends on the duration of the investment. If only one quarter passed, the annualized IRR would have been close to 100 percent, and the investor would be extremely satisfied. If, on the other hand, more than four years have elapsed, the IRR would be well below the investor's 12 to 18 percent expectation, and as such, the investor would be dissatisfied.

At one level the measurement of Equity Value Created seems basic. If you know how much and when equity was invested, as well as how much the equity is now worth, the calculation is straightforward. For example, see the table below:

Equity Investment = $100
Time Frame = 2 Years
Equity Value at End of Year 2 = $169
Equity IRR = 30%
Investor Reaction = Extremely Satisfied

The calculation depends on knowing the Equity Value at the beginning and end of the measurement period. The $100 is explicitly knowable at the launch of the company, and the $169 value would be explicitly known if the company was sold at the end of year two.[3] How, though, could this approach be applied on an ongoing basis during the investment cycle?

Approach to Measuring Equity Value Creation

The methodology outlined herein was designed to not rely on either (a) an exit or (b) the stock price. In contrast, the methodology measures the ongoing performance of a management team in its responsibility to grow Equity Value at a pace in excess of the equity hurdle rate. As such, it is applicable in all the following circumstances:

1. As a quarterly measurement of actual performance.
2. As a useful calculation for both private or public companies. If a company is public, the methodology does not rely on stock prices to isolate measurement from the external effects of the stock price, such as (a) macroeconomic environment,

---

3 If the company was public and the public price was seen as a sufficiently accurate view of intrinsic value, the calculation would also be straightforward. In fact, if the $100 of value grew to $169 of value over two years, the appreciation of stock price would yield the same 30 percent IRR result.

(b) perceived gaps in the Market Value relative to intrinsic value, or (c) random ebbs and flows in the stock price.
3. As a reliable measurement in determining value creation by a major division of a company.
4. As a way to measure forward-looking value creation implied by a forecast or budget.
5. Most importantly, as a methodology to assess management's performance and align incentives between management and investors.

The methodology outlines a practical annual and quarterly measurement that determines value creation performance and can drive value creation behavior.

Discussion of EBITDA Multiple

Central to this methodology is the use of EBITDA multiple as an approximate, albeit imperfect, indicator of the value of the business entity. Importantly, the use of an EBITDA multiple does not require putting undue weight on the absolute linkage between intrinsic value and the value as tabulated by multiplying EBITDA by an EBITDA multiple. In fact, the exact EBITDA multiple used in the calculation won't be material to the IRR calculation. For example, if a range of eight to twelve times EBITDA is viewed as an appropriate range for an industry sector, using ten times EBITDA will yield similar results as if eight or twelve were used.[4]

Instead, it relies on using a *static* EBITDA multiple to answer the following two questions:

---

4 Over time, the market is used to refine EBITDA multiples through a combination of public stock valuations and private M&A transactions. The important attribute is to apply like-for-like multiples across the periods of the measurement.

- Past performance: "On a like-for-like EBITDA multiple, how much Equity Value was created during the period, and what was the implied IRR?"
- Projected performance: "On a like-for-like EBITDA multiple, how much Equity Value will be created if the budget is achieved during the forecast period, and what will be the implied IRR?"

Equity Value Created Calculation

The equation for calculating how much Equity Value is created in a given period is straightforward.

1. Measure change in EBITDA
2. Multiply by the EBITDA multiple, which yields change in Enterprise Value
3. Add cash generated, which yields change in Equity Value Created

In table 1, we compare multiple scenarios that use an EBITDA multiple of ten with each having an identical change in EBITDA during the period. The amount of cash generated associated with the change in EBITDA determines the amount of Equity Value that was created or, if negative, destroyed.

Table 1: Equity Value Created				
	a	b	c	d
	Actual	a times 10	Actual	b + c
Scenario	Change in EBITDA	Change in Enterprise Value	Cash Generated	Equity Value Created
1	$10M	$100M	$0M	$100M
2	$10M	$100M	$50M	$150M
3	$10M	$100M	-$50M	$50M
4	$10M	$100M	-$200M	-$100M

To tabulate cash generated, subtract change in Net Indebtedness from change in equity invested.

Net Indebtedness is Gross Indebtedness less cash balance. If the company added to the debt during the period, Gross Indebtedness would rise, whereas if it paid down debt, Gross Indebtedness would lower. Net Indebtedness subtracts the cash balance from the Gross Indebtedness both at the beginning and end of the period, thereby reflecting the portion of the Enterprise Value that belongs to equity holders.

However, Net Indebtedness is also affected by change in equity investment, either as a result of new investments or distributions or cash dividends. The inclusion of change in equity in the calculation of cash generated ensures changes to cash balance caused by increases or decreases in equity investments do not affect Equity Value Created.[5]

---

5 If an acquisition is paid for in full or in part by issuing stock to the seller, the implied value of the equity grant should be treated as a new equity investment in this calculation.

Table 2 illustrates multiple examples that yield the same cash flow as scenario 2 above.

Table 2: Cash Flow Generated			
	a	b	c
	Actual	Actual	a - b
Scenario	Change in Equity Invested	Change in Net Indebtedness	Cash Generated
2a	$0M	-$50M	$50M
2b	$10M	-$60M	$50M
2c	-$20M	-$30M	$50M
2d	-$50M	$0M	$50M

Multiple of Invested Capital (MOIC) is an important measurement of value creation performance. The calculation is simple, with the numerator being the Equity Value at the end of the period (EEnding Equity Value) and the denominator being the equity value at the beginning of the period (beginning equity value).

Let's start with the denominator. Table 3 uses the variants of scenario 2 to show a range of beginning equity value.

Table 3: Beginning Equity Value			
a	b	c	d
Actual	a times 10	Actual	b - c
Scenario — Beginning EBITDA	Beginning Enterprise Value	Beginning Net Indebtedness	Beginning Equity Value
2a $20M	$200M	$100M	$100M
2b $15M	$150M	$125M	$25M
2c $30M	$300M	$100M	$200M
2d $90M	$900M	$300M	$600M

To calculate the numerator—the Ending Equity Value—simply add Equity Value Created from table 1 to Beginning Equity Value in table 3. MOIC can then be calculated by dividing the Ending Equity Value by the Beginning Equity Value, as shown in table 4.

Table 4: MOIC Achieved during the Period				
	a	b	c	d
	Table 3	Table 1	a + b	c / a
Scenario	Beginning Equity Value	Equity Value Created	End Equity Value	MOIC
2a	$100M	$150M	$250M	2.5 x
2b	$25M	$150M	$175M	7.0 x
2c	$200M	$150M	$350M	1.8 x
2d	$600M	$150M	$750M	1.3 x

The last step is to factor in the duration of time in which the value creation took place. The Equity IRR is estimated, and the level of investor satisfaction is assessed. Using 2d as the example, investor's assessment might range from dissatisfied to extremely satisfied depending on the time it took to create the value, as illustrated in table 5.

Table 5: Equity IRR Performance			
	a	b	c	
	Table 4	Actuals	Formula	
Scenario	MOIC	Years	Equity IRR	Investor's Assessment
2d(i)	1.3 x	0.25	100%	Extremely Satisfied
2d(ii)	1.3 x	1.00	25%	Satisfied
2d(iii)	1.3 x	2.00	13%	Neutral
2d(iv)	1.3 x	3.00	8%	Dissatisfied

Investor Alignment

Imagine a company that uses Equity IRR as its fundamental measurement tool. The following conversations would become commonplace:

INVESTOR: How is your business unit performing?

CEO: Very well. Our Equity IRR was 38 percent over the past two quarters and 32 percent over the past year.

INVESTOR: That's excellent! What are you forecasting to achieve over the next two quarters?

CEO: Well, it's at risk of dropping a bit, but it will be respectable. Twenty-five percent is the low end, and 30 percent is still within reach.

INVESTOR: I'll be very pleased anywhere in the 25–30 percent.

The savvy investor would focus on the excellent IRR historical performance and strong forecast. Performance against budget becomes far less relevant. No longer would budget preparation and approval be a contentious negotiation process. The absolute measurement of value creation would replace actual versus budget as the assessment of management's performance.

Summary

Management's responsibility is to create Equity Value at a pace that exceeds the investor's return threshold for an investment of a similar risk profile. As such, Equity Value creation and IRR must be measured. Moreover, incentive systems should be directly tied to the pace of equity value creation.

The Equity IRR value creation methodology provides the ability to measure Equity ValueEquity Value Created and Equity IRR. Better decision-making follows. Over time, the competency and culture of an organization center on delivering Equity IRRs at a rate that will satisfy investors.

MORE BANDWIDTH STORIES

Allstream / AT&T Canada

Allstream was the only competitive fiber platform that served nearly a dozen metropolitan areas stretching from Vancouver through Calgary, Edmonton, Toronto, Quebec City, and Montreal to Halifax in a pan-Canadian network. The company was formed from a combination of two businesses—Unitel Communications and Metronet Communications.

Unitel began in 1967 as CNCP Telecommunications, a joint venture of two of Canada's telegraph companies, one owned by Canadian National Railway and the other Canadian Pacific Railway. Telegraph was already dying fast, and in the late 1980s, as telecom boomed, Canadian Pacific acquired 48 percent ownership of the business, while the cable TV/ media giant Rogers Communications took a 30 percent stake. By 1990, Rogers had taken over the business entirely and rebranded it as Unitel Communications.

AT&T entered Canada in 1993 by acquiring a 20 percent stake in Unitel in exchange for US$31 million in cash and equipment valued at US$117 million. Over the next two years, Unitel struggled, leading Rogers to pass on exercising its right to acquire Canadian Pacific's 48 percent. Within a year, AT&T increased its ownership to 23 percent, the maximum then allowed under Canada's foreign ownership restrictions. The remaining two-thirds ownership shifted to three Canadian banks as an alternative to bankruptcy. Though AT&T's ownership remained capped at 33 percent, it gained management control of the entity.

However, this restriction was expected to be lifted within a few years in the early 2000s. In the late 1990s, Unitel was rebranded as AT&T Canada. Recall that AT&T CEO Michael Armstrong was undertaking an aggressive makeover of AT&T during the Bandwidth Boom. His quest for restoring AT&T's glory included AT&T's ambitious global Concert joint venture with British Telecom. Armstrong envisioned the Canadian portion of this global venture would be AT&T Canada.

Armstrong's ambition led him to broker a 1999 merger of AT&T Canada into the publicly traded MetroNet, with the resulting public entity assuming the AT&T Canada name. MetroNet was founded in 1995 in Calgary, with aspirations of being among the first local competitors to Bell Canada. In 1997 MetroNet went public and hired Brooks Fiber veteran D. Craig Young to be CEO, and his Brooks Fiber colleague and MFS veteran Mark Senda joined him as COO. In 1998 MetroNet acquired a national fiber network for US$131 million and purchased Rogers Telecom for US$690 million from Rogers Communications. These acquisitions positioned MetroNet as the largest competitor to Bell Canada.

AT&T Canada premerger shareholders would own 31 percent of the combined company, while MetroNet shareholders would own the remaining 69 percent. The transaction implied a Market Value of AT&T Canada of US$2.4 billion. In the complex transaction, AT&T would have the right to buy the shares held by MetroNet shareholders, expecting to exercise this right upon the lifting of foreign ownership restrictions that was expected to occur in the early 2000s.

During the meltdown, AT&T faced a multibillion-dollar obligation to fulfill its agreement with Unitel and buy the remainder. AT&T backed out, which sent AT&T Canada into bankruptcy. When AT&T Canada emerged from bankruptcy in 2003, it was rebranded as Allstream. AT&T sold its remaining interests in the Canadian company in 2004, writing off most of its investment.

In 2004 Manitoba Telecom, the incumbent telephone company that served the Manitoba province of Canada, acquired Allstream for US$1.3 billion. Mike Strople joined the company, now rebranded as MTS Allstream, in 2005 and eventually was named CEO.

"By 2011 it was clear that the Allstream business didn't fit with MTS, which led to a formal split of the businesses," Strople reflected when I shared with him a draft of *Bandwidth*. His mandate was to sell Allstream.

In 2013 Allstream entered an agreement to sell itself for US$520 million to Accelero Capital, which was a vehicle of the Egyptian tycoon Naguib Sawiris. Canada blocked the sale over concerns for national security. This left the company fluttering in the wind; Allstream was damaged goods.

"The unwinding of the Accelero purchase didn't feel good at the time, but that wasn't all bad. Allstream was consuming cash. MTS's CEO, not knowing what to do with the business and with zero clarity on who he could sell it to, gave me the freedom to turn it around. We made changes. In 2014 we were cash positive for the first time since the 1990s," Strople explained.

In 2015 Zayo purchased Allstream for $348 million, a bargain profitability multiple of only five times. Zayo remains the only fiber provider with a significant network across the US and Canada, leveraging its nine-thousand-mile network that serves all of Canada's major markets.

"The sale of Allstream to Zayo oddly made everybody happy. MTS was relieved to complete the sale, even at this reduced price; Zayo got a bargain," Strople added, emphasizing that the cash Zayo generated from Allstream in its first three years of ownership exceeded the purchase price.

"I expected the Allstream sale would be the end of my story. Instead I stayed at Zayo, which was probably the best part of the whole Allstream story for me," said Strople.

Strople became a valued partner to me, remaining at Zayo through our 2020 exit. He left in 2024 to become CEO of telMAX, a fast-growing Internet provider in Canada.

Enron Broadband and Dynegy Global Communications

In 1998, as the Bandwidth Boom Gold Rush was picking up steam, Enron disclosed plans to build a national fiber network to offer Bandwidth services. Enron announced partnerships with Touch America (then still named Montana Power) and WilTel 2 to build 1,600 miles of fiber in the Western United States. Enron's Oregon power company, Portland General, would contribute right-of-way for this joint build. Enron also announced plans to use its utility right-of-way to develop routes between Salt Lake City and Houston and between Houston and Miami. From there, swaps would be used to gain access to the Northeast and Midwest.

In early 1999 Enron announced it would partner with CapRock to build a one-thousand-mile Texas network connecting Amarillo, Dallas / Fort Worth, Waco, Austin, San Antonio, and Houston.

In 2000 Enron unveiled plans to pioneer the commodity trading of Bandwidth. In *Fortune* magazine's article titled "Enron Takes Its Pipeline to the Net," Enron president Jeffrey "Jeff" Skilling explained that the "Internet really isn't a network at all. . . . It actually comprises a bunch of separate networks." Enron viewed this as a complexity that would hinder the development of video and other Internet services. Enron would rescue the burgeoning Bandwidth industry by creating a system that would enable Bandwidth trading. "Like a market maker on Wall Street,

Enron will buy up pipeline space from other networks, maintaining an inventory that it can then mark up and sell to customers as needed," reported *Fortune*.

This was a telltale case of a solution looking for a problem. *Fortune* quoted Enron CEO Kenneth "Ken" Lay proclaiming, "I think we could become the preferred platform for e-commerce around the world." *Fortune* added, "If any old-world company could thrive in the Internet era, it's [Enron]."

The initiative was branded the Enron Intelligent Network.

Analysts applauded Enron, with one comparing them to the Great Bambino. "For Enron to say we can do Bandwidth trading is like Babe Ruth's saying, 'I can hit that pitcher,'" said Steven Parla, an energy securities analyst at Credit Suisse First Boston. "You tell him to get up there and take three swings. The risk is staggeringly low, and the potential reward is staggeringly high."

Wall Street was enamored, as the *Washington Post* reported in its January 2002 article titled "Broadband Strategy Got Enron in Trouble": "Asking few questions, investors sped to buy the stock. . . . Enron shares shot up from $40 in January to more than $70 in less than two months and went as high as $90 that summer."

WorldCom's John Sidgmore didn't get it. "Honestly, what possible expertise could Enron have to help in the communications industry? They have zero experience that I know of."

Enron executives began acting as though they agreed with Sidgmore's assessment. The *Washington Post* reported, "The stock price gains created a windfall for Enron executives and directors, who sold $924 million of company stock in 2000 and 2001. Kenneth D. Rice, who ran Enron Broadband Services, sold more than 1 million Enron shares for a total of $70 million in 2000 and the first half of 2001, according to regulatory and court filings. He left the company in August."

In May 2002 the *Wall Street Journal* published "Watson, Who Long Led Dynegy in Enron's Shadow, Steps Down."

> Almost from the time he helped found what became Dynegy in 1985, Chuck Watson worked in the shadow of a better-known company not far from him in Houston: Enron Corp.
>
> When Enron began ramping up its electricity trading in 1994, Dynegy followed with similar trades in 1995. After Enron took over an electric utility in 1997, Dynegy bought one two years later. When Enron launched an Internet-based commodities-trading platform in 1999, Dynegy was out with its own within months.
>
> And when Enron went into trading of capacity on high-speed broadband data lines, Dynegy again was right behind.

Dynegy, like Enron, was headquartered in Houston, Texas. Through a series of acquisitions, CEO Charles "Chuck" Watson grew Dynegy from a tiny regional energy company to a Fortune 500 enterprise.

Between 1999 and 2001, Dynegy's revenue grew from $15 billion to $42 billion. *Forbes* reflected on this burst in its May 2002 article titled "All the Reasons Chuck Watson Did Not Resign."

> This Enronesque pace would seem impossible from asset-based activities. This growth did not happen by accident. It certainly seems to have had something to do with Dynegy's share price rising from just under $9 to over $57 between April 1999 and April 2001. But the story behind the revenue growth has never been fully told.
>
> Watson . . . said, "Don't assume that financial complexity is wrong, evil or unnecessary. Our business is extraordinarily

intricate and we must employ unique and sophisticated financial arrangements to fulfill our dual responsibility to deliver
energy safely and manage financial risk."

Watson left out the part about the fraud, which led to prison time for
members of his accounting team.

Like Enron's management team, Watson saw an opportunity to use
the Bandwidth Boom to bolster Dynegy's financial performance. Dynegy
Global Communications was formed in 2000 when Dynegy acquired 80
percent of Colorado-based Extant for $152 million. Extant was founded
by Larry McLernon, who previously founded LCI International, which
was acquired by Qwest. Extant, it seemed, was McLernon's attempt to
remain relevant. It leased fiber from Level 3 and others to connect
twenty-eight cities as part of an ambitious plan to reach more than one
hundred cities by the end of 2003.

Dynegy was following its rival, Enron, off a cliff. "This is not a small
strategic step, it is a huge strategic step with a relatively small investment—
less than 10 percent of [Dynegy's] capital budget over the next few years,"
declared Watson as he announced Dynegy would spend $400 million to
complete Extant's network.

Soon thereafter, in September 2000, Dynegy's stock hit $59 a
share—its high-water mark.

Later in 2000, Dynegy jumped into the Europe Bandwidth market
by acquiring London-based iaxis for $200 million. iaxis operated an
8,750-mile Bandwidth network that connected forty data center facilities
across France, Germany, Italy, Spain, and the UK.

"This acquisition firmly establishes the platform for the expansion
of Dynegy's network connectivity and broadband communications
strategy into Europe and is an ideal complement to Dynegy's existing
US operations," Watson touted. "For a cost of less than $200 million,
Dynegy will have a significant presence in the European communications marketplace."

Dynegy doubled down in 2001 by partnering with another Colorado upstart, Telseon, to jointly build out networks in eighteen of the largest US markets. Dynegy committed $75 million and planned to combine Telseon's metro fiber with the intercity fiber associated with the Extant acquisition. John Kane, ICG's first CEO, was now Telseon's chief.

Pat Marburger, the president of Dynegy Broadband Trading and Origination, explained, "It's literally like taking a freeway with no off-ramps and adding a bunch of off-ramps, so traffic can flow into multiple locations. It gives us much greater reach into most major markets in the US." Dynegy positioned that they shaved twelve to eighteen months off their schedule and cut their capital outlay in half.

Watson believed his approach to the Bandwidth opportunity was more clever than that of his energy counterparts at Williams Companies and Enron Broadband. "To build fiber like Williams and Enron is extremely capital intensive. For them it's an asset play. We believe it's an intellectual play. We call what we're doing smart-build," said Watson. Like Enron, Watson planned to apply Dynegy's experience in energy commodities trading to Bandwidth.

In reality, Dynegy was buying struggling businesses that leased facilities from other carriers with hopes to resell commoditized Bandwidth in an increasingly overcrowded market.

———

Scott Yeager, Enron Broadband's senior vice president of strategic broadband and a former MFS colleague, led an Enron delegation to Level 3's Boulder headquarters. He described the Enron Intelligent Network and its Bandwidth trading ambitions. We were skeptical. When we persisted in asking questions about Bandwidth futures exchange, Bandwidth swaps, and accounting treatment, they would look at each other and say, "They just don't get it." I recall wondering what I was missing. Their

"Smartest Guys in the Room" tagline, which is the title of a fantastic book on the broader Enron debacle, was well earned.

Later, I visited Yeager at their headquarters in Houston, expecting him to introduce us to Enron's COO, Jeff Skilling, and CFO, Andrew S. Fastow. Showing signs of anxiety, Yeager left the room to fetch them only to return a half hour later saying they were busy. Before long, it became clear what was occupying their time.

———————

The year 2001 was not a good one for Enron. In August Jeff Skilling quit, with Chairman Ken Lay assuming the CEO title. Enron reported a big third-quarter loss. In October the SEC opened a formal investigation, with reports that Enron executives were enriched through undisclosed partnerships being part of the probe. Andrew Fastow was replaced. Enron reduced previously reported earnings. In November Dynegy signed a letter of intent to rescue Enron through a merger that would value Enron at $8 billion. Dynegy advanced $1.5 billion to Enron, which was backed by Enron's oil pipeline assets. Just three weeks later, as the full magnitude of Enron's situation became increasingly clear, Dynegy pulled out of the intended merger. Just ten days before Dynegy's pull out was made public, Lay's wife, Linda, sold 500,000 Enron shares owned by their foundation. On December 2, with its stock trading well below $1 per share, Enron filed for bankruptcy.

In January 2002 the *Forbes* article titled "Shell Game" described the elaborate partnerships Enron used to mask its financial performance, while Lay and Skilling sold millions of dollars' worth of shares. *Forbes* cited an accounting professor describing the scheme. "It's like somebody sat down with the rules and said, 'How can we get around them? They structured these things to comply with the letter of the law but totally violated the spirit."

Enron's Bandwidth trading junket played a major role in Enron's deception. *Forbes* elaborated,

> In June 2000, for example, Enron sold $100 million worth of [Dark Fiber]. The "buyer" was a partnership run by Fastow called LJM2 (the acronym reportedly comes from the initials of his wife and children), set up in 1999 to trade assets with Enron. On that deal, Enron booked a $67 million profit, a significant piece of the $318 million gross profit the company reported for the broadband business in 2000. LJM2 later sold $40 million of the Dark Fiber to what Enron refers to as "industry participants," and the remainder to another Enron-related partnership for $113 million in December. What's curious is that the value of the fiber ostensibly increased 53% between June and December—during the same time that, in open markets at least, the value of Dark Fiber plunged by 67%. LJM2 reaped a $2.4 million profit from the fiber trade, contributing to the $30 million of undisclosed gains the LJM partnerships delivered to Fastow."

An analyst told the *Washington Post*, "[Enron executives] were great spin doctors. They had the answers. The answers were lies."

Insiders also knew it was a fraud. "From a common-sense standpoint, we knew it was smoke and mirrors," a former Enron Bandwidth trader told the *Washington Post*. A Global Crossing participant called it "the Houston poker game."

Enron and Qwest used the swap game to help each other hit financial targets. In early 2002 the *New York Times* reported the two companies "raced to complete the transaction as the third quarter was ending in September. Enron and Qwest valued the transaction at more than $500 million." Executives close to the deal told the *New York Times* that "the swap helped Qwest soften a deteriorating situation in profit and revenue at the end of last year's third quarter. The deal also allowed Enron . . . to avoid recording a huge loss by selling an asset whose value had plummeted on the open market." A former Enron executive said Qwest's pitch was "We will overpay for the assets, and you will overpay me on the contract."

Multiple executives from Enron Broadband Services, including my former colleague Scott Yeager, were indicted on charges of conspiracy, fraud, insider trading, and money laundering. NBC News reported testimony from the 2005 trial about the software that Enron touted as enabling Bandwidth trading: "'It was referred to as the secret sauce or pixie dust,' former director of product engineering Shawna Meyer told jurors. . . . 'Sprinkle it around or add some secret sauce to the network and it would solve all the problems that we had.'"

On July 7, 2004, Lay was indicted on eleven counts of securities fraud, wire fraud, and making false and misleading statements. The trial commenced in early 2006. In May the jury found Lay guilty on six counts of conspiracy and fraud. While awaiting sentencing, Lay died from a heart attack while at his home in Aspen, Colorado.

Skilling was convicted of federal felony charges and was found guilty of conspiracy, insider trading, making false statements, and securities fraud. He was sentenced to twenty-four years in prison and fined $45 million. He served twelve after multiple appeals resulted in a reduced sentence of fourteen years.

CFO Fastow was sentenced to a six-year prison sentence and spent five years behind bars. His wife, Lea Weingarten, who worked at Enron and participated in the fraud, was sentenced to twelve months.

Yeager from Enron was acquitted on six fraud counts in an indictment that charged that to inflate the value of Enron's stock, he deceived the public about the value of Enron's project to develop a nationwide fiber communications system. The jury hung on the remaining insider trading and money laundering counts. The Supreme Court, in a 6–3 vote, issued a ruling that barred a retrial on these two remaining counts.

tw telecom purchased Enron's fiber assets as part of a competitive auction in early 2004. The acquired 4,200-mile network served twenty-one states, with a concentration in Oregon.

A similar fate to Enron's played out at Dynegy. By late 2001 Watson used Dynegy's third-quarter earnings call to paint a rosy picture, declaring that the industry had touched bottom and demand would pick up. "The tragic events of September 11 means more video conferencing than before," Watson predicted. He added, "There will be a permanent change in the way of doing business after what happened."

Watson saw Enron's late 2001 collapse as an opportunity. The *Wall Street Journal* explained,

> Then when Enron began to crumble amid disclosures of accounting irregularities, Mr. Watson pounced, offering last November to buy out his longtime rival and mentor for about $9 billion. He soon dropped the takeover bid, though, and began stressing how different Dynegy was from Enron. His company was a sturdy alternative to Enron built upon hard assets, not accounting monkey-business, he said.

Well, Dynegy was not sturdy. In early 2002 the SEC launched an investigation of Dynegy's energy trading activities and an off–balance-sheet entity—code-named Project Alpha—that was used to evade taxes. In May 2002 Dynegy's board terminated Watson.

Strangely, the board characterized the departure as a "constructive termination" instead of a "for cause" firing. As such, Watson was paid $33 million in severance—far in excess of the $7 million he would have made had they retained him for the remaining eight months of his employment contract. Watson was already wealthy via a sale of $250 million of Dynegy stock in 2000.

Dynegy characterized Watson's departure as a response to "an industry that is changing more rapidly and profoundly than ever before . . . new opportunities . . . new challenges [and] events of the past year have hurt the credibility of this sector and eroded investor confidence."

Unlike with Enron, Dynegy's Bandwidth excursion was not a factor in the SEC investigation. However, it did contribute to the financial duress. In June 2002 Dynegy disclosed an asset write-down of $450 million attributed to its Bandwidth dalliance.

Dynegy sold its European Bandwidth business in January 2003. "This is a strong first step in Dynegy's strategic commitment to divest its investment in communications and focus on our core energy businesses," said Dynegy CEO Bruce Williamson, Watson's successor.

Three months later Dynegy sold the US Bandwidth business. "The completion of this transaction will end our participation in the communications sector, enabling us to focus entirely on the energy business." At the time of the sale, Dynegy's network served forty-four markets with sixteen thousand miles of fiber.

"[Dynegy] no longer has to pour money into it," a Raymond James analyst told *Midland Daily News.* "You can sell anything, it's the price that's the question. I'll be very interested to see what they got for it and if they were paid in cash." My guess is they received very little for either the US or the European business. Dynegy was satisfied to just slam the door shut on its Bandwidth business.

On the day Dynegy announced its exit from Bandwidth, its share price closed at $2.49 a share, down 95 percent since its 2000 entry into the Bandwidth business.

Three Dynegy executives were indicted by the SEC for improperly boosting cash flow and reducing taxes. US attorney Michael Shelby cited Jamie Olis, Gene Shannon Foster, and Helen Christine Sharkey for "withholding the truth about Dynegy's true fiscal condition from the SEC, shareholders and the public." Olis, Foster, and Sharkey were members of Dynegy's tax and risk management group.

The SEC determined that Dynegy used inappropriate accounting transactions to bolster cash flow by $300 million and cut taxes by $79 million. Dynegy paid the SEC $3 million to settle the investigation without admitting wrongdoing. Foster and Sharkey pleaded guilty and agreed to cooperate with the SEC investigators. Foster ended up with fifteen months behind bars, a $1,000 fine, and three years' probation. Sharkey served thirty days in prison and was required to pay a $10,000 fine immediately. Olis was found guilty and received a twenty-four-year sentence, later resentenced to six years.

Watson, despite being CEO when the fraud led to a massive overstatement of financial performance, was never charged.

Lightower

M/C Partners was focused on more than just Zayo. In 2007, the year of Zayo's launch, M/C partnered with Wachovia Corporation to purchase the US subsidiary of the international power company National Grid for $290 million. The asset base was a 1,100-mile fiber network and 350 wireless towers in the Northeast. The new entity was named Lightower Fiber Networks.

In 2008 Lightower sold its tower assets to SBA. Lightower then made two acquisitions that put it on a pathway to be a meaningful rival to Zayo. They acquired DataNet Communications Group and KeySpan Communications. The fiber network tripled to 3,600 miles, including Boston, Albany, Manhattan, Long Island, New Jersey, and Connecticut. It featured unique fiber routes that connected transatlantic fiber landing stations to Manhattan.

John R. Galanti started Datanet Communications Group in 2000 with the backing of Quadrangle Capital Partners. In 2007 I talked to Galanti about Zayo acquiring his company. I should have been more aggressive. His price seemed high to me relative to the quality of assets and the geography. With the benefit of hindsight, the price was very reasonable, and the assets would have bolstered Zayo's early collection of assets. I should have jumped on the opportunity, especially since Datanet provided a big boost to Lightower. I remember the deals I misplayed just as much as I recall my successful acquisitions.

In 2009 Rob Shanahan, the co-founder and former CEO of Conversent Communications, took over for Doug Wiest as Lightower's CEO. Shanahan remained CEO for the last eight years of Lightower's impressive journey.

Lightower continued its regional rollup strategy. In May 2010 the company purchased Veroxity Technology Partners, which added two thousand miles of fiber in Boston and Connecticut and a customer base that was nicely concentrated in the financial services vertical.

Next up in 2010 was Lexent Metro Connect, which added 200 buildings and 150 miles of fiber concentrated in the financial exchanges and data centers of New York City and New Jersey. Lexent featured low-latency routes between these financial trading locations. Ray La Chance, another one of our industry's good guys and a true New Yorker, built Lexent over six years.

In early 2011 Lightower followed up with NSTAR Communications, which consisted of the fiber assets of the NSTAR energy conglomerate. Some 280 miles of fiber along with 225 buildings were added in the Boston metropolitan area.

In early 2011 Lightower perhaps had a more valuable set of assets than Zayo, as the revenue mix was more weighted on infrastructure, and the assets were concentrated in the uniquely important Northeast region. Perhaps Lightower instead of Zayo was destined to be a leader of the consolidation. We each had the opportunity to make a bold move—Sidera Networks was shopping itself to both of us. Shanahan took the bait.

The prior name of Sidera Networks was RCN Business Solutions, a subsidiary of the Cable TV company Residential Communications Networks Corporation (RCN). The backstory of RCN intersected with MFS.

When I joined MFS in the early 1990s, Kiewit was exploring a parallel venture to MFS. Before joining MFS, I participated in Ameritech's study of the Cable TV industry. Kiewit's analysis focused on leveraging MFS's fiber assets to provide service to consumers. The result of this initiative was America's first facilities-based provider (overbuilder) of bundled telephone, Cable TV, and Internet service. The leader of this initiative was David McCourt, who joined Kiewit when MFS purchased his Boston company—Corporate Communications Network—in the early 1990s.

McCourt began his career as a probation officer in Washington, DC, but shifted early on to launch a construction company named McCourt Cable Systems. Like Steve Garofalo of MFN, McCourt learned about fiber networks by designing and building networks for others, particularly Cable TV providers.

McCourt was a charismatic guy, but he rubbed a lot of people, including me, the wrong way. He came across as arrogant, only wanting to spend time with those whom he considered power brokers. Similar to a lot of the grand storytellers of the Bandwidth Boom, he focused far more on the boldness of the strategy than the execution details.

McCourt observed Crowe's progress with MFS and made his case that Kiewit should sponsor him in a residential communications strategy. His vision was to offer a comprehensive set of communications services aimed at residential customers that included voice, Internet, and Cable TV services.

To execute his strategy, McCourt convinced Kiewit to back him in the acquisition of the public company C-TEC Corporation, which operated both phone and Cable TV properties in the Northeast. His plan was to use C-TEC as a platform to expand service into adjacent geographies using MFS fiber. After years of jiggering, he carved out RCN as the overbuilder competitor to both the telephone and Cable TV company. His strategy was aimed at densely populated areas along the Boston to Washington, DC, corridor, where he could target high-rise locations.

In 1996 RCN acquired Liberty Cable, a company providing television services to forty thousand subscribers in multitenant New York City buildings through wireless antennae. RCN's thesis was to replace the antennae with fiber and offer the full set of communications services.

McCourt raised $1.7 billion in 2002 to fund a mosaic of other acquisitions that were at best loosely related to his strategy. To buy time during the meltdown, RCN generated $245 million by selling its original C-TEC New Jersey cable properties. However, McCourt couldn't put off the inevitable.

In July 2004 RCN declared bankruptcy, and McCourt resigned as CEO. Level 3's 21.6 percent ownership became worthless.

On December 31, 2004, RCN emerged from bankruptcy. A total of $1.2 billion in debt was converted to equity valued at approximately $700 million. Peter Aquino replaced McCourt as CEO.

A year later RCN created the subsidiary RCN Business Solutions, coinciding with its acquisition of Con Edison Communications from the power utility Consolidated Edison. This bolstered its fiber network in New York City. The following year, 2007, RCN acquired NEON Communications, a 4,800-mile regional intercity network that connected data centers and carrier hotels from Maine to Virginia. Another name change followed, this time to RCN Metro Optical Networks.

ABRY Partners, a Private Equity firm in Boston, had close relationships with Zayo's investors Charlesbank and M/C. They proposed that ABRY sponsor Zayo's acquisition of RCN, combining RCN Metro Optical Networks and selling off the core RCN Cable TV business. Though intriguing to me, I was concerned with the complexity of the transaction. ABRY decided to proceed on its own.

In 2010 ABRY acquired RCN for $1.2 billion. They rebranded RCN Metro Optical Networks as Sidera Networks and separated it from RCN.

Later that year, Sidera made two acquisitions that reinforced its transition to a Bandwidth Infrastructure provider. They acquired Cross Connect Solutions in Philadelphia and Long Island Fiber Exchange.

I continued to converse with ABRY about merging Sidera into Zayo, but I did not have the conviction to proceed forward. Sidera had interesting assets in the valued Northeast corridor, but much of its network was intertwined with the Cable TV network of RCN. Its revenue base included a high proportion of smaller enterprise customers.

In December 2012 Sidera announced a merger into Lightower. ABRY remained an investor. Berkshire Partners, another Boston investment firm, became a new investor in the combined company. A $2 billion value was placed on the combination of Lightower and Sidera.

Berkshire was also a firm that knew Zayo well. Whenever I visited Boston, I would bounce between M/C, Charlesbank, ABRY, and Berkshire. As with ABRY, we didn't find a way to bring Berkshire on as an investor despite our mutual interest.

———

Lightower's final acquisition was Fibertech. Stephen Garofalo of MFN created the Dark Fiber product. However, John Purcell, the CEO of Fibertech, showcased how to navigate the Dark Fiber business model to create enormous stakeholder value.

Fibertech's origin story starts with Rochester Telephone Corporation, which was founded in 1920 as a merger of Rochester Telephonic Exchange and Rochester Telephone Company. In 1995 the company listed on the NYSE and rebranded itself as Frontier Corporation. Global Crossing bought Frontier in 1999 and sold it to Citizens Utilities Company in 2001. Citizens changed its name to Frontier Communications.

Purcell joined Rochester Telephone in 1965 after graduating from Le Moyne College. He remained there for over thirty years, slowly working his way to lead Frontier's regulatory group as vice president during the period of rapid deregulation. Purcell became vital to Frontier's open market plan, which was the first time a telephone company sought to introduce competition in the local telephone market.

Frontier was a good environment for developing entrepreneurial talent in the otherwise stale telecom industry. Many of its executives led competitive fiber companies, including Rob Shanahan of Lightower, David Rusin of AFS, and Arunas Chesonis of PaeTec.

In 2000 Purcell left Frontier to team up with his Rochester Tel colleague, Frank Chiaino, and launch Fibertech. Their simple strategy was to be the underlying Dark Fiber infrastructure for voice, Internet, and managed service providers.

As part of the commercial transactions, Fibertech's customers funded much of the capital expenditures to build out the networks. Fibertech, in turn, provided Dark Fiber to the customers, so they didn't need to buy optronics. Not much outside capital was needed from investors. Fibertech's profitability margins were an extraordinarily high 75 percent. A handful of sales professionals, product managers, fiber engineers, and network operators were all that was needed. The financial and operational systems were simple. After all, it takes little effort to send out invoices to a few dozen customers that leased fiber under fixed-rate, long-term contracts. Revenue and customer churn was nonexistent because Fibertech's customers needed the fiber to survive.

Unlike almost every other fiber company launched in 2000, Fibertech was unaffected by the meltdown. Purcell and Chiaino steadily built their Dark Fiber business. They sold it to Lightower in 2015 for $1.9 billion. Though I loved FiberTech's business model, the $1.9 billion price tag was steep for me, especially considering the lower-tier markets that FiberTech served. But it turned out to be a good acquisition for Lightower.

Bolstered by the 2015 acquisition of Fibertech, Shanahan and his investors were ready for their well-deserved payday. Crown Castle was destined to be their exit; Lightower was the largest Bandwidth pure play and fit neatly into Crown's strategy. However, I wasn't going to make it easy on Crown. I battled Crown all the way to the finish line, topping its $6.9 billion bid with a $7.1 billion Hail Mary. They matched my bid and completed the purchase.

Shanahan and the Lightower team produced an exceptional outcome for their investors. Rob, you were a worthy and friendly rival. Well done.

McLeodUSA

Clark McLeod was a junior high math teacher in 1970s Cedar Rapids, Iowa, when MCI won a landmark legal battle with AT&T. The ruling opened the floodgates to competition for long-distance voice traffic, and McLeod made an unexpected career pivot. In 1981 he formed TeleconnectUSA to compete with AT&T. Renamed TelecomUSA, the company grew to be the fourth-largest, long-distance provider, and McLeod sold it in 1990 to MCI for $1.25 billion.

Earning $50 million from the sale, McLeod turned to philanthropy. But he later told *Forbes* he had surmised he "was better at creating jobs for people than giving money away." And so McLeod jumped back into the game, launching McLeodUSA, with an aspiration to compete against the Baby Bells. "The 1980s were 'a bake sale' compared to now," McLeod said to *Forbes*. He added, "There are more disruptive trends going on than in any other business."

McLeod emphasized that his strategic focus would be plain old telephone service, "POTS" in old Bell System jargon. "The other [local competitors] didn't identify the opportunity; they all looked at the Internet. Is the Internet controlled by a monopoly? Hell no, there are thousands of providers. What is the protected monopoly? Local services."

In describing his superregional strategy to the *Wall Street Journal*, McLeod explained, "There's a real market to serve customers in the second-and-third-tier markets on a regional basis. . . . Nobody else in the country is doing this. The bigger phone companies are either too focused on merging or on going international."

In 1996 McLeodUSA raised $240 million in a Nasdaq IPO priced at $20 per share.[1] Salomon Smith Barney executed the IPO. As expected,

---

1 Adjusting for a stock split that occurred a few years later, the IPO stock price was $3.33 per share.

Salomon analyst Jack Grubman launched coverage with a buy recommendation and a twelve-month price target of $40. "McLeod represents one of the truly great business models that will be executed in the new era of telecom. [McLeodUSA will be] one of the best return vehicles in what will be a high-return segment of the telecom industry."

In 1997 McLeodUSA acquired Consolidated Communications, a one-hundred-year-old rural phone company based in Mattoon, Illinois. CEO Richard Lumpkin was the great-grandson of the founder, Iverson A. Lumpkin, and the fourth Lumpkin to run the company. The $400 million price tag consisted of $155 million in cash, $60 million in assumed debt, and 8.5 million shares of McLeodUSA stock, then valued at $24.50 per share.

"These are markets McLeodUSA itself would have likely entered, thus a potential competitor has been eliminated," Grubman wrote when praising the acquisition. In the same year, McLeodUSA formed an alliance with AT&T to offer local phone service to business customers in thirty markets. The stock rose to $35.75.

In 1999 McLeodUSA purchased Ovation Communications for $500 million, which provided services in Milwaukee. McLeod once again offered a mix of cash ($141 million), assumed debt ($83 million), and stock (5.1 million shares). Ovation's CEO, Tim Devine, a former MFS colleague of mine, said, "We have carved out a niche as a super-regional [local phone firm] in the Midwest."

The Ovation acquisition included Phone Michigan,[2] which operated in Flint, Michigan. Additionally, McLeodUSA funded the ongoing three-hundred-mile expansion into Detroit.

Later in 1999 McLeodUSA bought Access Communications, which served Arizona, Boise, Las Vegas, New Mexico, Oregon, Salt Lake City, and Washington State. The $250 million price consisted of $50 million in cash, $97 million in assumed debt, and 1.9 million

---

2 Ovation acquired BRE Communications, which did business as Phone Michigan.

shares valued at $101 million. At this point McLeod's stock had reached $53 per share.

McLeodUSA caught the attention of Teddy Forstmann of the famed leveraged buyout firm Forstmann Little & Co. McLeod told the *Washington Post* in 2002,

> [Teddy] looked me in the eye and said, "How much funding do you need to execute on your existing plan?"
>
> "A billion dollars," McLeod answered.
>
> "That's what we want to put into the company," Forstmann said, and it was done.

In 1999 Forstmann Little invested $1 billion for a 12 percent stake, which implied a value of over $10 billion, including debt. The price per share was $36.50, a 15 percent premium to McLeodUSA's most recent stock price. As the twenty-first century was about to begin, McLeod, the erstwhile junior high math teacher, was armed and dangerous and ready to go national.

In January 2000 McLeodUSA acquired Splitrock for $2.1 billion. This gigantic purchase price for a company with little revenue and few customers was curious, especially since Clark McLeod was on the board. Roy Wilkens, the former CEO of WilTel, was also a Splitrock board member. He joined McLeodUSA's executive team to lead the newly created data services business unit.

Later in 2000, McLeodUSA purchased CapRock for $535 million, adding a 5,200-mile fiber network in the Southeast, along with a strategic relationship with AT&T, to its platform.

McLeod wasn't alarmed by the growing tension in the Bandwidth markets. In fact, he welcomed it as an opportunity. In January 2001, when most telecom providers were shut out of the markets, McLeod raised $750 million in high-yield debt. "McLeod has seen this mess before: Wall Street loves telecom deregulation, so it throws billions at bad businesses. A bloodbath ensues. The survivors remake their markets. And he gets rich," explained *Forbes* in its March 2001 article titled "Conqueror in the Carnage."

Despite the bravado, McLeodUSA couldn't escape the quicksand. In May 2001 McLeodUSA lowered its revenue guidance by $100 million. Its stock price, which was already down sharply from its 2000 high, fell 31 percent to $6.77. McLeod announced it was abandoning its national strategy and returning to its focus on its thirty-one-thousand-mile fiber network that served twenty-five states.

In July McLeodUSA drew $175 million of its secured debt facility.

By August 2001 McLeodUSA's stock traded below $3 a share, down 90 percent from the $53 price at the time of the Splitrock acquisition. Salomon's Grubman explained that the company "had some missteps in the last year and a half, most notably the ill-advised acquisition of Splitrock." He sidestepped his own glowing praise for the deal when it had been announced and that Salomon had advised McLeodUSA in support of the $2.1 billion price.

Forstmann Little invested an additional $100 million in August to stabilize the company. Forstmann also agreed to forgo $175 million of dividends, over five years, due to it from its year 2000 $1.0 billion investment. Steve Gray was named CEO, with Clark McLeod retaining only his chairman position. Teddy Forstmann became a member of McLeodUSA board's executive committee.

"Wall Street is all about greed and fear," Forstmann explained to the *New York Times*. He added, "These guys were taken on the greed ride and then on the fear ride. We're now going on the reality ride. This will either make me look very smart or very dumb." Dumb turned out to be the answer.

In October an additional $200 million of the secured debt was drawn. McLeodUSA drew $175 million of its secured debt facility, leaving $342 million available under its $1.3 billion facility.

The stock continued its free fall, dropping below $1 a year by November. McLeodUSA wrote off $2.5 billion, attributing most of it to the Splitrock acquisition.

In November 2002 the *Washington Post* published "A CEO's Lesson: What Goes Up . . ."

> Before their money vanished with their faith, Clark E. McLeod's investors could sometimes find him just by glancing skyward. It was spectacular to stand there on the edge of a cornfield, crane the neck and look up at his cream-colored jet hurtling by. A state-of-the-art Gulfstream G-5, with sleeper cabins and a flight attendant, glinted in the sun, the $39.5 million chariot of a CEO who had joined the ranks of American technology demigods.
>
> A snapshot of an epoch: the great American ride, not long before the crash.
>
> "He's up there," someone would say, and there would come gruff murmurs of awe from men seldom amazed by anything.
>
> A self-made billionaire who had risen from hard days when he puttered around in a sputtering Volkswagen, McLeod seemed above them, somehow, and this they didn't mind—in the way lucky people seldom resent a man they regard as a benefactor. McLeod's luck seemed to float down to his hometown, so even when he exercised stock options to the tune of about $98 million in February 2000, who could begrudge him when middle-income investors in his company, McLeodUSA and other telecoms sometimes made $50,000 in a day, and riches seemed to be their reward for simply believing? And he was one of their own, after all, a former schoolteacher who grew up just three miles away in the town of Marion, in the middle-class home of his educator parents.

Speed and ascent, ambition and acquisition: They were the paramount ethics during the 1990s in technology and telecommunications, where a titan's climb seemed limitless and his fall improbable, where no one had forecast the spiral coming. Up in the pantheon of American chief executives, McLeod pushed hard, and harder still—too fast, as he later acknowledged—and on borrowed wings at that, far beyond a safe and rational point.

The *Washington Post* described the life that McLeod created for himself.

At Clearwater Farm, where McLeod and his wife, Mary, raised thoroughbred dressage horses, cars stopped and dreamers furtively spilled out to stare, able to see the horses roaming in meadows near McLeod's recently built equestrian arena, with exotic Teflon-coated, dust-free gravel.

Inside, his friends reported with awe, a mirror nearly the length of a football field offered a rider a flattering view of himself atop his steed.

Those not fortunate enough to be among his official visitors queued up nearby to glimpse slivers of his house. It sat behind white pines and evergreens he had ordered uprooted and trucked in from out of state, only heightening the sense of a great man's removal. Tucked out of sight were man-made ponds he had stocked with bass and koi, near a 20-foot waterfall.

He lived like Gatsby.

In 2000 McLeod self-published a book called *This Way Up*. The *Washington Post* pointed out the hypocrisy.

Integrity was high on his list, he wrote, but it was also the value that people "fail at most." He pledged to lead the way to success.

"Integrity is doing what's right. Always. . . . Integrity requires saying what needs to be said, when it needs to be said, even when it's not easy to say it."

The (shareholder) lawsuits charged that McLeod had not said what needed to be said when it was not easy to say it—that he had failed to be candid when the company began careening, and that when he did speak, he and others materially misrepresented the company's ills.

In December McLeodUSA sold the wholesale dial-up access business to Level 3 for $55 million and associated liabilities. This was Splitrock's primary revenue stream, which was associated with the access services provided to Prodigy.

Also in December, McLeodUSA warned it might file for Chapter 11. A prepackaged restructuring was agreed to with Forstmann Little. The plan involved McLeodUSA selling its directory publishing business, with Forstmann Little agreeing, if no alternative buyer emerged, to acquire the business for $535 million. Yell Group of Britain stepped up and bought the business for $600 million.

The proceeds were used to buy back the $2.9 billion of bond debt for $670 million cash and 15 percent of McLeodUSA post-recapitalization stock. The $140 million revolving credit line and the $1.3 billion in bank debt would be reduced by a few hundred million dollars with proceeds from asset sales.

Forstmann Little invested an additional $175 million, and it also agreed to convert its $1 billion existing preferred stock into common stock. After restructuring, Forstmann Little owned 58 percent of McLeodUSA and a total of 30 percent of the restructured stock would be granted to the public shareholders. In April 2002 McLeodUSA emerged from bankruptcy and returned to Nasdaq.

Forstmann explained his logic in a note to his investors: "By eliminating all this excessive debt and getting to an appropriate capital structure, we believe that the value of this company can be improved to the point that will at least allow us to get even on our total investment." Well, that isn't how the game played out.

Before its sale to McLeodUSA in 1997, Consolidated Communications had been in the Lumpkin family for one hundred years, with four generations of Lumpkins serving as the CEOs. In July 2002 Richard Lumpkin, the fourth CEO, repurchased Consolidated with the backing of Spectrum Equity Investors and Providence Equity. Lumpkin paid $271 million for the asset he sold five years earlier for $400 million, basically equivalent to being paid $130 million to take a five-year sabbatical. Great-granddad would have been proud.

McLeodUSA used the proceeds from the Consolidated sale to reduce its bank debt from $960 million to $715 million.

In the late 1980s, Ted Forstmann's golf partner asked him what it meant to be taken over by a buyout firm. "It means the barbarians are at the gates" was Forstmann's memorable reply.

Barbarians at the Gates: The Fall of RJR Nabisco, written by Bryan Burrough and John Helyar in 1989 and considered one of the best business books ever, featured Forstmann of Forstmann Little battling Henry Kravis of KKR for control of the tobacco and food conglomerate that owned Camel, Newport, Oreos, and Fig Newtons. KKR prevailed.

Forstmann founded Forstmann Little in 1978. Before his dallies in *Bandwidth*, Forstmann earned a whopping 58 percent compounded annual return on investment. The firm's most successful investment was Gulfstream Aerospace, which Forstmann acquired in 1990 for $800 million. Forstmann appointed himself chairman and CEO and, over his seven-year reign, turned Gulfstream into one of the most successful jet manufacturers. In 1999 he sold Gulfstream to General Dynamics for $5.6 billion, earning more than $5 billion in return for the firm's investors and public shareholders.

Chris A. Davis was on Forstmann Little's Gulfstream turnaround team, where she was CFO and a board member. In August 2001 she joined McLeodUSA's newly created turnaround team—code named Gulfstream II. In April 2002 she was named chairman and CEO, replacing McLeod and Gray. Akin to The Who's 1989 "The Kids Are Alright" reunion tour, Forstmann's band was back together and ready to rock.

One year later Davis put a positive spin on McLeod's restructuring: "We've done a lot of refocusing and rebuilding of the company and significantly improved the balance sheet." However, the top and bottom-line financial performance of the business continued to struggle.

Two years passed and the financial challenges continued. In the second quarter of 2005, McLeodUSA reported $160 million in revenue, which was down from $192 million from the same quarter of 2004. McLeod reported a gigantic loss of $268 million for the quarter. The stock price, which traded below $1 for the previous twelve months, was down to $0.02 a share, causing a delisting from Nasdaq. McLeodUSA warned of a second bankruptcy. McLeodUSA also disclosed it had been trying to sell itself off but found no buyers.

In August Davis resigned as CEO. McLeodUSA hired Alvarez and Marsal, and Alvarez put a chief restructuring officer in charge. In October 2005, three years after the first bankruptcy, McLeodUSA filed a petition for its second bankruptcy.

Forstmann's $1.175 billion investment was worthless.

In January 2006 McLeodUSA emerged from bankruptcy. Its $677 million of debt was reduced to $73 million, and with $23 million of cash on its balance sheet, its net debt was now $50 million. Secured junior bank debt was converted into equity of the restructured entity.

Royce Holland, the former COO of MFS and, more recently, the former CEO of Allegiance Telecom, was named the new CEO. "We will be focused on small and medium sized business customers, particularly those that have a footprint matching our 25-state network."

In March 2007, one year later, McLeodUSA filed to take the company public on Nasdaq with plans to sell $172.5 million in common stock. Instead, McLeodUSA agreed to sell to PAETEC Holding for $492 million, plus the assumption of $65 million of debt. PAETEC was sold to Windstream in 2011 for $2.3 billion. In 2019 Windstream filed for Chapter 11 bankruptcy.

Holland presumably had at least 5 percent of the equity, which means he made more than $25 million for this two-year effort. Nice job, Royce!

McLeodUSA and its executives were involved in several lawsuits and investigations.

McLeod agreed to pay $4.4 million to settle a New York lawsuit associated with spinning. The judge called the practice "a sophisticated form of bribery." The practice of spinning, which involved numerous Bandwidth executives, involved bankers allocating shares of hot IPOs to executives as a reward for using their banking services.

McLeodUSA consented to an SEC cease and desist order regarding the misrepresentation of Bandwidth swap transactions. Revenue and EBITDA were overstated by $11.2 million and $7.1 million, respectively, in 2001, which had the effect of portraying the company as growing rapidly when, in fact, it was in decline. The company made accounting errors in the recording of the transactions in its financial statements. Additionally, management was not transparent on the positive yet unstainable lift that the swaps gave to McLeodUSA's financial performance.

The state of Connecticut, which was an investor in Forstmann Little, sued Forstmann Little for making reckless investments. "I have no intention of sitting on the sidelines while any money manager takes Connecticut's pension funds and places them in jeopardy," said the state's treasurer. "Forstmann's actions raise serious concerns regarding both their motives and the professional conduct that is expected of a fiduciary."

A Connecticut jury found that Forstmann Little had mishandled state pension funds. However, they also determined Connecticut State wasn't entitled to collect any damages associated with its $95 million in losses. The rationale was that the state acquiesced to Forstmann Little for too long before complaining about it.

A class action lawsuit was filed in 2002 that claimed McLeod, Gray, and Davis misled shareholders about McLeodUSA's financial performance. The claim was settled by McLeodUSA's insurance company for $30 million. This tiny amount meant that an investor would receive a paltry nine cents per share.

Only in a tragedy as big as the Bandwidth Bust does the story of McLeodUSA, CapRock, and Splitrock not make it into the main body of the book.

Spread Networks

Spread Networks was made famous by Michael Lewis's book titled *Flash Boys*. In 2009 Dan Spivey led the construction of an 827-mile fiber network that connected New York and Chicago in the shortest conceivable path. The only goal was to reduce the transmission time for data from seventeen to thirteen milliseconds. By doing so, specialized hedge fund traders would gain an information advantage over other investors.

These traders used this edge to see order activity before trades were executed. They were able to front run these orders by buying or selling at prices slightly better than the orders about to hit the exchanges. The term high-frequency trading meant they make an enormous number of trades to exploit this information advantage. In the process the hedge fund traders became billionaires.

The use of Spread's fastest route gave traders this advantage. So long as Spread sold it to only a handful of traders, the lucky customers would have a surefire way to generate enormous gains. In exchange these customers would pay Spread a premium for these fast routes.

Though this sounds wrong, it wasn't illegal.

Eventually, through regulation and alternative technology of microwaves, the premium available to Spread began to decline.

Spivey, a trader on the Chicago Board Options Exchange, was the brainchild behind Spread. He was hired by a New York hedge fund to devise a low-latency arbitrage strategy that would identify small discrepancies between pricing activity in New York and Chicago. He realized that if the algorithm had access to the fastest path, the opportunity would be almost limitless. Spivey convinced James L. "Jim" Barksdale, the original CEO of Netscape, to back him. Barksdale put up over $200 million of the cash he earned from Netscape's sale to AOL in 1999.

With the help of Barksdale's son David, the three of them executed the plan—Spivey as Chairman, David as CEO, and Barksdale as the bankroll with the powerful rolodex. In August 2010 the route was up and being used by the traders.

I visited regularly with Spivey and David in the mid-2010s, including as they started to attend Metro Connect in Miami. Spivey was reluctant to engage in any acquisition discussions, but David showed interest. They knew their original contracts with lucrative pricing were soon to expire. For certain, the renewals would be at lower pricing because of lower-latency microwave routes, but how much lower was impossible to know. Spread, the lowest-latency, all-fiber route, would still command a premium.

Spread was uniquely valuable to Zayo in additional ways, as the route filled a critical hole in our intercity fiber network. Also, I loved the way Spread understood the unique value of its routes and its skill at pricing service in a way that reflected value. As an example, Spread created different tiers of service simply by splicing unnecessary fiber reels onto some routes to slow down the speed. This allowed Spread to sell these slower routes at a lower price than the company's cherished fastest path.

I saw the opportunity for Zayo to create unique routes by leveraging the combination of the Spread Network and other Zayo assets. Spread would allow Zayo to further strengthen its position as a valued strategic supplier to the financial services community. With all this in mind, we agreed on a price of $127 million in a 2018 acquisition.

This was a win-win deal. The Barksdales and Spivey had already made a great return on their investment, so this was all gain for them. Spread by Zayo became a valuable part of Zayo's business offerings.

Both Spivey and the Barksdales lived in Jackson, Mississippi. I visited them when we closed the deal, and they treated me to a most enjoyable closing dinner.

TelCove / Adelphia

TelCove was originally a business unit under the Pennsylvania Cable TV provider Adelphia Communications. John Rigas and his brother Gus founded Adelphia in 1952. In the late 1990s, Adelphia, with over five million subscribers, was the sixth-largest Cable TV provider.

In 1991 Adelphia created a Bandwidth business unit branded as Hyperion Telecommunications. In 1998, as the Gold Rush was gaining momentum, Hyperion went public on Nasdaq, though Adelphia retained 79 percent ownership. Hyperion's value soared to well over $1 billion. The next year, the name was changed to Adelphia Business Solutions.

The Bust hit Adelphia hard. Needing cash, it raised $1.7 billion when it spun out its 79 percent ownership in Adelphia Business Solutions in late 2001. Four months later Adelphia Business Solutions declared bankruptcy. Two months later, in June 2002, Adelphia also filed for bankruptcy.

During the Adelphia bankruptcy, Rigas and his son Timothy Rigas, who was Adelphia's CFO, were accused of using $2.3 billion of Adelphia assets for their own purposes. They were also misrepresenting Adelphia's financial challenges. The father and son were found guilty of conspiracy, securities fraud, and bank fraud. Rigas entered prison at the age of eighty-two and was released nine years later at the age of ninety-one. Timothy Rigas served twelve years behind bars.

When Adelphia Business Solutions emerged from bankruptcy in 2004, it rebranded to TelCove. Adelphia and TelCove dropped lawsuits against one another, with Adelphia agreeing to pay TelCove $60 million and to purchase $39 million in services over five years. Bob Guth, who spent the prior eight years at Adelphia, was named TelCove's CEO. Soon thereafter, Guth did some bottom feeding, acquiring East Coast fiber assets of Chicago-based Exelon for $49 million. In 2006 TelCove sold to Level 3 for $1.2 billion, consisting of $637 million of Level 3 shares, $445 million in cash, and $155.5 million in assumed debt. At the time of the acquisition, TelCove served seventy markets on the East Coast with a fiber network of 22,000 miles.

Velocita

Koch Industries originated as an energy business and owned substantial ROW along oil pipelines. It, too, saw an opportunity to leverage these assets by participating in the fiber boom. In 1999 Koch partnered with entrepreneur John Warta to launch PF.Net.

Warta had already launched two fiber companies. In 1988 he founded Electric Lightwave (ELI), which was backed by Citizens Utilities. Warta left ELI in 1991. His explanation to me was that a new chairman was named at Citizens, which caused "everything to fall apart." Warta described a deal that he orchestrated in which he presold twelve of forty-eight fibers on a route between Vegas and Phoenix, and the customer would pay 75 percent of the cost. As part of the deal, Warta would personally own 10.5 percent of the value of the route—an arrangement that was put in place when he first brought Citizens into ELI as his partner. This didn't sit well with the new chairman; his solution was to replace Warta with a Citizens executive.

As mentioned in the tw telecom section, Warta founded GST Telecom in 1994. While at GST, Warta developed a relationship with Koch Industries. Warta said he petitioned the GST board to allow Koch to collaborate on a new venture, which was named PF.Net. Joining Koch as financial backers were AT&T spinoff Lucent Technologies and investment firm Odyssey Partners. Together, they invested $700 million.

"We're excited about the partnership," said a Koch spokesperson. "We think [Warta] will bring a great knowledge of the telecom business to this venture."

PF.Net's strategy was described in the article "Warta Ventures Bags $700M" published by the *Portland Business Journal* in late 1999.

> PF.Net has dubbed itself the company that is building "the network to the net." Its strategy is to assemble a national communications network that will feature the latest internet transmission technology. The goal is to build a so-called end-to-end fiber optic network that will span the country and allow for the fast transmission of information via the internet. While other telecom players are assembling advanced networks in specific regions, companies such as PF.Net are betting that a unified national network will be more attractive to prospective customers such as phone companies, internet service providers and corporations that would buy or lease space on the network.

This strategy was far from unique. Level 3, Qwest, WilTel 2, and Broadwing were building similar networks.

However, Warta had a card hidden up his sleeve. Warta told me he spent a year in Morristown, New Jersey, where AT&T was headquartered, and convinced AT&T executive Frank Ianna that "these guys at

Level 3 are gonna kill you." Warta claimed that he influenced Ianna to select three upstarts to build AT&T a new nationwide fiber network. The largest portion would be awarded to PF.Net. In late 1999 AT&T selected PF.Net as one of three fiber companies to build out its next-generation intercity fiber network.

PF.Net's investors must have reached a conclusion that the opportunity was sufficiently big enough that they needed seasoned leadership. They brought on a high-profile chairman who was very familiar to AT&T—Annunziata—following his short reign at Global Crossing.

PF.Net also hired a CEO who had ties to AT&T—Kirby "Buddy" G. Pickle. After he left AT&T, Pickle became an executive at MFS and, more recently, the COO of Teligent, the company headed up by AT&T veteran Alex Mandl. Pickle left Teligent because its business model wasn't gaining traction. In 2000 the stock price fell from a high of $100 per share to below $15. The *Wall Street Journal* shared an analyst's explanation of why Pickle, who was seen as Mandl's successor, was departing Teligent. "Mandl has taken on a personal mission to turn around some of Teligent's recent misfortunes, so if Buddy wanted to move up he had to go elsewhere."

Warta rationalized why Pickle was brought in to replace him as CEO. "My original contract actually expired last December, but the folks from Odyssey Partners asked me to extend the agreement, and so did AT&T. With our new team coming on board, I will step back and let them run things."

Pickle rebranded PF.Net to Velocita Communications. AT&T climbed aboard as an investor, bringing the total amount raised to $1 billion. The *Washington Post* published "The Optimist" in late 2000.

> Buddy Pickle believes 2001 will be Act I in a dramatic turn-around in the telecommunications sector.
>
> Pickle can't afford to be pessimistic. He's leading one of the most ambitious telecommunications start-ups ever, at a time

when most broadband providers have been ditched by Wall Street.

Still, Pickle is an indefatigable optimist.

"The people who are going to make money are the guys like us," who are building the "new economy" equivalent of the railroad," Pickle said. "It's not like two years ago, when ideas got money, even foolhardy ones." No matter what the economy does, "the great ideas will continue to be worked on."

The Velocita network was envisioned to be over eighteen thousand miles with connections to more than fifty major cities, all operational by early 2002. With the backing of anchor tenant AT&T, and the ability to leverage the AT&T ROW, what could go wrong? Everything turned out to be the answer, as detailed in Chapter V, "The Bust of the Bandwidth Bubble."

By early 2002 Velocita completed 80 percent of its fourteen-thousand-mile fiber network. Before becoming operational, it ran out of money. In June 2002 it declared bankruptcy.

"Bankruptcy is just another step in that process," explained Pickle. "It is another step in conjunction with business plans to get rid of overhead and other expenses associated with the services business." To make sure I follow Pickle, I'll paraphrase his business plan: Use other people's money to build out a business, declare bankruptcy so you don't have to return their investment, then continue on. That sounds like a Dill Pickle.

Bankruptcy, it turned out, was the final step for Velocita. AT&T purchased the assets for $37 million in late 2002. Koch, Warta, and the other investors lost nearly all their $1 billion investment.

GLOSSARY OF
FINANCIAL TERMS

Indebtedness means the aggregate value of the bank debt, loans, and other liabilities of a company. In this book we will not differentiate between Gross Indebtedness and Net Indebtedness. The difference between the two is the company's cash and cash equivalents subtracted from gross to derive net.

Market Value means the value of all the equity ownership of a company. The expressions market capitalization, market cap, and equity value have nearly identical meanings. I will use the term **Enterprise Value** in this book to Market Value plus Net Indebtedness.

Profitability is the expression I use in this book for the financial term EBITDA, which stands for earnings before interest, taxes, depreciation, and amortization. EBITDA captures the amount of cash that a business is generating or burning before accounting for capital expenditures, interest payments, and taxes.

A **public company** has shares that are traded in open markets. **Going public** is the process of a **private company** becoming a public company, which culminates in an initial public offering (**IPO**). The two top stock exchanges that list public company stocks are the New York Stock Exchange (**NYSE**) and **Nasdaq**. The Securities and Exchange Commission (**SEC**) is an agency of the federal government that protects investors against market manipulation.

Public Equity are investors whose focus is publicly traded stocks. **Private Equity** (**PE**) are investors whose focus is established private companies. **Venture Capital** (**VC**) are investors who focus on start-ups.

Multiple of Invested Capital or MOIC is the ratio derived by dividing the dollars that are returned to the investor by the dollars invested by the investor. For example, if an investor invested $100 and received $500 in return, the MOIC would be calculated by dividing $500 by $100, resulting in a multiple of five. **Absolute Return** is the increase in value, or gain, received on an investment. For example, $400 would be the Absolute Return for a $100 investment that returned $500 to the investor.

Mergers and Acquisitions (**M&A**) is the process when two companies combine into one. If the acquired company is a public company, the majority of shareholders are required to approve the transaction through a **shareholder vote**. The board of directors of a public company are obligated to ensure that the company is sold for the highest possible price. To ensure it is meeting its responsibility, the board typically insists the buyer allows for a window of time (**go-shop period**) for other bidders to offer a higher price. If another bidder emerges and successfully outbids the original buyer, a fee is typically owed to the original buyer (**breakup fee** or **termination fee**) to compensate them for losing the deal. A higher fee means the buyer is less likely to pull out. Most M&A is the result of a constructive dialogue between a buyer and seller. However, some M&A is driven by a buyer who is not welcomed by the seller (**hostile takeover**), as the seller prefers an alternative outcome. The fiduciary obligations of a board of directors of a public company are used by the buyer to force the seller to consider the buyer's hostile offer.

Bankruptcy means the value of a company is less than its **Net Indebtedness**. When this happens, the ownership rights to the company shift to the debt holders. A bankrupt company will file for protection, which is referred to as **declare bankruptcy,** and receive protection from creditors while it sorts through how to resolve the **bankruptcy**. A company will elect to file under either the **Chapter 7** or **Chapter 11 Bankruptcy Code.** Under **Chapter 7,** a company is typically liquidated with its assets sold to the highest bidder, with proceeds used to pay back creditors. Under **Chapter 11 Bankruptcy,** a company is reorganized and continues to operate, with creditors assuming some or all the post-bankruptcy ownership of the company. Sometimes, when a company is heading toward bankruptcy, the management team negotiates a transaction with investors that represents a proposed resolution. The pre-packaged (**pre-pack**) transactions are presented to the bankruptcy judge in hopes of accelerating the bankruptcy process.

LIST OF NAMES

The names of individuals and companies listed below were not explicitly named in the book, in order to declutter the story and streamline the narrative. They are included here for those interested in the minute details of the Bandwidth adventure in its entirety.

Chapter I: The Storyteller

George Lynch was my Level 3 supervisor at Illinois Bell. Colleagues referred to Lynch as George Lunch, as vendors were required to treat him for a meal whenever they wanted to meet with him.

Dan's colleagues at Ameritech were Andrew Morely, Dan Foreman, Dave Donalik, Greg Smitherman, Jason Lee, Jason Weller, Jeff White, Lynn Refer, Mitch Moore, and Tom Toulton. Moore, Morely, and Refer would join Dan at either MFS or Level 3.

Chapter II: The Inventors of the Internet Backbones

BBN—Leo Beranek

Digital Equipment Corporation (DEC) was the computing company that developed the prototype PDP-1 for BBN.

The device BBN developed for DARPA that evolved into modern-day Internet routers was called the **Interface Message Processor (IMP)**.

The other two government packet networks that could not interface with ARPANET were called **Packet Radio net** and **SATNET**.

BBN's first IP-based router was the T/20. Its successors were the T/200 and the C/30, a micro-programmable, medium-speed packet processor.

UUNET—John Sidgmore and Rick Adams

Science Application International Corporation was Rick Adams's first employer.

Dan Lynch, UUNET board member, "They are just like us . . ."

Mike O'Dell, UUNET's chief scientist, "The project was swallowing cash . . ."

The nonprofit that Adams was part of was called the **Usenix Association**.

PSINet—Bill Schrader

NYSERNET was the name of the nonprofit that Schrader created to connect New York State's universities, research institutions, medical centers, and libraries.

The original name of PSINet was **Performance Systems International**.

Matrix Partners, **Sigma Partners**, and **Amerindo** were PSINet's three sponsors.

The nonprofit trade association that PSINet co-founded with UUNET was named **Commercial Internet eXchange**.

In 1998 **USinternetworking Inc.** made the offer to buy the controlling interest in PSINet.

Chapter III: The Renegades That Laid the Fiber

Teleport

Trenton Borden R. Putnam, commissioner of New Jersey's Department of Commerce and Economic Development, "'the amount of data that will be flying around . . .'"

Guy F. Tozzoli, director, Port Authority World Trade Department, "It's like an airport . . ."

MFS—Jim Crowe and Royce Holland

Institutional Communications Company was the name of the DC start-up that was also constructing a local fiber network to bypass the Baby Bells.

Broadwing—Ralph Swett

Broadwing's largest microwave acquisition was the 1988 purchase of Michigan-based **ALC Communications.**

In 1997 and 1998, Broadwing's acquisitions included Network Long Distance and Telecom One of Chicago; web hosting company SmartNAP; system integrators the Data Place, Network Evolutions, and ntr.net; and Coastal Telephone Company.

WilTel—Roy Wilkens

Richard Klugman, analyst, Merrill Lynch, "It's an excellent fit . . ."

XO—Craig McCaw

Twin City Cablevision was the name of the Cable TV provider in Centralia, Washington, that Craig McCaw's father left to his family estate.

AboveNet—John Kluge and Steve Garofalo

Alan J. Gottesman, analyst, Unterbridge & Towbin, "There's not anything left . . ."

Chapter IV: The Bandwidth Boom Gold Rush

WorldCom—Bernie Ebbers

Advantage Company, a seventeen-year-old Nashville company, was the public company WorldCom merged with to become a publicly traded company.

Resurgens Communications Group was the third company in the three-way merger between WorldCom and Metromedia Communications Corporation.

Richard Klugman, analyst, Merrill Lynch, "It's an excellent fit . . ."; "If BT's shares hold up . . ."

Blake Bath, analyst, Lehman Brothers, "This is a bomb, a major bomb . . ."

Scott Cleland, analyst, Legg Mason Precursor Group, "Telecom is about scale, scale, scale . . ."

Eric Strumingher, analyst, PaineWebber, "[WorldCom/Sprint] would be the largest . . ."

Kevin Dundon and **John Scarano** were the two members of the senior team that Dan consulted about how they were going to use IP technology to send voice, data, and video over the same lines with lower costs.

Richard Klugman, analyst, Donaldson, Lufkin & Jenrette, "We believe that as in past deals . . ."

Simon Flannery, analyst, Morgan Stanley, "We believe WorldCom's current valuation . . ."

Richard Klugman, analyst, Jefferies & Company, "'There was some rumbling in the investment community . . ."

Level 3—Jim Crowe

Jeffrey Kagan, analyst, Kagan Telecom Associates, "Crowe struck gold last time out . . ."

Salvatore Muoio, analyst, S. Muoio & Co., "It's going to be a fantastic company . . ."

Josh Howell, Crowe's right hand for Corporate Marketing at Level 3

Ike Eliott, engineer who patented the Softswitch at Level 3

Rob Hagans, lead engineer, Level 3, who led the team that rewrote XCom's software

Kevin Dundon, Level 3 colleague who took Shawn Lewis (XCom) out with Dan for drinks

Kevin Dundon and **John Scarano** were the two members of the senior team that Dan consulted about how they were going to use IP technology to send voice, data, and video over the same lines with lower costs.

Global Crossing—Gary Winnick

David Lee, president, Global Crossing, "It was like doing business . . ."

Jack Scanlon, first Global Crossing CEO, former senior Motorola executive, "A chance to build a global . . ."

Racal Telecom was the UK Bandwidth provider Global Crossing acquired for $1.65 billion in cash in 1999.

IXNet was the private company Global Crossing acquired a minority ownership position in that did not own a fiber network. IPC Communications was the owner of IXNet.

Istar Internet is the Canadian ISP PSINET exchanged $25 million of its stock for.

PSINet—Bill Schrader

Lawrence Winkler, treasurer, PSINet, "We are now set up for 2000 . . ."

Harold S. Wills, president and COO, PSINet, "Some day, they will buy us and we will . . ."

AT&T—Robert E. Allen and C. Michael Armstrong

Michael J. Balhoff of securities firm Legg Mason Wood Walker, "As best as I can tell . . ."

Mark Cooper, research director, Consumer Federation of America, "The forces of evil unleashed . . ."

Anne Bingaman, former antitrust chief, Justice Department, "Creates huge competitive problems . . ."

Walter Y. Elisha, board member, AT&T, "He's a bright guy, but . . ." and "We are now searching . . ."

Stuart Conrad, analyst, Deutsche Morgan Grenfell, "From AT&T's perspective . . ."

John Hodulik, analyst, Lazard Frères & Co., "It makes the most sense . . ."

Jeffrey Kagan, analyst, Kagan Telecom Associates, "This is the best news . . ."

Steve Allen, division manager, AT&T Network Services, "AT&T is responding to unprecedented . . ."

Stephen R. Leeolou, co-founder and CEO, Vanguard, "We believe that this . . ."

Genuity—Charles Lee

James L. Freeze, SVP and CSO, Genuity, "It's not about just having great . . ."

Philippe Guglielmetti, chairman and founder, Integra, "The market had concerns . . ."

XO—Craig McCaw and Dan Akerson

Jose Romero, telecom analyst, Safeco Asset Management Co., "With Nextlink . . ."

Jay Sherwood, senior analyst, RS Mid-Cap Opportunities Fund, "They are one of the few . . ."

AboveNet—John Kluge and Steve Garofalo

Nicholas "Nick" Tanzi, president and COO, MFN (AboveNet), "MFN plans to extend . . ."

Chapter V: The Bandwidth Bust

Larissa Herda, CEO, tw telecom, "Unfortunately, we see a slowdown in business . . ."

Aryeh Bourkoff, analyst, UBS Warburg, "All the companies in the emerging telecom . . ."

WorldCom—Bernie Ebbers

Richard Klugman, telecommunications analyst, Jefferies & Company, "I've been following . . ."

Mark Abide, bookkeeper, WorldCom, "This is worth looking into from an audit . . ."

Howard J. Schilit, founder, Center for Financial Research and Analysis, "The kiboshing . . ."

Ken Johnson, spokesman, House Energy and Commerce Committee, "Ebbers was aware that . . ."

Vincent Wright, bus driver and juror for Ebbers trial, "In order to find the defendant guilty . . ."

Aran Nulty, elementary school teacher and juror number ten on Ebbers trial, "The most difficult thing . . ."

Salina Strong, juror number four on Ebbers trial, "I think Ebbers pretty much . . ."

Qwest—Joe Nacchio and Dick Notebaert

Solomon Dennis "Sol" Trujillo, CEO, U S West, "No way we will engage . . ."

Troy A. Eid, United States attorney for Colorado, "Convicted felon Joe Nacchio . . ."

Global Crossing—Gary Winnick and John Legere

Robin Wright, executive, Global Crossing, "Ends up being the plug . . ."
Randall Lee, director, SEC's Los Angeles office, "Global Crossing's senior executives . . ."
Janet Mahoney, former Global Crossing employee and plaintiff, "This is not justice . . ."

WilTel 2—Howard Janzen and Jeff Storey

Renee Christian, spokeswoman, Williams Communications, "The goal is to come out . . ."

PSINet—Bill Schrader

Ian P. Sharp, board member, PSINet, "Was more interested in building . . ."
Drake Johnstone, analyst, Davenport & Company, "The risk of bankruptcy . . ."

Genuity

Richard Klugman, analyst, Jefferies & Co., "It sounds like they're anticipating . . ."
Susan Kraus, spokeswoman, Genuity, "This is absolutely coincidental . . ."

Touch America—Bob Gannon

Frank Morrison Jr., former state Supreme Court justice, "We're trying to salvage . . ."
Carl Anderson, retiree, Montana Power, "I get pretty emotional . . ."

Monica Boehmer, change management administrator, "You can see a company . . ."

360 Networks—Greg Maffei

Sam Jadallah, former head of worldwide sales, Microsoft, "When Maffei bets, he bets big . . ."

GT Group Telecom was the Canadian fiber provider 360networks acquired after emerging from bankruptcy.

AT&T—C. Michael Armstrong

Ken McGee, analyst, Gartner Group, "It's hard to escape the feeling . . ."

Oscar Castro, manager, Montgomery Global Communications fund, "If Comcast gets away . . ."

Level 3—Jim Crowe

Richard Rolnik, head of high-yield telecom research, BNP Paribas, "Level 3's bonds are . . ."

Tom Friedberg, analyst, Brean, Murray & Co., "This is creative financial . . ."

Jack Grubman—Sarbanes–Oxley

Elliot Dorbian, former broker, Salomon, "When Grubman said wonderful . . ."

Steve Holwerda, portfolio manager, Ferguson, Wellman, Rudd, Purdy, and Van Winkle, "It's buyer beware . . ."

Jacob H. Zamansky, securities lawyer, "Jack Grubman is the king . . ."

Helen Scott, professor of securities law, New York University, "Does it look crummy?"

Eric Efron, co-portfolio manager, USAA Aggressive Growth Fund, "Whenever [Grubman] picks . . ."

Chapter VI: The Phoenix Rises from the Ashes

Broadwing—David Huber

Jeffrey Kagan, analyst, Kagan Telecom Associates, "Times change . . . [fiber] companies are selling . . ."

tw telecom—Larissa Herda

Bruce Roberts, analyst, Dillon, Read & Co. Inc. in New York, "By almost any measure . . ."

Mark Gershien was Herda's boss who wanted her to remain in Chicago.

Chapter VII: The Reconsolidation

Adam Quinton, analyst, Merrill Lynch, "What you have been seeing is the gradual reconsolidation . . ."

Level 3—Jim Crowe and Jeff Storey

Greg Maffei, CEO, 360networks, "I hope Level 3 is amazingly successful . . ."

Colby Synesael, analyst, Merriman Curhan Ford & Co., "Over the last two years he did a lot . . ."

Colby Synesael, analyst, Merriman Curhan Ford & Co., "There's a lot of disorganization . . ."

Colby Synesael, analyst, Cowen Securities, "Of all the companies that I cover . . ."

Jonathan Chaplin, telecom investment analyst, JPMorgan Chase, "Given that most of the issues were operations-related . . ."

Lumen—Glen Post and Jeff Storey

Daniel R. "Dan" Hesse, CEO, Sprint, "As CEOs go . . ."
George Cummings, CEO, Progressive Bank in Monroe, Louisiana, "During the wintertime . . ."

Appendix 3

Enron Broadband & Dynegy Global Communications

Douglas Carmichael, accounting professor, Baruch College in Manhattan, "It's like somebody sat down with the rules . . ."
Carol Coale, analyst, Prudential Securities, "[Enron executives] were great spin doctors . . ."
Rudy Sutherland Jr., former broadband trader, Enron, "From a common-sense standpoint . . ."
Jon Kyle Cartwright, analyst, Raymond James, "[Dynegy] no longer has to pour money into it . . ."

McLeodUSA

Denise L. Nappier, treasurer, State of Connecticut, "I have no intention of sitting . . ."

Velocita

Mary Beth Jarvis, spokesperson, Koch Industries, "We're excited about the partnership . . ."
Bo Fifer, analyst, Deutsche Bank Alex. Brown, "Mandl has taken on a personal mission . . ."

BIBLIOGRAPHY

Chapter II

Beranek, Leo. "A Culture of Innovation: Insider Accounts of Computing and Life at BBN." 2011. PDF file. https://walden-family.com/bbn/bbn-print2.pdf.

Bloomberg News, "Company News; PSINet Rejects Offer from USInternetworking." *New York Times*, January 28, 1995.

"Cincinnati Bell Is Set to Name Company Formed by IXC Deal." *Wall Street Journal*, November 15, 1999.

Flynn, Laurie. "Technology; Internet Server Takes a Big Step." *New York Times*, February 5, 1995.

"IXC and PSINet Agree on 20-Year Network Deal." *New York Times*, July 24, 1997.

Lewis, Peter H. "Company News; Microsoft's Next Move Is on Line." *New York Times*, January 13, 1995.

Lewis, Peter H. "Uunet and MFS Plan to Merge as Internet Meets Fiber Optics." *New York Times*, May 1, 1996.

Mallaby, Sebastian. *The Power Law: Venture Capital and the Making of the New Future*. Penguin Publishing Group, 2022.

Norris, Floyd. "Market Watch; How to Make $34 Million in a Year." *New York Times*, May 28, 1995.

Schiesel, Seth. "An Iconoclast Goes It Alone; on the Net PSINet's Schrader Is a Force to Reckon With." *New York Times*, March 13, 2000.

Schiesel, Seth. "GTE Discloses 3 Big Deals in Growth Bid." *New York Times*, May 7, 1997.

Schrader, William. "William Schrader Professional Profile." LinkedIn. Accessed November 20, 2023.

Swisher, Kara. "Anticipating the Internet." *Washington Post*, May 5, 1996.

"Uunet Posts a Loss After Acquisition." *New York Times*, January 30, 1996.

Zuckerman, Laurence. "Internet Access Service Is Extending Its Reach." *New York Times*, March 13, 1995.

Zuckerman, Laurence. "AT&T Expected to Buy Stake in an Internet Access Provider." *New York Times*, July 12, 1995.

Zuckerman, Laurence. "Innovator Is Leaving the Shadows for the Limelight." *New York Times*, July 17, 1995.

Chapter III

Andrews, Edmund L. "Company News; AT&T Completes Deal to Buy McCaw Cellular." *New York Times*, September 20, 1994.

Andrews, Edmund L. "The Local Call Goes Up for Grabs." *New York Times*, December 29, 1991.

"AT&T's Sale of Wireless Stake Would Be US's Biggest IPO Ever." *Wall Street Journal*, December 6, 1999.

Berger, Marilyn. "John W. Kluge, Founder of Metromedia, Dies at 95." *New York Times*, September 8, 2010.

"Billionaire Boys." *Forbes*, June 6, 2013.

Blumenstein, Rebecca. "How the Fiber Barons Plunged the US into a Telecom Glut." *Wall Street Journal*, June 18, 2001.

Burgess, John. "Big vs. Small Long-Distance Fight Heats Up." *Washington Post*, February 12, 1989.

"Citigroup Retires the Salomon Name." *Los Angeles Times*, April 7, 2003.

Corr, O. Casey. "Money from Thin Air." *Seattle Times*, April 4, 1993.

Cox Enterprises. "History." Last modified July 10, 2024. https://www.coxenterprises.com/about-us/history.

"Craig McCaw's Cosmic Ambition—the Dyslexic Visionary Who Fathered Cellular Looks to Launch an Even Bigger Industry." *Fortune*, May 26, 1996.

"Craig McCaw—the Wireless Wizard of Oz." *Forbes*, June 22, 1998.

Denton, Jon. "Firm Combines Fiber Optics with Pipelines." *Oklahoman*, September 6, 1990.

Dresser, Michael. "ACSI Moves Quietly." *Baltimore Sun*, October 3, 1995.

Dwyre, Bill. "One Word for Phil Anschutz: Private." *Los Angeles Times*, October 10, 1999.

"Fiber Optics Pulls Politics Underground." *Chicago Tribune*, May 8, 1989.

Fisher, Miles. "Mike Milken—Philanthropist, Financier and Chairman of the Milken Institute." *Coffee with the Greats*. Season 1, Episode 8. Aired December 2020.

Gieschen, Frederik. "How Craig McCaw Pioneered the Wireless Age." *The Alchemy of Money*, June 10, 2021.

Griffin, Greg. "A Look at Joe Nacchio." *Denver Post*, December 23, 2005.

Hardy, Quentin. "Craig's Higher Calling." *Forbes*, June 12, 2000.

Hardy, Quentin. "We're on a Collision Course." *Forbes*, June 12, 2000.

Hardy, Stephen. "All Fiber Is Local." *Lightwave + Broadband Technology Report*, January 31, 1999.

Henderson, Nell. "Parts Worth More Than a Whole Metromedia." *Washington Post*, July 1, 1986.

Kiewit Corporation. "Our Story." Last modified July 10, 2024. https://www.kiewit.com/.

"IXC Communications Walking the Talk." Channel Futures, March 31, 1998.

"IXC Unveils Nationwide High-Speed Data Services." *HPC wire*, February 7, 1997.

Kupfer, Andrew. "Craig McCaw's Cosmic Ambition." *Fortune*, May 1996.

Landler, Mark. "Rich, 82, and Starting Over." *New York Times*, January 5, 1997.

Lippin, Steven. "WorldCom Reaches Pact to Buy MFS in $14.4 Billion Stock Deal." *Wall Street Journal*, August 26, 1996.

Madigan, Sean. "e.spire Seeks Bankruptcy Protection." *Washington Business Journal*, March 22, 2001.

Mari, Ruth. "Staten Island Teleport Is Seen as a Link to Future." *New York Times*, October 17, 1982.

McGill, Douglas C. "Teleport on Staten I. Envisioned as City's Link to a Bright Future." *New York Times*, September 10, 1983.

"MCI to Deploy Fully Digital Network." United Press International, August 21, 1990.

Mehta, Stephanie N. "Why AT&T Called Teleport CEO for Aid with Local-Phone Plans." *Wall Street Journal*, January 9, 1998.

"Merger Rumors Surround Qwest." *Forbes*, June 9, 1999.

"Metromedia Company History." Funding Universe. Accessed January 9, 2024. http://www.fundinguniverse.com/company-histories/metromedia-company-history/.

MFS Communications. "Form S-1 Registration Statement under the Securities Act of 1933." October 10, 1995. https://www.sec.gov/Archives/edgar/data/794323/0000912057-95-008579.txt.

Morris, Kathleen, and Steven V. Brull. "Phil Anschutz: Qwest's $7 Billion Man." Bloomberg News, December 7, 1997.

Mote, David. "LCI International," International Directory of Company Histories. Encyclopedia.com. Accessed July 12, 2024. encyclopedia.com/books/politics-and-business-magazines/lci-international-inc.

Myerson, Allen R. "LDDS Offers $2 Billion for Wiltel Network." *New York Times*, May 5, 1994.

Naik, Gautam. "MFS Enters Ameritech's Turf by Providing Local Services." *Wall Street Journal,* May 23, 1996.

"Nextlink Raises $258 Million in Its Initial Stock Offering." *Wall Street Journal*, September 27, 1997.

O'Reilly, Brian, and Ann Harrington. "Billionaire Next Door Philip Anschutz May Be the Richest American You've Never Heard Of." CNN Money, September 6, 1999.

"Qwest Follows Growth Strategy with Acquisition of Denver ISP." *Wall Street Journal*, October 2, 1997.

Ramirez, Anthony. "The Media Business; Cable TV Concerns to Own Local Telephone Company." *New York Times*, March 5, 1992.

"Roy Wilkens." Alumni of Influence: Missouri University of Science and Technology. Accessed November 28, 2023. https://influence.mst.edu/2016/honorees/wilkens/.

Salamie, David E. "The Anschutz Company." International Directory of Company Histories. Encyclopedia.com. Accessed February 1, 2024. https://www.encyclopedia.com/books/politics-and-business-magazines/anschutz-company.

Sandberg, Jared. "Qwest Sees Itself Rising to Top with Its High-Speed Network." *Wall Street Journal*, December 24, 1996.

Schiesel, Seth. "At Qwest, a Chance to Become Really Big." *New York Times*, June 15, 1999.

Schiesel, Seth. "BellSouth Agrees to Acquire 10% of Qwest." *New York Times*, April 20, 1999.

Schiesel, Seth. "Qwest Set to Acquire LCI for $4.4 Billion in Stock." *New York Times*, March 10, 1998.

Schiesel, Seth. "The Markets: Market Place; Merger Deal with Plenty of Intrigue." *New York Times*, July 22, 1999.

"Sewer Rats and Billionaires." *Forbes*, April 3, 2000.

Sorkin, Andrew Ross. "McKinsey & Co. Isn't All Roses in a New Book." *New York Times*, September 2, 2013.

Telephone World. "The History of Sprint." Last modified September 24, 2023. https://telephoneworld.org/long-distance-companies/the-history-of-sprint/.

"Teleport Using IPO Cash to Buy Out Continental." *Multichannel News*, May 13, 1996.

Temes, Judy. "Wall Street's Week: IPOs Hot, but Some Stocks Are Not: Companies Ignore Turbulence as Investors Snap Up New Issues." *Crain's New York Business*, November 2, 1997.

Weeks, Linton. "John Kluge's Winning Hand." *Washington Post*, July 8, 2003.

"Western Union to Sell Subsidiary." *New York Times*, March 27, 1985.

Wilkens, Roy. "Roy Wilkens Professional Profile." LinkedIn. Accessed November 28, 2023.

Williams Companies. "Our History." Last modified June 28, 2024. https://www.williams.com/our-company/our-history/.

Wulfing v. Kansas City Southern Industries, 842 S.W.2d 133 (Mo. Ct. App. 1992).

Chapter IV

"A WorldCom Deal for MCI Could Be the Birth of a Titan." *Wall Street Journal*, October 1, 1997.

Altafiber. "Broadwing Delivers Record Revenue and EBITDA for First Quarter." April 19, 2001. https://investor.cincinnatibell.com/news/press-releases/press-release-details/2001/Broadwing-Delivers-Record-Revenue-and-EBITDA-for-First-Quarter/default.aspx.

"AT&T Agrees to Acquire TCI, Creating a Telecom Behemoth." *Wall Street Journal*, June 24, 1998.

"AT&T Goes for the Last Mile." *Forbes*, October 15, 1998.

"AT&T to Buy Velocita Assets." *Wall Street Journal*, November 8, 2002.

Baker, M. Sharon. "Maffei's Firm Eyes Seattle." *Puget Sound Business Journal*, March 19, 2000.

Baker, Sharon. "McCaw's Nextlink on a Roll." *Puget Sound Business Journal*, November 22, 1998.

Barrett, Larry. "Microsoft CFO Greg Maffei Quits." ZDNET, December 21, 1999.

Bartash, Jeffry. "Global Crossing Sells Local Business." MarketWatch, July 12, 2000.

"Bell Atlantic and Metromedia Fiber Network Announce Strategic Agreements." Verizon News, October 7, 1999.

Berman, Dennis K. "Telecom Investors Envision Potential in Failed Networks." *Wall Street Journal*, August 14, 2003.

Blumenstein, Rebecca, and Joann S. Lublin. "Armstrong Pays $110 Billion in Effort to Transform AT&T." *Wall Street Journal*, November 9, 1999.

Blumenstein, Rebecca, John R. Wilke, Nicole Harris, and Deborah Solomon. "WorldCom May Call Off Takeover of Sprint as US Sues to Halt It." *Wall Street Journal*, June 28, 2000.

Blumenstein, Rebecca. "AT&T to Buy Vanguard Cellular in a Deal Valued at $900 Million." *Wall Street Journal*, October 6, 1998.

Blumenstein, Rebecca. "AT&T to Offer Concert Services of British Telecommunications." *Wall Street Journal*, October 8, 1998.

Blumenstein, Rebecca. "Global Crossing Cosseted Executives as Pink Slips Loomed, Stock Fell." *Wall Street Journal*, February 21, 2002.

Blumenstein, Rebecca. "Qwest Wins U S West for $35 Billion, Ending Duel with Global Crossing." *Wall Street Journal*, July 19, 1999.

Braga, Michael. "WorldCom Defends Deal with Intermedia." *Tampa Bay Times*, October 19, 2000.

Branigin, William. "Gary Winnick, Global Crossing Founder at Center of Epic Rise and Fall, Dies at 76." *Washington Post*, November 14, 2023.

Bredemeier, Kenneth. "PSINet Deal Would Double Its Size." *Washington Post*, March 22, 2000.

"British Telecom to Buy MCI to Form Communications Giant." *Wall Street Journal*, November 3, 1996.

"British Telecom, MCI Defend Revised Merger Agreement." *Wall Street Journal*, August 22, 1997.

Bulkley, Kate. "Corporate Profile: Global Crossing—His Company Is Two Years Old. But It's Already Worth Pounds 20bn." *Independent*, November 17, 1999.

Burrough, Bryan and John Helyar. *Barbarians at the Gate: The Fall of RJR Nabisco*. HarperPerennial, 1991.

Cardwell, Diane. "Robert E. Allen, 81, Dies; Led an AT&T in Transition." *New York Times*, September 13, 2016.

Cauley, Leslie, and John J. Keller. "A Single Phone Call Ends AT&T-SBC Plan to Merge." *Wall Street Journal*, June 30, 1997.

Cauley, Leslie. "Armstrong's Vision of Cable Empire at AT&T Unravels on the Ground." *Wall Street Journal*, October 18, 2000.

Cauley, Leslie. "AT&T to Enter Local Market with $11 Billion Teleport Deal." *Wall Street Journal*, January 8, 1998.

Cauley, Leslie. "Comcast to Acquire MediaOne in $48.63 Billion All-Stock Deal." *Wall Street Journal*, March 23, 1999.

Chien, C. A. "C. Michael Armstrong 1983—Biography." Business Biographies. Reference for Business.com. Accessed February 23, 2024. https://www.referencefor-business.com/biography/A-E/Armstrong-C-Michael-1938.html.

"Cincinnati Bell to Acquire IXC Communications." *Fiber Optics Online*, 1999.

"Command Center: Peter Shaper." *Milsat Magazine,* September 2009.

"Company News; Forstmann Little to Invest $850 Million in Nextlink." *New York Times*, December 9, 1999.

"Company News; LDDS Communications to Acquire ACC for Stock." *New York Times*, March 22, 1994.

"Company News; Long-Distance Phone Merger Adds Third Company." *New York Times*, February 26, 1993.

"Company News; Nextlink Acquires Rest of Internext for $220 Million." *New York Times,* December 8, 1999.

"Company News; Nextlink to Expand via Wireless Licenses." *New York Times,* January 15, 1999.

"Company News; PSINet Agrees to Buy Data Service Provider." *New York Times*, August 24, 1999.

"Company News; PSINet to Acquire Web Outsourcing Services Company." *New York Times*, March 23, 2000.

"Company News; PSINet to Buy IStar Internet for $25 Million." *New York Times*, November 11, 1997.

"Company News; Williams Communications and Parent to Trade Assets." *New York Times*, March 1, 2001.

Coster, Helen. "The Fastest Corporate Downfalls." *Forbes*, July 14, 2000.

Darwin, Jennifer. "Telecom Firm Splitrock Bids to Make a Connection with Investors." *Houston Business Journal*, July 18, 1999.

Desloge, Rick. "Brook's Money Tree." *St. Louis Business Journal*, April 16, 2000.

Dougal, April S. "LDDS-Metro Communications," International Directory of Company Histories. Encyclopedia.com. Accessed January 24, 2024. https://www.encyclopedia.com/books/politics-and-business-magazines/ldds-metro-communications-inc.

Ewing, Terzah. "Williams Sees Its Fiber-Optic Lines as the Fuel for Growth in the Future." *Wall Street Journal*, November 21, 1997.

"Fathers of Invention: How Level 3 Worked Its Way to the Main Floor." *USA Today*, April 1, 1998.

"Global Crossing Closes Offer." *New York Times*, June 21, 1999.

"Global Crossing Completes Acquisition of Racal Telecom." Business Wire, November 24, 1999.

Goodman, Peter S. "MCI Wins Bidding for Sprint." *Washington Post*, October 4, 1999.

Goodman, Peter S., and Renae Merle. "End of Its Merger Run Led to WorldCom's Fall." *Washington Post*, June 29, 2002.

"GTE Internetworking Changes Name to Genuity." Verizon News, April 6, 2000.

Hagerty, James R. "Alex Mandl, Who Gave Up a Top AT&T Job to Lead a Startup, Has Died at Age 78." *Wall Street Journal,* April 1, 2022.

Hardy, Quentin. "Craig's Higher Calling." *Forbes*, June 12, 2000.

"Harris Corporation to Acquire CapRock Communications, a Global Provider of Mission-Critical, End-to-End Managed Communications Services." PR Newswire, May 21, 2010.

"Head of AT&T; Business Unit Quits After Just 8 Months." *Los Angeles Times*, February 25, 1999.

Heinzl, Mark. "Gregory Maffei's Telecom Revolution Ended with 360networks's Meltdown." *Wall Street Journal*, September 10, 2001.

Holson, Laura M. "Fast Times at Global Crossing; Enjoying the Spotlight While Building an Upstart in Telecommunications." *New York Times*, October 12, 2000.

Holson, Lauren M. "Fast Times at Global Crossing; Enjoying the Spotlight While Building an Upstart in Telecommunications." *New York Times*, October 12, 2000.

Holson, Lauren M. "Fast Times at Global Crossing; Revolving Door: 3 Chief Executives Quit in 3 Years." *New York Times*, October 12, 2000.

Hutchison Whampoa. "Hutchison Whampoa and Global Crossing Complete Telecom Joint Venture in Hong Kong." January 12, 2000. http://www.hutchison-whampoa.com/en/media/press_each.php?id=18.

"In His Own Words: Ebbers Takes the Stand." *Wall Street Journal*, March 2, 2005.

"Is Qwest the Best?" *Forbes*, June 6, 2001.

Jones, Kathryn. "Company News; LDDS Buying Provider of Global Phone Service." *New York Times*, August 2, 1994.

"Justice Approves WorldCom Deal; C&W to Buy MCI Internet Assets." *Wall Street Journal*, July 15, 1998.

Keller, John J. "Armstrong Boosts Stock Price, but AT&T Has a Ways to Go." *Wall Street Journal*, January 16, 1998.

Keller, John J. "Armstrong Boosts Stock Price, but AT&T Has a Ways to Go." *Wall Street Journal*, January 16, 1998.

Keller, John J. "Ex-MFS Managers Plan to Build Global Network Based on Internet." *Wall Street Journal*, January 20, 1998.

Keller, John J. "Jumping Off Sidelines, GTE Makes a Cash Bid for MCI." *Wall Street Journal*, October 16, 1997.

Keller, John J. "Walter, Allen Never Grew Close; Chief Aired Concerns to Board." *Wall Street Journal*, July 18, 1997.

Keller, John J., and Bryan Gruley. "Talks between AT&T, SBC Spark a Major Commotion." *Wall Street Journal*, May 28, 1997.

Keller, John J., and Steven Lipin. "Jumping Off Sidelines, GTE Makes a Cash Bid for MCI." *Wall Street Journal*, October 16, 1997.

Koo, Carolyn. "Global Crossing to Buy IXnet, IPC." TheStreet, February 22, 2000.

Kreifeldt, Erik. "Global Crossing Acquires Frontier for $11.2 Billion, Claims Largest Fiber Network." Fiber Optic Online, March 17, 1999.

Landler, Mark. "After 9 Months, AT&T President Quits under Pressure." *New York Times*, July 17, 1997.

"Letter." *Wall Street Journal*, April 22, 1999.

"Level 3 Agrees to Acquire XCOM in $165 Million Bid for Business." *Wall Street Journal*, April 6, 1998.

Lewis, Shawn. "Shawn Lewis Professional Profile." LinkedIn. Accessed August 1, 2024.

Lipin, Steven, and John J. Keller. "MCI-BT Merger Is Uncertain in Wake of World-Com's Bid." *Wall Street Journal*, October 6, 1997.

Lipin, Steven, and John J. Keller. "WorldCom Launches MCI Bid to Rival British Telecom's Pact." *Wall Street Journal*, October 1, 1997.

Lipin, Steven, and Leslie Cauley. "WorldCom Reaches Pact to Buy MFS in $14.4 Billion Stock Deal." *Wall Street Journal*, August 26, 1996.

Lipin, Steven, Nicole Harris, and Rebecca Blumenstein. "MCI WorldCom to Buy Sprint in Record $115 Billion Takeover." *Wall Street Journal*, October 5, 1999.

Lipin, Steven. "Analysts Applaud AT&T's Plan to Buy Teleport for $11.3 Billion." *Wall Street Journal*, January 9, 1998.

Lockard, Meredith. "McLeodUSA Buys Splitrock for $2.1 Billion in Stock." *Fiber Optics Online*, January 11, 2000.

Lublin, Joann S., and Quentin Hardy. "Motorola's Jack Scanlon Quits To Run Start-Up Global Crossing." *Wall Street Journal*, April 1, 1998.

"MCI Accepts WorldCom Bid in a $37 Billion Stock Deal." *Wall Street Journal*, November 10, 1997.

McKnight, John. "Sierra Pacific Communications and Touch America to Build 750-Mile, High-Speed, Fiber-Optic Network between Salt Lake City and Sacramento." ElectricNet, May 5, 2000.

"McLeodUSA Acquires CapRock Communications." *Fiber Optics Online*, October 4, 2000.

McMillan, Dan. "Warta Ventures Bags $700M." *Portland Business Journal*, November 19, 1999.

Mehta, Stephanie N. "Metromedia Fiber Network Plans to Buy AboveNet for $1.76 Billion." *Wall Street Journal*, June 24, 1999.

Mehta, Stephanie N. "Why AT&T Called Teleport CEO for Aid with Local-Phone Plans." *Wall Street Journal*, January 9, 1998.

Mehta, Stephanie N. "WorldCom Quietly Completes $37 Billion Acquisition of MCI." *Wall Street Journal*, September 15, 1998.

"Merged Company Named." *New York Times*, November 16, 1999.

"Metromedia Fiber Network Completes Merger with SiteSmith." Premises Networks, February 8, 2001.

"Michael Milken." *Forbes,* August 5, 2024.

Mills, Mike. "Heir Is Unknown, Undaunted." *Washington Post*, October 23, 1996.

Montana Associated Technology Roundtables. "Montana's Power Failure—High-Tech Meltdown Meant Doom for Touch America." August 10, 2003. https://matr.net/news/montanas-power-failure-high-tech-meltdown-meant-doom-for-touch-america/.

Myerson, Allen R. "LDDS Offers $2 Billion for Wiltel Network." *New York Times*, May 5, 1994.

Myerson, Allen R. "LDDS to Purchase Wiltel for $2.5 Billion." *New York Times*, August 23, 1994.

Naik, Gautam. "BT, MCI to Talk with Bidders; Possibly Seeking 3-Way Accord." *Wall Street Journal*, October 23, 1997.

Naik, Gautam. "Builder of MFS Decides to Sell His Company within an Hour." *Wall Street Journal*, August 27, 1996.

"Net-Services Firm Genuity to Buy France's Integra for $113.7 Million." *Wall Street Journal*, June 1, 2001.

"New AT&T Chairman Got Big Stock Deal." *New York Times*, March 27, 1998.

Noguchi, Yuki. "Deal to Give WorldCom Control of Digex." *Washington Post*, September 6, 2000.

Noguchi, Yuki. "The Optimist." *Washington Post,* December 24, 2000.

Nuzum, Christine. "Lawyers Give Final Appeals to Jury in Ex-WorldCom CEO's Criminal Case." *Wall Street Journal*, March 4, 2005.

O'Brien, Timothy. "A New Legal Chapter for a 90's Flameout." *New York Times*, August 15, 2004.

Olavsrud, Thor. "With Fanfare, Genuity Unveils Black Rocket." Internet News, September 19, 2000.

"Phone Giants Merge." *Chicago Tribune*, August 27, 1996.

"Pickle Resigns at Teligent to Join Fiber-Optic Firm." *Wall Street Journal,* September 12, 2000.

"PSINet Drops as Its Fourth-Quarter Loss Nearly Doubles." *New York Times*, February 22, 2000.

"PSINet Expects Fourth-Quarter Revenue to Nearly Double." *New York Times*, January 10, 2000.

"PSINet to Trade Services for Equity." *New York Times*, February 24, 2000.

"Qwest Completes Purchase of U S West." *New York Times,* July 3, 2000.

Rehak, Judith. "Mid-Caps May Beat Both David and Goliath : Nextlink:Spiced with Variety (folo)." *New York Times*, September 11, 1999.

Richards, Bill. "For Montana Power, a Broadband Dream May Turn Out to Be More of a Nightmare." *Wall Street Journal*, August 22, 2001.

Romero, Simon, and Geraldine Fabrikant. "The Rise and Fall of Global Dreams." *New York Times*, March 3, 2002.

Romero, Simon. "Global Crossing Memo Indicates Early Warning of Downfall." *New York Times*, October 1, 2002.

Romero, Simon. "Technology; PSINet Faces Cash Squeeze and May Seek Bankruptcy." *New York Times*, April 4, 2001.

Romero, Simon. "WorldCom and Sprint End Their $115 Billion Merger." *New York Times*, July 14, 2000.

Schiesel, Seth, and Simon Romero. "Technology; WorldCom: Out of Obscurity to under Inquiry." *New York Times*, March 13, 2002.

Schiesel, Seth. "AT&T to Pay $11.3 Billion for Teleport." *Wall Street Journal*, January 9, 1998.

Schiesel, Seth. "From a Supplier of Gas Comes a Digital Pipeline." *New York Times*, January 12, 1998.

Schiesel, Seth. "GTE Discloses 3 Big Deals in Growth Bid." *New York Times*, May 7, 1997.

Schiesel, Seth. "MCI-Worldcom Match: How They'll Fit, or Won't." *New York Times*, October 3, 1997.

Schiesel, Seth. "SBC to Acquire Stake in Unit of Williams." *New York Times*, February 9, 1999.

Schiesel, Seth. "Technology; Genuity to Introduce First Major Internet Service Offerings." *New York Times*, September 19, 2000.

Schiesel, Seth. "The Markets: Market Place; 2 Big Names Said to Drop Idea of Bidding for WorldCom." *New York Times*, February 6, 2002.

Schiesel, Seth. "The Markets: Market Place; Merger Deal with Plenty of Intrigue." *New York Times*, July 22, 1999.

Schiesel, Seth. "The Markets: Market Place; Qwest Communications Stumbles When, Engaged to U S West, It Falls for Deutsche Telekom." *New York Times*, March 7, 2000.

Solomon, Deborah and Nikhil Deogun. "Digex Ignores Minority Shareholders by Backing Intermedia's Merger Deal." *Wall Street Journal*, September 8, 2000.

Solomon, Deborah. "Genuity Unit of GTE Plans Stock Offering." *New York Times*, May 25, 2000.

Solomon, Deborah. "Global Crossing Completes Network, but Faces a Challenging Environment." *Wall Street Journal*, June 22, 2001.

Solomon, Deborah. "WorldCom, Intermedia Agree to Revise Terms of Deal in Bid to Settle Digex Suit." *Wall Street Journal*, February 16, 2001.

Spagat, Elliot. "Williams Communications Awarded Executives Bonuses to Repay Loans." *Wall Street Journal*, May 4, 2002.

Speedcast, 2016. "Speedcast to Acquire Harris CapRock for US$425M." November 2, 2016. https://www.speedcast.com/newsroom/press-releases/2016/speedcast-to-acquire-harris-caprock-for-us425m/.

Steinberg, Brian, and Shawn Young. "Level 3's Wall Street Debut Attracts Believers, Doubters." *Wall Street Journal*, April 1, 1998.

Stremfel, Michael, and Anthony Palazzo. "Rise and Fall of Global Pipe Dream." *Los Angeles Business Journal*, February 17, 2002.

Sullivan, Tom. "Surge in Internet Optimism Fuels Decade's Biggest Junk-Bond Deal." *Wall Street Journal*, April 23, 1998.

"The $20 Billion Crumb." *Forbes*, April 19, 1999.

"Time Warner, Broadwing, Change Names after Billions in Losses." Bloomberg.

"Touch America Completes Fiber-Optic Link From Spokane to Boise." *Fiber Optics Online*, June 7, 1999.

US Department of Justice. "Evaluation of the United States Department of Justice." Last accessed July 12, 2024. https://www.justice.gov/atr/evaluation-united-states-department-justice-29.

US Securities and Exchange Commission. "Form 10-Q Quarterly Report Pursuant to Section 13 or 15(d) of the Securities Exchange Act of 1934 for the Quarterly Period Ended September 30, 1999." https://www.sec.gov/Archives/edgar/data/10 63880/000095013499009400/0000950134-99-009400-d1.html.

Vanderwater, Bob. "Williams Unit Goes Public." *Oklahoman*, October 3, 1999.

"Velocita to Build Second Phase of AT&T's Next Generation Network." *Fiber Optics Online*, February 15, 2001.

White, James A. and Paul Ingrassia. *Corporate History*. Cincinnati, Cincinnati Bell, 1973.

"Williams, Montana Power, Enron Units Plan Venture." *Wall Street Journal*, September 8, 1997.

"WorldCom and Brooks Fiber Announce Merger." Verizon News Archives, October 1, 1997.

"WorldCom's CEO Got a Bonus of $10 Million Despite Tough 2000." *Wall Street Journal*, April 2, 2001.

Worldwide Fiber. "Pacific Fiber Link L.L.C., Merger to Form Worldwide Fiber Networks, Inc." March 2, 2000.

Ziegler, Bart. "Giant Merger Could Produce Bumper Crop of Surplus CEOs." *Wall Street Journal*, May 28, 1997.

Chapter V

"3 Ex-Officials of Global Crossing Are Fined in SEC Settlement." *New York Times*, April 13, 2005.

"360networks Acquires Touch America." *Photonics Spectra*, January 19, 2004.

"360networks Agrees to Sale of Canadian Assets for $205M." *Puget Sound Business Journal*, May 26, 2004.

"360networks to Buy Group Telecom." CBC News, November 18, 2002.

"AT&T Spins Off Its Wireless Unit Today." *Los Angeles Times,* July 9, 2001.

Barakat, Matthew. "XO Warns Bankruptcy Is Imminent." *Midland Daily News*, March 28, 2002.

Bellamy, Clayton. "Williams Comm. CEO Blames Tech. Meltdown for Share Drop." *Wall Street Journal*, April 26, 2002.

Berman, Dennis K. "Ex-Global Crossing Official Says 'Roundtripping' Skewed Results." *Wall Street Journal*, February 6, 2002.

Berman, Dennis K. "Global Crossing Stayed Silent on Early Accounting Warning." *Wall Street Journal*, February 5, 2002.

Berman, Dennis K. "Telecom Investors Envision Potential in Failed Networks." *Wall Street Journal*, August 14, 2003.

Berman, Dennis K., Phillip Day, and Henry Sender. "Global Crossing Files for Chapter 11, Plans to Reorganize with Asia Firms." *Wall Street Journal*, January 29, 2002.

Blumenstein, Rebecca. "WorldCom's CEO Ebbers Resigns amid Board Pressure over Probe." *Wall Street Journal*, April 30, 2002.

Branigin, William. "Gary Winnick, Global Crossing Founder at Center of Epic Rise and Fall, Dies at 76." *Washington Post*, November 14, 2023.

Braude, Nicholas P. "WorldCom Gets Court Approval to Sell Ranch Seized from Ex-CEO." *Wall Street Journal,* July 2003.

Brin, Dinah Wisenberg. "Williams Communications Aims to Emerge from Bankruptcy Quickly." *Wall Street Journal*, April 24, 2002.

Cauley, Leslie. "Armstrong's Vision of Cable Empire at AT&T Unravels on the Ground." *Wall Street Journal*, October 18, 2000.

"Company News; Leucadia Offers $365 Million for Rest of Wiltel." *New York Times*, May 16, 2003.

"Corvis Closes Focal Purchase." LightReading, September 2, 2004.

Connell, Megan. "The Fall of Enron and the Creation of the Sarbanes–Oxley Act of 2002." La Salle University Digital Commons, 2017.

Crewsell, Julie, and Nomi Prins. "The Emperor of Greed." CNN Money, June 24, 2002.

"Defiant Joe Nacchio Lashes Out Against Those Who Sent Him to Prison." CBS News, May 28, 2014.

Deogun, Nikhil, and Deborah Solomon. "AT&T Board Approves New Plan to Split Company's Core Businesses." *Wall Street Journal*, October 25, 2000.

Deogun, Nikhil, and Deborah Solomon. "Comcast Makes Offer for AT&T Broadband." *Wall Street Journal*, July 9, 2001.

Deogun, Nikhil, and Deborah Solomon. "Surprising Comcast Offer May Give AT&T Breakup an Unexpected Push." *Wall Street Journal*, July 10, 2001.

Dooley, John. "For Some Telecom Firms, Cost Cuts Haven't Fixed Credit." *Wall Street Journal*, September 11, 2001.

Dreazen, Yochi J. "Comcast Deal Will Likely Win Approval but Internet Access May Be Sticking Point." *Wall Street Journal*, December 21, 2001.

Eckert, Deborah. "Williams Comm. Creditors Seek Info on Former Parent, Executive Perks." *Wall Street Journal*, June 5, 2002.

Eichenwald, Kurt. "For WorldCom, Acquisitions Were Behind Its Rise and Fall." *New York Times*, August 8, 2002.

Eichenwald, Kurt. "Lawyer Says Ex-Chief of WorldCom Knew of Problems." *New York Times*, July 12, 2002.

Ewalt, David M. "Qwest's Last Stand." *Forbes*, April 22, 2005.

"Ex-WorldCom Executive Sullivan Pleads Guilty to Conspiracy." *Wall Street Journal*, June 9, 2004.

"As Shares Sank, Some Executives Shed Costly Toys." *New York Times*, March 24, 2002.

Feder, Barbaby J., and David Leonhardt. "Turmoil at WorldCom: The Finance Executive; from Low Profile to No Profile." *New York Times*, June 27, 2002.

Feldman, Amy, and Joan Caplin. "Is Jack Grubman the Worst Analyst Ever?" CNN Money, April 25, 2002.

"Former Qwest Executives Indicted for Fraud." *New York Times*, February 25, 2003.

"Former WorldCom CFO Sullivan Receives Five-Year Prison Sentence." *Wall Street Journal*, August 12, 2005.

Frosch, Dan. "Closing Arguments Begin in Trial of Ex-Qwest Chief." *New York Times*, April 11, 2007.

Frosch, Dan. "Ex-Chief at Qwest Found Guilty of Insider Trading." *New York Times*, April 20, 2007.

"Genuity Is Left on the Launching Pad." *Forbes*, June 6, 2013.

Gilpin, Kenneth N. "Qwest Discloses US Attorney Is Investigating the Company." *New York Times*, July 10, 2002.

Gilpin, Kenneth N. "WorldCom Changes Its Name and Emerges from Bankruptcy." *New York Times*, April 20, 2004.

"Global Crossing Memos Released." *New York Times*, September 20, 2002.

"GM CEO Ed Whitacre to Step Down Sept. 1." CBS News, August 12, 2010.

Gold, Russell. "Worst 1-Year Performer: Williams." *Wall Street Journal*, March 10, 2003.

Goodman, Peter S., and Renae Merle. "End of Its Merger Run Led to WorldCom's Fall." *Washington Post*, June 29, 2002.

Griffin, Greg. "The Case against Joe Nacchio." *Denver Post*, April 11, 2007.

Hays, Constance L. "Prosecuting Martha Stewart: The Overview; Martha Stewart Indicted by US on Obstruction." *New York Times,* June 5, 2003.

Hendricks, David. "Retired AT&T Chief Ed Whitacre Set to Lead GM." *SFGate,* June 24, 2009.

"International Business; 360networks Misses Payment." *New York Times*, June 16, 2001.

"Interview: Cynthia Cooper, Sherron Watkins, Coleen Rowley." *Time,* December 30, 2002.

Jander, Mary. "Corvis & Broadwing: Together at Last." LightReading, February 25, 2003.

Kapner, Suzanne. "AT&T and British Telecom Will Shut Down Joint Venture." *New York Times*, October 17, 2001.

Kennedy, Peter. "360net Buying Dynegy's Network." *Globe and Mail*, April 1, 2003.

Kerven, Anne. "MFN Lands $611 M, Moves Execs, Lowers Guidance." EEWorld Online, October 1, 2001.

King, Carolyn. "US Court Confirms 360networks Plan of Reorganization." *Wall Street Journal*, October 3, 2002.

Kohn, David. "Who Killed Montana Power?" *60 Minutes*. CBS News, February 6, 2003.

Kolesnikov, Sonia. "ST Telemedia to Buy Global Crossing Stake." United Press International, April 30, 2003.

Landler, Mark. "For Salomon, as Adviser, Millions Plus Revenge." *New York Times*, October 3, 1997.

Latour, Almar, and Dennis K. Berman. "Verizon Takeover of MCI Approved for $6.8 Billion." *Wall Street Journal*, February 14, 2005.

Latour, Almar, and Shawn Young. "Boards of SBC and AT&T Approve $16 Billion Deal." *Wall Street Journal*, January 31, 2005.

Latour, Almar, and Shawn Young. "Ebbers Denies He Knew about WorldCom's Fraud." *Wall Street Journal*, March 1, 2005.

Latour, Almar, and Shawn Young. "Ebbers Offers Wave of Denial on the Stand." *Wall Street Journal*, March 2, 2005.

Latour, Almar, and Shawn Young. "Ebbers Trial Gets Under Way In US Court." *Wall Street Journal*, January 26, 2005.

Latour, Almar, and Shawn Young. "Next Call for Ebbers: A Verdict." *Wall Street Journal*, March 4, 2005.

Latour, Almar, and Shawn Young. "WorldCom Ex-Finance Chief Says He Told CEO of Accounting Issues." *Wall Street Journal*, February 9, 2005.

Latour, Almar, Shawn Young, and Li Yuan. "Ebbers Convicted in Massive Fraud." *Wall Street Journal*, March 16, 2005.

Latour, Almar, Susan Pulliam, and Shawn Young. "WorldCom Testimony Heats Up." *Wall Street Journal*, February 8, 2005.

Latour, Almar. "Meet the New TV Guy." *Wall Street Journal*, November 23, 2004.

"Level 3 Acquires CorpSoft for $89M." LightReading, February 25, 2002.

"Level 3 Communications Will Buy Back Up to $1.8 Billion in Long-Term Debt." *Wall Street Journal*, September 11, 2001.

"Level 3 Plans to Buy Software Distributor Company Insists Purchase Fits within Plans." *Denver Post*, February 26, 2002.

"Level 3 Reports Jump in Sales but Posts Wide Loss on Charges." *Wall Street Journal*, January 26, 2001.

"Level 3 Shopping Spree Stalls." *Forbes*, July 29, 2002.

Ling, Connie. "Level 3 Exits Asia with Sale of Assets to Reach Venture." *Wall Street Journal*, December 20, 2001.

"Ma Bell's New Offspring." CNN Money, October 25, 2000.

Martinez, Barbara. "Analyst's Report on Level 3 Sparks Questions on Timing." *Wall Street Journal*, April 2, 1999.

"MCI Emerges from Bankruptcy." NBC News, April 20, 2004.

Meisner, Jeff. "Another Spin at 360." *Puget Sound Business Journal*, September 15, 2002.

Meisner, Jeff. "Even without Maffei, 360 Hungrily Hunts Deals." *Puget Sound Business Journal*, July 14, 2005.

Miller, Rich. "Digital Realty Acquires AboveNet Centers." *DataCenter Knowledge*, October 2, 2006.

Morgenson, Gretchen. "Telecom, Tangled in Its Own Web." *New York Times*, March 24, 2002.

Morgenson, Gretchen. "Telecom's Pied Piper: Whose Side Was He On?" *New York Times*, November 18, 2001.

"Net-Services Firm Genuity to Buy France's Integra for $113.7 Million." *Wall Street Journal*, June 1, 2001.

Noguchi, Yuki. "Cogent to Buy US Division of PSINet." *Washington Post*, February 27, 2002.

Nuzum, Christine, and Chad Bray. "Prosecutors Urge Jurors to Convict 'Corrupted' Ex-WorldCom CEO." *Wall Street Journal*, March 3, 2005.

Nuzum, Christine, and Susan Pulliam. "WorldCom's Ex-Controller Deals Blow to Ebbers's Defense." *Wall Street Journal*, January 28, 2005.

Nuzum, Christine. "Ex-WorldCom Chairman, Whistle-Blower Say Ebbers Didn't Know about Fraud." *Wall Street Journal*, February 25, 2005.

Nuzum, Christine. "Former WorldCom Controller Says He Used Improper Accounting." *Wall Street Journal*, February 5, 2005.

Nuzum, Christine. "Former WorldCom Executive Myers Says Ebbers Feared Being Wiped Out." *Wall Street Journal*, January 31, 2005.

Nuzum, Christine. "Former WorldCom Finance Chief: Ebbers Ordered to Hide Information." *Wall Street Journal*, February 11, 2005.

Nuzum, Christine. "Metromedia Fiber Files for Bankruptcy Protection." *Wall Street Journal*, May 20, 2002.

Nuzum, Christine. "Prosecutor Claims Ebbers Told Lies to Inflate WorldCom Stock Price." *Wall Street Journal*, January 26, 2005.

Nuzum, Christine. "Two Challenge Ebbers's Defense in Testimony in WorldCom Case." *Wall Street Journal*, February 3, 2005.

O'Brien, Timothy L. "WorldCom to Exit Bankruptcy and Change Name to MCI." *New York Times*, April 14, 2003.

Pacelle, Mitchell, and Dennis Berman. "Level 3 to Pay $242 Million for Most of Genuity Assets." *Wall Street Journal*, November 29, 2002.

Pacelle, Mitchell, and Elliot Spagat. "Level 3 Makes Bold $1.1 Billion Bid to Buy Williams Communications." *Wall Street Journal*, July 24, 2002.

Padgett Jackson, Tim. "The Rise and Fall of Bernie Ebbers." *Time*, May 13, 2002.

Pham, Alex, and David Colker. "Global Crossing Ex-Execs to Pay $324 Million." *Los Angeles Times*, March 20, 2004.

"Playing Defense." *Wall Street Journal*, March 4, 2005.

Powell, Rob. "AboveNet Emerges from Coma." Telecom Ramblings, May 16, 2008.

Powell, Rob. "AboveNet Keeps Chugging." Telecom Ramblings, August 10, 2009.

Powell, Rob. "AboveNet Splits!" Telecom Ramblings, August 4, 2009.

Powell, Rob. "Industry Spotlight: AboveNet's Bill LaPerch." Telecom Ramblings, March 11, 2010.

Pulliam, Susan, and Deborah Solomon. "How Three Unlikely Sleuths Exposed Fraud at WorldCom." *New York Times*, October 30, 2002.

Pulliam, Susan. "Top WorldCom Executives, Analyst Feel Congressional Heat in Hearing." *Wall Street Journal*, July 9, 2002.

Quigley, Kelly. "Focal Emerges as Private Company." *Crain's Chicago Business*, July 8, 2003.

"Reactions Mixed to Buffett's Level 3 Deal Some Wonder If Telecom Field Still Too Risky." *Denver Post*, July 9, 2002.

Ripley, Amanda. "Cynthia Cooper: The Night Detective." *Time*, December 30, 2002.

Robinson, Doug and Lee Benson. "Ian Cumming Lived Life with Little Fanfare and Died the Same Way." *Deseret News,* February 7, 2018.

Robinson, Rick. "Williams Pays Debt for Wil-Com Facilities." *Oklahoman,* March 30, 2002.

Romero, Simon, and Geraldine Fabrikant. "Deal's Loose Ends Are Tied to Global Crossing." *New York Times,* February 21, 2002.

Romero, Simon, and Seth Schiesel. "The Fiber Optic Fantasy Slips Away." *New York Times,* February 17, 2002.

Romero, Simon. "Qwest Announces Accounting Flaws." *New York Times,* July 29, 2002.

Romero, Simon. "Technology; Global Crossing Warns on Revenue and Appoints a Chief." *New York Times,* October 5, 2001.

Sandberg, Jared, Rebecca Blumenstein, and Shawn Young. "WorldCom Internal Probe Uncovers Massive Fraud." *New York Times,* June 26, 2002.

Schiesel, Seth, and Simon Romero. "Genuity Faces Bankruptcy as Verizon Ignores an Option." *New York Times,* July 26, 2002.

Schiesel, Seth. "2 Companies Agree to Buy Control of Global Crossing." *New York Times,* August 10, 2002.

Scinta, Christopher. "Metromedia Fiber Sees Enterprise Value of $230M–$330M." *Wall Street Journal,* May 22, 2003.

Searcey, Dionne. "Former Qwest CEO Joseph Nacchio: Tales from a White-Collar Prison Sentence." *Wall Street Journal,* September 30, 2013.

Searcey, Dionne. "SBC Decides to Stick with AT&T Brand Name." *Wall Street Journal,* October 27, 2005.

"Shareholders Sue Metromedia." LightReading, August 23, 2001.

Sheng, Ellen, Nick Baker, and Shawn Young. "MCI Emerges from Chapter 11 Protection, Pledges New Products, Partnership Deals." *Wall Street Journal,* August 21, 2004.

Simonson, Sharon. "Equinix Buys AboveNet's Data Center in San Jose." *Silicon Valley Business Journal,* December 12, 2004.

Solomon, Deborah, and Nikhil Deogun. "Amid Steep Business Declines, AT&T Unveils Breakup Plan." *Wall Street Journal,* October 26, 2000.

Solomon, Deborah, Kara Scannell, and Mark Heinzl. "360networks to Announce Backing from W.L. Ross." *Wall Street Journal,* October 2, 2002.

Solomon, Deborah. "AT&T Board Rejects Bid by Comcast for Cable Unit." *Wall Street Journal,* July 19, 2001.

Solomon, Deborah. "AT&T CEO Tries to Hold on to an Icon Loaded with Debt, Hit by Competition." *Wall Street Journal,* November 16, 2001.

Solomon, Deborah. "AT&T Names David Dorman to Be New Chairman, CEO." *Wall Street Journal,* July 18, 2002.

Solomon, Deborah. "How Upstart Qwest's Merger with Baby Bell Led to Trouble." *Wall Street Journal*, April 2, 2002.

Solomon, Deborah. "Woes of Level 3, Williams Raise Fears of Growing Telecom Sector Shakeout." *Wall Street Journal*, January 30, 2002.

Spagat, Elliot. "Williams Communications Awarded Executives Bonuses to Repay Loans." *Wall Street Journal*, May 4, 2002.

Spagat, Elliot. "Williams Communications to Delay Interest Payments, Extend Deadline." *Wall Street Journal*, April 2, 2002.

Stern, Christopher. "Global Crossing Chairman Resigns." *Washington Post*, December 31, 2002.

"Sullivan Pleads Not Guilty." *New York Times*, April 23, 2003.

"Technology Briefing | Telecommunications: Executive Buys Stake in Metromedia Fiber." *New York Times*, July 18, 2003.

"Technology; PSINet Suffers Losses, Weighs Independence." *New York Times*, November 3, 2000.

Temasek. "Our Portfolio." Last modified June 28, 2024. https://www.temasek.com.sg/en/our-investments/our-portfolio.

"The Rise and Fall of Bernie Ebbers." *Forbes*, June 6, 2013.

Timmons, Heather. "Spitzer Raises the Heat on Citigroup." Bloomberg.

US Securities and Exchange Commission. "Order Instituting Cease-and-Desist Proceedings, Making Findings, and Imposing a Cease-and-Desist Order Pursuant to Section 21C of the Securities Exchange Act of 1934." Administrative Proceeding, Accounting and Auditing Enforcement. Release No. 2231. File No. 3-11891. April 11, 2005.

"Verizon and Genuity Get Cozier." LightReading, May 3, 2002.

"Verizon Global, Genuity Solutions Cancel Metromedia Fiber Pacts." *Wall Street Journal*, April 7, 2002.

Vuong, Andy. "CenturyLink Completes Purchase of Qwest." *Denver Post*, April 1, 2011.

Vuong, Andy. "Ed Mueller: The Man Who Made the Call to Sell Qwest." *Denver Post*, June 4, 2010.

Vuong, Andy. "Judge: Greed Toppled Ex-CEO." *Denver Post*, July 27, 2007.

Wei, Lingling, and Judith Burns. "WorldCom's Former CFO Sullivan Pleads Guilty to Federal Charges." *Wall Street Journal*, March 3, 2004.

Weil, Jonathan. "Now, Telecoms' Valuations Are Called into Question." *Wall Street Journal*, June 25, 2001.

"When Buffett Buys, a Bottom Must Be Near." *Forbes*, July 8, 2002.

"WilTel Names Chief Executive." *Wall Street Journal*, November 1, 2002.

"WorldCom Hearing with House Financial Services Committee." CNN, July 8, 2002.

"WorldCom Trial Star Witness Admits He Lied." *New York Times*, February 17, 2005.

"Yesterday's Man." *Economist*, May 2, 2002.

Young, Shawn, Almar Latour, and Susan Pulliam. "Ebbers Lawyer Paints Sullivan as Chronic Liar." *Wall Street Journal*, February 17, 2005.

Young, Shawn, Almar Latour, and Susan Pulliam. "Linking Ebbers to the Fraud at WorldCom Proves Difficult." *Wall Street Journal*, February 18, 2005.

Young, Shawn, and Almar Latour. "Ebbers Feels Pinches on Finances." *Wall Street Journal*, February 23, 2005.

Young, Shawn, and Christine Nuzum. "Sullivan Says Ebbers Ducked Merger Talks to Keep Veil on Fraud." *Wall Street Journal*, February 11, 2005.

Young, Shawn, and Evan Perez. "WorldCom's Sullivan Garnered Respect from Wall Street Crowd." *Wall Street Journal*, June 27, 2002.

Young, Shawn. "Analysts Expect Qwest to Lower Forecast as Telecom Slump Begins to Hit Home." *Wall Street Journal*, December 13, 2001.

Young, Shawn. "Ebbers Defense Claim Is Backed by Testimony of Two Witnesses." *Wall Street Journal*, February 25, 2005.

Young, Shawn. "Genuity, Shunned by Verizon, Scrambles amid Credit Crunch." *New York Times*, July 26, 2002.

Young, Shawn. "Metromedia Will Restate Its Results after Auditor Finds Internal Flaws." *Wall Street Journal*, April 18, 2002.

Young, Shawn. "Qwest Chief Defends Calpoint Contract as Deal Fuels Worries about Revenue." *Wall Street Journal*, October 1, 2001.

Young, Shawn. "Qwest Predicts Charge on Its Assets In 2002 of $20 Billion to $30 Billion." *Wall Street Journal*, April 2, 2002.

Young, Shawn. "Sullivan Says He Told Ebbers of Cost-Hiding Moves." *Wall Street Journal*, February 10, 2005.

Zipern, Andrew. "Technology Briefing: Internet; PSINet Hires Adviser as Shares Drop." *New York Times*, March 20, 2001.

Zuckerman, Gregory, and Deborah Solomon. "Telecom Debt Debacle Could Lead to Losses of Historic Proportions." *Wall Street Journal*, May 11, 2001.

Chapter VI

"A Nation Challenged; Telecom Concern Trims Outlook." *New York Times*, September 26, 2001.

Arango, Tim. "How the AOL-Time Warner Merger Went So Wrong." *New York Times*, January 10, 2010.

Backover, Andrew. "Chief of Embattled ICG Resigns." *Denver Post*, August 23, 2000.

Backover, Andrew. "ICG's New Team Sees a Data-Services Future." *Denver Post*, May 15, 2000.

Beauprez, Jennifer. "ICG Gets Capital to Fuel Growth." *Denver Post*, February 29, 2000.

Casselman, Ben. "Fiber-Optics Pioneer Says: Buy Florida, Sell Maryland." *Wall Street Journal*, December 15, 2006.

Cauley, Leslie. "Cincinnati Bell Plans to Acquire IXC for $2.2 Billion Plus $1 Billion in Debt." *Wall Street Journal*, July 22, 1999.

"Ciena Corp. Built on Dreams, Risks Decision." *Baltimore Sun,* August 16, 1998.

"Cincinnati Bell to Acquire IXC Communications." *Fiber Optics Online.* 1999.

"Communications: The Next Wave." *Forbes*, October 6, 1997.

De La Merced, Michael J. "Level 3 to Buy tw telecom for $5.7 Billion." *New York Times*, June 16, 2014.

Desloge, Rick. "Xspedius Executives Dial up Big Payouts from Time Warner Deal." *St. Louis Business Journal,* October 1, 2006.

Dresser, Michael. "ACSI Moves Quietly." *Baltimore Sun,* October 3, 1995.

Dugan, Ianthe Jeanne, and Ariana Eunjung Cha. "AOL to Acquire Time Warner in Record $183 Billion Merger." *Washington Post*, January 11, 2000.

Ernst, Steve. "GST Anticipates Profitability for First Time." *Puget Sound Business Journal,* June 6, 1999.

Gara, Antoine. "Verizon to Buy XO Communications' Fiber Business for $1.8B from Billionaire Carl Icahn." *Forbes*, February 22, 2016.

"GST Activates Network in Houston." *Portland Business Journal*, December 7, 1999.

"GST Announces Completion of Fiber Optic Link." *Portland Business Journal,* November 20, 1997.

"GST in Chapter 11." *Puget Sound Business Journal,* May 17, 2000.

"GST Telecom Inc." *Bloomberg.* Accessed August 5, 2024. https://www.bloomberg.com/profile/company/58762Z:US?embedded-checkout=true.

Hardy, Quentin. "Weaving the Perfect Net." *Forbes*, July 3, 2000.

Henry, Shannon. "A New Breed of Local Telcos." *Washington Technology,* August 29, 1997.

https://www.sec.gov/Archives/edgar/data/1111634/000089742308000098/exhibit991.htm.

https://www.sec.gov/Archives/edgar/data/1111634/000114420411040735/v228407_prem14c.htm.

Hudson, Kris. "ICG's Decline Hangs Up Liberty Media, Investors." *Denver Post,* November 12, 2000.

"ICG Goes to Venture Capitalists." Channel Futures, November 30, 2004.

"ICG Out of Bankruptcy." *Denver Business Journal*, October 10, 2002.

"ICG to Be Acquired." *Denver Business Journal*, July 19, 2004.

"In re ICG Communications, Inc. Securities Litigation." Bernstein Litowitz Berger & Grossmann LLP. Court: United States District Court for the District of Colorado. Case Number: 00-cv-1864. Class Period: 12/09/1999–09/18/2000. https://www. blbglaw.com/cases-investigations/icg-communications-inc.

Jackson, Margaret. "Brooks Fiber Alums Form Xspedius." *St. Louis Business Journal*, October 13, 2002.

Jander, Mary. "Corvis & Broadwing: Together at Last." LightReading, February 25, 2003.

Lee, 'Consella A. "Xspedius Gets Bankruptcy Court OK to Buy e.spire for $68 Million." *Wall Street Journal*, June 5, 2002.

Leib, Jeffrey. "Telecom Going Public." *Denver Post*, May 12, 1999.

Lipscombe, Paul. "Morrison & Co. Finalizes FiberLight Acquisition." *Data Center Dynamics,* April 25, 2023.

Markoff, John. "Fiber-Optic Technology Draws Record Stock Value." *New York Times*, March 3, 1997.

McMillan, Dan. "'Discharged' Warta Slams GST in Lawsuit." *Portland Business Journal,* March 28, 1999.

McMillan, Dan. "GST Suit Says Bosses Ripped Company Off." *Portland Business Journal,* November 1, 1998.

McMillan, Dan. "Warta's GST Retirement a Big Surprise to Many." *Portland Business Journal,* June 21, 1998.

Mildenberg, David. "New Telecom Lining up for Entry Here." *Charlotte Business Journal*, June 6, 2005.

Mills, Mike. "e.spire Rings Up Sales Against the Bells." *Washington Post*, May 24, 1998.

"Mpower to Acquire ICG's California Customers, Network." *Converge! Network Digest*, October 21, 2004.

Nocera, Joe. "The Kettle? The Pot Says He's Black." *New York Times*, August 29, 2008.

Noguchi, Yuki. "Icahn Suggests Another Plan for XO." *Washington Post*, March 27, 2002.

Noguchi, Yuki. "XO Emerges from Bankruptcy Protection." *Washington Post*, January 16, 2003.

Noguchi, Yuki. "XO Ends Deal with Forstmann for $25 Million." *Washington Post*, October 14, 2002.

Pearlstein, Steven. "The Puzzling Allure of David Huber." *Washington Post*, January 7, 2005.

Potter, Beth. "Level 3 Spree Grabs ICG." *Denver Post*, April 17, 2006.

Powell, Rob. "Icahn Makes His Move on XO." Telecom Ramblings, July 10, 2009.

Powell, Rob. "Is Icahn about to Take XO Private?" Telecom Ramblings, October 17, 2008.

Powell, Rob. "More Monkey Business by Icahn at XO." Telecom Ramblings, July 31, 2009.

Raabe, Steve. "2 Suits Accuse ICG of Financial Misconduct." *Denver Post*, September 23, 2000.

Raabe, Steve. "ICG Names Another CEO." *Denver Post*, September 27, 2000.

"Randall Curran '76 Joins FTI Consulting as Senior Managing Director." Depauw University News & Media, July 31, 2004.

Reardon, Marguerite. "Level 3 to Buy Broadwing for $1.4 Billion." CNET, October 17, 2006.

Richtel, Matt, and Andrew Ross Sorkin. "Verizon Agrees to Acquire MCI for $6.6 Billion, Beating Qwest." *New York Times*, February 14, 2005.

Romero, Simon. "Technology Briefing: Deals; Time Warner to Buy GST for $690 Million." *New York Times*, September 22, 2000.

Rose, Frank. "Surviving the Fiber-Optic Fire Sale." *Wired*, November 1, 2002.

Sherer, Paul M. "ICG Files for Chapter 11 Protection, Lines Up $350 Million Credit Line." *Wall Street Journal*, November 15, 2000.

Sherer, Paul M. "ICG Sees Major Investors Bail Out, Signaling Shakeout for Telecom Sector." *Wall Street Journal*, September 20, 2000.

Sherer, Paul M., and Gary McWilliams. "A Brash Provider of Internet Services Became Unplugged; Chapter 11 Looms." *Wall Street Journal*, November 13, 2000.

Sidel, Robin, and Dennis K. Berman. "Broadwing Agrees to Sell Bulk of Broadband Unit to Venture." *Wall Street Journal*, February 25, 2003.

"Technology Briefing: Telecommunications." *New York Times*, April 29, 2003.

"Telecomeback." *Forbes*, January 21, 2002.

"The Inside Story of ICGs Collapse." *eWEEK*, June 4, 2001.

Thomas Jr., Landon. "Bryan Hits the Wall: Handsome Investor's $60 Million Goes Poof! as He Takes Down Malone with Him." *New York Observer*, October 2, 2000.

"Time Warner Buys Enron Fiber." LightReading, January 25, 2004.

"Time Warner Telecom to Change Name." Channel Futures, March 12, 2008.

Toscano, Paul. "Portfolio's Worst American CEOs of All Time." CNBC, April 30, 2009.

"TWTC Buying Xspedius for $531M." *Denver Business Journal,* July 28, 2006.

US Securities and Exchange Commission. "Form 10-K Annual Report Pursuant to Section 13 or 15(d) of the Securities Exchange Act of 1934. Time Warner Telecom

Inc." Commission File Number 0-30218. For the Fiscal Year Ended December 31, 1999.

US Securities and Exchange Commission. "Time Warner Telecom to Acquire Xspedius Communications for $531.5 Million." Filed by Time Warner Telecom Inc. Pursuant to Rule 425 Under the Securities Act of 1933. Subject Company: Time Warner Telecom Inc. Commission File No. of Subject Company: 0-30218. July 27, 2006.

US Securities and Exchange Commission. SEC Filing, April 28, 2009. https://www.sec.gov/Archives/edgar/data/1111634/000089742309000265/exhibit991.htm.

US Securities and Exchange Commission. SEC Filing, August 1, 2008.

US Securities and Exchange Commission. SEC Filing, July 15, 2011.

US Securities and Exchange Commission. SEC Filing, June 12, 2008. https://www.sec.gov/Archives/edgar/data/1111634/000089742308000090/xocm13da1.htm.

US Securities and Exchange Commission. SEC Filing, August 4, 2009. https://www.sec.gov/Archives/edgar/data/1114634/000089742309000265/xoholdings13da3.htm.

"XO Communications KOs Allegiance Bidders." *Forbes*, February 13, 2004.

"XO Hires GlobalX-er as CEO." LightReading, April 28, 2003.

Chapter VII

"American Tower Corporation Announces Agreement to Acquire Global Tower Partners." Business Wire, September 6, 2013.

Ante, Spencer E., and Joann S. Lublin. "Qwest Deal Is One of CenturyTel Chief's Riskiest Moves." *Wall Street Journal*, April 23, 2010.

Avery, Greg. "Intrigue Surrounds Departure of Level 3's O'Hara." *Denver Business Journal*, March 16, 2008.

Benoit, David. "Corvex Builds Stake in CenturyLink." *Wall Street Journal*, May 8, 2017.

Berman, Dennis K. and Shawn Young. "Level 3 Adds to Telecom Network with $1 Billion Deal for TelCove." *Wall Street Journal*, May 1, 2006.

Buckley, Sean. "Zayo Acquires Allstream for $348M, Gains Instant Canadian Fiber Presence." *Fierce Network*, November 23, 2015.

Buckley, Sean. "Zayo Acquires Optic Zoo Networks for $25M, Deepens Canadian Fiber Footprint." *Fierce Network*, January 23, 2018.

"CenturyLink Buying Qwest with Stock." *Forbes*, April 22, 2010.

"CenturyLink Transforms, Rebrands as Lumen." Lumen News Details, September 14, 2020.

"CenturyTel and EMBARQ Agree to Merge." Lumen News Releases, October 27, 2008.

Chuang, Tamara. "Colorado's Level 3 Communications Is No More, as CenturyLink Closes $30B Purchase." *Denver Post*, November 1, 2017.

Chuang, Tamara. "Faster Internet Feeds Boulder's Zayo: IPO Jumps 16 Percent." *Denver Post*, October 17, 2014.

Cologix. "Stonepeak Successfully Completes $3.0 Billion Equity Recapitalization of Portfolio Company Cologix." April 6, 2022. https://cologix.com/news/stonepeak-successfully-completes-3-0-billion-equity-recapitalization-portfolio-company-cologix/.

"Colony Capital Acquires Digital Bridge Holdings for $325 Million and Announces Planned Strategic Initiatives to Become the Premier Platform for Digital Infrastructure and Real Estate." Business Wire, July 25, 2019.

"Colony Capital Acquires Digital Bridge Holdings for $325 Million and Announces Planned Strategic Initiatives to Become the Premier Platform for Digital Infrastructure and Real Estate." Business Wire, July 25, 2019.

De La Merced, Michael J. "Zayo to Buy AboveNet for $2.2 Billion to Extend Fiber Network." *New York Times*, March 19, 2012.

DeGrasse, Martha. "ExteNet Systems Acquired by Digital Bridge." *RCR Wireless News*, July 23, 2015.

Del Deo, Nick, Craig Moffett, and Cathy Yao. "Dark Fiber and Towers: Batman vs. Superman, or 'You're No Jack Kennedy.'" MoffettNathanson, October 14, 2014.

FitzGerald, Drew, and Joann S. Lublin. "Telecom CEO's Deal-Making Puts Louisiana Town at Center of Internet." *Wall Street Journal*, November 2, 2016.

FitzGerald, Drew, and Joshua Jamerson. "CenturyLink to Buy Level 3 Communications for $25 Billion." *Wall Street Journal*, October 31, 2016.

FitzGerald, Drew, and Will Feuer. "This CEO Just Bet $1 Million on Herself." *Wall Street Journal*, November 8, 2023.

FitzGerald, Drew. "Level 3 to Buy tw telecom for $5.7 Billion." *Wall Street Journal*, June 16, 2014.

Gregorian, Dareh. "Jury Finds Trump Friend Tom Barrack Not Guilty of Foreign Lobbying and Lying to FBI." NBC News, November 4, 2022.

Gutierrez, Carl. "Level 3 Drops amidst CFO Mystery." *Forbes*, October 15, 2007.

Hagerty, James R. "Walter Scott Jr., Business Partner of Warren Buffett, Dies at Age 90." *Wall Street Journal*, September 27, 2021.

"James Q. Crowe, Telecommunications Pioneer, Dies at 74." PR Newswire, July 7, 2023.

"Jeff Storey to Succeed Jim Crowe as Level 3 CEO after 'Fast Search.'" *Denver Post*, April 11, 2013.

Kosman, Josh. "NY Fiber Firm Buy in Works." *New York Post*, June 15, 2011.

Le Maistre, Ray. "How Zayo Spent $3.7B on Acquisitions." LightReading, July 4, 2014.

"Lee Jobe, Career Telecom Exec, Dies." Channel Futures, February 27, 2010.

"Level 3 Closes Genuity Purchase at Lower Price." *Wall Street Journal*, February 5, 2003.

"Level 3 Communications to Buy Rival WilTel." *Wall Street Journal*, November 1, 2005.

"Level 3 President's Departure Surprises." *Denver Post*, March 10, 2008.

"Level 3 to Acquire Looking Glass." *Wall Street Journal*, June 6, 2006.

"Level 3 to Buy Broadwing in Deal Worth $1.4 Billion." *Wall Street Journal*, October 17, 2006.

"Lumen Announces CEO Transition." PR Newswire, September 13, 2022.

Marshall, Barbara. "Meet Wellington's Royal Couple of Polo." *Palm Beach Post*, May 11, 2012.

Mukherjee, Supantha. "AboveNet CEO Keen to Make His First Acquisition." Reuters, October 21, 2011.

Noland, Kim. "Failure to Launch." *Wall Street Journal*, October 29, 2007.

Ovide, Shira. "Level 3-Global Crossing Deal: Reaction Roundup." *Wall Street Journal*, April 11, 2011.

Powell, Rob. "AMP Capital Announces Entry into a Definitive Agreement to Purchase Everstream." Telecom Ramblings, March 8, 2018.

Powell, Rob. "Antin to Buy FirstLight Fiber." Telecom Ramblings, February 21, 2018.

Powell, Rob. "Brookfield Matches Macquarie Again for Cincinnati Bell." Telecom Ramblings, March 4, 2020.

Powell, Rob. "Crown Castle Announces Agreement to Acquire Sunesys." Telecom Ramblings, April 30, 2105.

Powell, Rob. "Crown Castle Makes Another Fiber Move, Buys FPL Fibernet." Telecom Ramblings, November 1, 2016.

Powell, Rob. "Digital Colony Powers Beanfield Buy of Aptum's Metro." Telecom Ramblings, May 7, 2020.

Powell, Rob. "Enel Buys into Ufinet." Telecom Ramblings, June 28, 2018.

Powell, Rob. "EQT Adds to Fiber Portfolio, Buys inexio." Telecom Ramblings, October 2, 2019.

Powell, Rob. "EQT: IP-Only and GlobalConnect to Merge." Telecom Ramblings, November 22, 2019.

Powell, Rob. "Jeff Storey Resurfaces at Level 3." Telecom Ramblings, December 5, 2008.

Powell, Rob. "Level 3 Closes Global Crossing Deal, Plans Reverse Stock Spit." Telecom Ramblings, October 4, 2011.

Powell, Rob. "Lumos and Spirit Are Now Segra." Telecom Ramblings, January 15, 2019.

Powell, Rob. "Macquarie to Acquire Bluebird, Swap with Uniti." Telecom Ramblings, January 15, 2019.

Powell, Rob. "Ridgemont Moves into Cross River Fiber." Telecom Ramblings, August 13, 2014.

Powell, Rob. "SSE Enterprise Telecoms Sells 50% Stake to Infracapital." Telecom Ramblings, December 21, 2018.

Powell, Rob. "Stonepeak Closes Deal with Lumen, Launching Cirion." Telecom Ramblings, August 2, 2022.

Powell, Rob. "Stonepeak to Acquire Lumen's Latin American Business." Telecom Ramblings, July 26, 2021.

Powell, Rob. "Telia to Sell Telia Carrier." Telecom Ramblings, October 6, 2020.

Powell, Rob. "TPG to Acquire Both RCN and Grande from ABRY." Telecom Ramblings, August 15, 2016.

Powell, Rob. "Wilcon Sells to Crown Castle." Telecom Ramblings, April 17, 2017.

Rooney, Ben. "CenturyTel, Qwest in $10.6 Billion Merger." CNN Money, April 22, 2010.

Rose, Frank. "Surviving the Fiber-Optic Fire Sale." *Wired*, November 1, 2002.

Samtani, Hiten. "How Marc Ganzi Bet Colony Capital's Future on the Next-Gen Economy." *Real Deal*, May 17, 2021.

SDC Capital Partners. "SDC Capital Partners Closes Minority Investment in Bandwidth Infrastructure Group." October 16, 2020. https://www.sdccapitalpartners.com/news/newsinfo/sdc-capital-partners-closes-minority-investment-in-bandwidth-infrastructure-group.

Shumsky, Tatyana. "CFO Exit Dings CenturyLink's Market Value." *Wall Street Journal*, September 25, 2018.

Solomon, Deborah. "Woes of Level 3, Williams Raise Fears of Growing Telecom Sector Shakeout." *Wall Street Journal*, January 30, 2002.

"Stonepeak Infrastructure Partners Completes Transaction to Acquire Majority Interest in euNetworks." Business Wire, January 18, 2018.

Tan, Gillian, Kiel Porter, and Nabila Ahmed. "Zayo Is Said to Attract Blackstone-Stonepeak Group Interest." Bloomberg News, November 18, 2018.

"Telecom Zayo Set for Launch." *Denver Post*, August 29, 2007.

USA Today US Edition, October 16, 2014.

Vuong, Andy. "Former Level 3 CEO Crowe Plots Next Move: 'I Don't Use the "R" Word.'" *Denver Post*, August 23, 2013.

Vuong, Andy. "Level 3's CFO Will Step Down." *Denver Post*, October 15, 2007.

"Zayo Closes Acquisition of Electric Lightwave." Zayo Press Release, March 1, 2017.

"Zayo Closes Acquisition of Spread Networks." Zayo Press Release. February 28, 2018.

"Zayo Completes 360networks Acquisition." *Denver Post*, December 2, 2011.

"Zayo Completes Transition to a Private Company." Zayo Press Release, March 9, 2020.

Appendix 3: More Bandwidth Stories

"$60M Deal Will Give Them Liberty Cable." *New York Daily News,* April 1, 1996.

"Adelphia Changes Hyperion Name." *Pittsburgh Business Times,* July 2, 1999.

"Adelphias Hyperion Unit Files for $150m IPO." *Multichannel News,* April 5, 1998.

"All The Reasons Chuck Watson Did Not Resign." *Forbes,* May 29, 2002.

"Allstream Sale to Accelero Rejected over 'National Security.'" CBC, October 7, 2013.

"AT&T to Buy Velocita Assets." *Wall Street Journal,* November 8, 2002.

Barboza, David, and Barnaby J. Feder. "Enron's Many Strands: The Transactions; Enron's Swap with Qwest Is Questioned." *New York Times*, March 29, 2002.

Barrionuevo, Alexei. "Enron Chiefs Guilty of Fraud and Conspiracy." *New York Times,* May 25, 2006.

Barrionuevo, Alexei. "Ex-Enron Chief Is Sentenced to 24 Years." *New York Times,* October 23, 2006.

Behr, Peter. "Broadband Strategy Got Enron in Trouble." *Washington Post,* January 1, 2002.

Beltran, Eamon. "McLeodUSA to Sell Unit for $271M, Will Reduce Bank Loans." *Wall Street Journal,* July 22, 2002.

"Berkshire Partners to Lead $2 Billion Merger of Lightower, Sidera." Reuters, December 27, 2012.

Berman, Dennis K. "Paetec Aims to Be a Force with McLeod Deal." *Wall Street Journal,* September 17, 2007.

Blumenstein, Rebecca, and Solange De Santis. "AT&T to Merge Canadian Unit, MetroNet in $2.4 Billion Accord." *Wall Street Journal,* March 5, 1999.

Brown, Matthew L. "Lightower Closes Veroxity Acquisition." *Worcester Business Journal,* September 7, 2010.

Brown, Matthew L. "Westborough's RCN Taken Private, Name Changed to Sidera." *Worcester Business Journal,* September 13, 2010.

Buckley, Sean, "Lightower Snaps up Fibertech in $1.9B All-Cash Deal." *Fierce Network,* April 27, 2015.

Buckley, Sean. "Lightower Completes Its $2 Billion Merger with Sidera." *Fierce Network*, April 11, 2013.

Buckley, Sean. "Sidera Networks Bolsters Colocation Footprint with Cross Connect Solutions Acquisition." *Fierce Network*, November 22, 2010.

Buckley, Sean. "Sidera Networks to Acquire Long Island Fiber Exchange." *Fierce Network*, December 13, 2010.

Buckley, Sean. "Zayo Completes Spread Networks Acquisition, Creates New Low-Latency Division." *Fierce Network,* March 1, 2018.

Buckley, Sean. "Zayo Gets New York-to-Chicago Low-Latency Route with $127M Spread Networks Deal." *Fierce Network,* November 28, 2017.

Burns, Johnathan. "Bankruptcies Continue to Rise in Telecom Service Segment." *Wall Street Journal,* February 1, 2002.

Burrough, Bryan, and John Helyar. *Barbarians at the Gate: The Fall of RJR Nabisco.* Harper & Row, 1989.

Butler, Brandon. "Lightower, NSTAR in Pact." *Worcester Business Journal,* January 5, 2011.

Cherney, Elena. "AT&T Canada Reaches Deal with Its Creditors." *Wall Street Journal*, October 16, 2002.

Cimilluca, Dana. "A Publicly Traded McLeodUSA: Take III." *Wall Street Journal,* March 1, 2007.

Clinton, Chad. "McLeodUSA Draws $200M under $1.3 Billion Credit Line." *Wall Street Journal,* November 19, 2001.

"Company News; Adelphia Communications to Spin Off Unit and Cut Debt." *New York Times,* November 10, 2001.

"Company News; AT&T to Complete Purchase of AT&T Canada Shares." *New York Times*, October 5, 2002.

"Company News; AT&T to Use Mcleodusa Network for Service in 30 Cities." *New York Times*, October 31, 1997.

"Company News; McLeodUSA in Acquisitions to Expand Service in West." *New York Times,* June 3, 1999.

"Company News; Rogers Says It Will Not Pursue Stake in Unitel." *New York Times*, April 20, 1995.

"Con Ed Sells to RCN." LightReading, December 6, 2005.

Coyle, Brian. "McLeodUSA Files IPO to Sell up to $172.5M in Common Stock." *Wall Street Journal,* March 23, 2007.

Cummins, Chip, Jathon Sapsford, Paul Beckett, and Thaddeus Herrick. "Watson, Who Long Led Dynegy in Enron's Shadow, Steps Down." *Wall Street Journal,* May 29, 2002.

Darwin, Jennifer. "Dynegy Signs $75 Million Broadband Deal." *Houston Business Journal,* May 20, 2001.

Davis, Christina. "Hardened Salesman Turned Hard-Charging CEO I Rob Shanahan President and CEO, Lightower Fiber Networks." *Worcester Business Journal,* March 29, 2010.

De Santis, Solange. "MetroNet Nimbly Moved from Obscurity to a Partnership with Powerful AT&T." *Wall Street Journal,* March 8, 1999.

De Santis, Solange. "Unitel Expected to Change Name to AT&T Canada." *Wall Street Journal,* September 9, 1996.

Deogun, Nikhil, and Stephanie N. Mehta. "Forstmann Little to Invest $1 Billion for 12% Stake in Telephone Carrier." *Wall Street Journal,* August 30, 1999.

"Dynegy Exits Telecom." LightReading, January 23, 2003.

"Dynegy Sees Improved Broadband Performance and Continued Energy Growth." *POWERGRID International,* October 15, 2001.

"Dynegy Sells Comms to 360networks." LightReading, March 31, 2003.

"Dynegy Sells European Telecom Business." *Midland Daily News,* January 22, 2003.

"Dynegy to Acquire European Broadband Outfit." *Houston Business Journal,* November 7, 2000.

"Enron Is Latest Company with Plans for a National Fiber-Optic Network." *Wall Street Journal,* July 20, 1998.

"Enron, CapRock to Build Fiber Optic Network." *Houston Business Journal,* February 4, 1999.

"Enron's Rise and Fall." *Wall Street Journal,* December 5, 2001.

"Ex-Chief of McLeod in $4.4 Million Settlement." *New York Times,* July 31, 2006.

Farnsworth, Clyde H. "AT&T Sets Unitel Deal in Canada." *New York Times,* January 8, 1993.

Feltner, Kerry. "Leaders at Fibertech Pass the Baton." *Rochester Business Journal,* August 14, 2015.

Feltner, Kerry. "Telecom Industry Veteran John Purcell Dies." *Rochester Business Journal,* April 10, 2017.

Fisher, Daniel. "Shell Game." *Forbes,* January 7, 2002.

Fox, David A. "Adelphia Business Solutions Files Chapter 11." *Nashville Post,* March 27, 2002.

Grant, Peter. "RCN Will Sell Cable Systems in New Jersey for $245 Million." *Wall Street Journal,* August 28, 2002.

Hardy, Quentin. "Conqueror in the Carnage." *Forbes,* March 5, 2001.

Heinzl, Mark. "Manitoba Telecom to Buy Allstream." *Wall Street Journal,* March 19, 2004.

Herron, Jeremy. "McLeodUSA Exits Bankruptcy, Names Royce Holland CEO." *Wall Street Journal,* January 10, 2006.

Hopkins, Marc. "McLeodUSA Files for Ch. 11 in Chicago, Gets $50 Million in DIP Financing." *Wall Street Journal,* October 31, 2005.

Keller, John J. "McLeod to Expand Holdings with Phone Carrier Purchase." *Wall Street Journal,* June 16, 1997.

Kirkpatrick, David. "Enron Takes Its Pipeline to the Net." *Fortune,* January 24, 2000.

Knight, Chris. "Lightower Fiber Networks to Acquire Lexent Metro Connect." Pamlico Capital, September 14, 2010.

Kovatch, Karen. "Battle Lines Are Drawn, but Who Will Open Fire?" *Pittsburgh Business Times,* May 26, 1997.

Larson, Eugénie. "AT&T Acquires Velocita." LightReading, November 11, 2002.

"Lea Fastow Freed from Halfway House." NBC News, July 8, 2005.

Leahy, Michael. "A CEO's Lesson: What Goes Up . . ." *Washington Post,* November 9, 2002.

Leahy, Michael. "Boom and Bust: Telecom's Trajectory a Symbol of an Era." *Seattle Times,* November 25, 2002.

"Level 3 Announces $1 Billion TelCove Buy." *New York Times,* May 1, 2006.

"Level 3 Announces $1 Billion TelCove Buy." *New York Times,* May 1, 2006.

"Level 3 to Buy Dial-Up Asssets of McLeodUSA for $55 Million." *Wall Street Journal,* December 7, 2001.

Levensohn, Michael. "DataNet Investors Gain from Buy-In." *Times Herald-Record,* September 4, 2004.

Lewis, Michael. *Flash Boys.* W. W. Norton & Company, March 31, 2014.

"Lightower Fiber Acquires DataNet Communications Group and KeySpan Communications Corp." Business Wire, March 6, 2008.

McKenna, Maeve. "McLeodUSA Acquires CapRock Communications." *Fiber Optics Online,* October 4, 2000.

"McLeod Ends Restructuring, Starts Trading." *Wall Street Journal,* April 18, 2002.

"McLeod Executives Resign, Company Restructures." *Springfield Business Journal,* August 18, 2005.

"McLeodUSA and Consolidated to Merge in $400 Million Deal." *New York Times,* June 17, 1997.

"McLeodUSA CEO Sees Better Days Ahead." *Wall Street Journal,* February 22, 2006.

"McLeodUSA Inc. Warns Bankruptcy Possible after Delisting." *Wall Street Journal,* June 20, 2005.

"McLeodUSA Names New Chairman." CNET, April 24, 2002.

"McLeodUSA Posts $3.11 Billion Loss Due to Huge Restructuring Charge." *Wall Street Journal,* November 15, 2001.

"McLeodUSA Shares Plummet 31% after Firm Lowers Its Projections." *Wall Street Journal,* May 4, 2001.

McMillan, Dan. "Ex-GST Boss Founds New Telecom Firm." *Portland Business Journal,* June 6, 1999.

McMillan, Dan. "Warta Ventures Bags $700M." *Portland Business Journal,* November 14, 1999.

Mehta, Stephanie N. "McLeodUSA Unveils Plan to Purchase SplitRock in $1.75 Billion Stock Deal." *Wall Street Journal,* January 7, 2000.

Meier, Barry. "Corporate Conduct: The Overview; 2 Guilty in Fraud at a Cable Giant." *New York Times,* July 9, 2004.

Mullins, Robert. "AT&T, McLeod Bring Big Resources to Local Phone Race." *Milwaukee Business Journal,* January 24, 1999.

Murphy, Kate, and Alexei Barrionuevo. "Fastow Sentenced to 6 Years." *New York Times,* September 27, 2006.

"National Grid Wireless Changes Name to Lightower." *Converge! Network Digest,* August 16, 2007.

"National Grid: M/C Venture Partners and Wachovia Capital Partners to Buy National Grid Wireless US for $290 Million." Reuters, August 9, 2007.

"No AT&T Stake in AT&T Canada." *Wall Street Journal,* November 7, 2002.

Noguchi, Yuki. "PF.Net Communications to Change Name." *Washington Post,* December 21, 2000.

Noguchi, Yuki. "The Optimist." *Washington Post,* December 24, 2000.

Norris, Floyd. "Dynegy Chief Is Much Richer for Being Forced Out." *New York Times,* May 30, 2002.

Paterik, Stephanie. "Dynegy Moves to Allay Worries with $2 Billion Liquidity Plan." *Wall Street Journal,* June 25, 2002.

Perin, Monica. "Dynegy Fulfills Broadband Goal with Agreement to Acquire Extant." *Houston Business Journal,* August 6, 2000.

Peters, Jeremy W., and Simon Romero. "Enron Founder Dies before Sentencing." *New York Times,* July 5, 2006.

Philyaw, Jason. "McLeodUSA in Recapitalization Agreements with Forstmann." *Wall Street Journal,* December 10, 2001.

"Pickle Resigns at Teligent to Join Fiber-Optic Firm." *Wall Street Journal,* September 12, 2000.

Powell, Rob. "Windstream Files Chapter 11." Telecom Ramblings, February 25, 2019.

Powell, Rob. "Windstream Pounces, to Acquire PAETEC." Telecom Ramblings, August 1, 2011.

"Private Equity Firm Acquires Broadband Provider." *New York Times,* March 5, 2010.

"RCN Closes Neon Deal." LightReading, November 13, 2007.

"RCN Set to Emerge from Chapter 11." *Converge! Network Digest,* December 7, 2004.

"RCN to acquire NEON Communications Group." *Lightwave,* June 25, 2007.

"RCN's McCourt to Quit CEO Post, Remain Chairman." *Wall Street Journal,* July 21, 2004.

"Rebuilt McLeodUSA Faces Funding Questions." *Wall Street Journal,* April 4, 2003.

Rifkin, Glenn. "John J. Rigas, Cable TV Magnate Who Pillaged His Company, Dies at 96." *New York Times,* September 30, 2021.

"Rogers Wireless to Drop AT&T Brand." *Wall Street Journal,* December 2, 2003.

Roush, Matt. "McLeod Dials up New Presence in Local Market." *Crain's Detroit Business,* October 18, 1999.

Santo, Brian. "Crown Castle Gets 'Crown Jewel' Lightower for $7.1B." *Fierce Network,* July 19, 2017.

"SBA Communications Corporation Agrees to Acquire Light Tower Wireless LLC." GlobeNewswire, July 22, 2008.

Scannell, Kara. "Forstmann Little Details Plan to Restructure McLeodUSA." *Wall Street Journal,* December 4, 2001.

Scannell, Kara. "Forstmann Little Will Give McLeod $100 Million More." *Wall Street Journal,* August 2, 2001.

"Settlement Proposed in McLeodUSA Shareholder Lawsuits." *Wall Street Journal,* September 20, 2006.

Sorkin, Andrew Ross. "A Bold Gambler Ups the Ante despite Long Odds." *New York Times,* August 2, 2001.

Sorkin, Andrew Ross. "Business; Will He Be K.O.'d by XO? Forstmann Enters the Ring, Again." *New York Times,* February 24, 2002.

Steiner, Christopher. "Wall Street's Speed War." *Forbes,* September 9, 2010.

Stevens, Matt, and Matthew Haag. "Jeffrey Skilling, Former Enron Chief, Released after 12 Years in Prison." *New York Times,* February 22, 2019.

Taub, Steven. "Ex-Dynegy Finance Pair Get Prison Terms." *CFO Magazine,* January 6, 2006.

"Technology Briefing | Telecommunications: Adelphia Pact with Former Unit Approved." *New York Times,* March 24, 2004.

"Technology Briefing | Telecommunications: RCN Files for Bankruptcy Protection." *New York Times,* May 28, 2004.

"Technology Briefing | Telecommunications: RCN Plans to Exit Bankruptcy." *New York Times,* August 24, 2004.

"telMAX Announces Michael Strople as CEO." PR Newswire, January 3, 2024.

"Theodore Forstmann, 1940–2011: The Pioneer of Private Equity." *The Week,* January 8, 2015.

"Three Dynegy Executives Indicted for Conspiracy and Fraud." *New York Times,* June 12, 2003.

"Time Warner Buys Enron Fiber." LightReading, January 25, 2004.

Tsau, Wendy. "AT&T Canada's Restructuring Plan Wins Court Approval." *Wall Street Journal*, February 26, 2003.

"United States v. Scott Yeager." US Department of Justice Criminal Division. https://www.justice.gov/criminal/criminal-vns/case/united-states-v-scott-yeager.

US Securities and Exchange Commission. "McLeodUSA Incorporated." Securities Exchange Act of 1934, Release No. 50385. Accounting and Auditing Enforcement, Release No. 2103. Administrative Proceeding, File No. 3-11662. September 15, 2004.

"Utility Owner Sells Stake in Fiber Optic Venture." *New York Times*, April 22, 2004.

"Vancouver-Based Electric Lightwave Sells for $1.4 Billion." *The Oregonian*, November 30, 2016.

"Velocita Seeks Bankruptcy Protection, Exits Telecom Services." *Washington Post,* June 5, 2002.

"Velocita to Build Second Phase of AT&T's Next Generation Network." *Fiber Optics Online*, February 15, 2001.

"Witness: Enron Broadband Software 'Pixie Dust.'" NBC News, April 21, 2005.

"Zayo to Acquire Allstream." Business Wire, November 23, 2015.

INDEX